D0915441

MAKING
MICE

MAKING MICE

STANDARDIZING ANIMALS FOR AMERICAN BIOMEDICAL RESEARCH, 1900–1955

KAREN A. RADER

PRINCETON UNIVERSITY PRESS
PRINCETON AND OXFORD

Published by Princeton University Press, 41 William Street,
Princeton, New Jersey 08540
In the United Kingdom: Princeton University Press, 3 Market Place,
Woodstock, Oxfordshire OX20 1SY

Library of Congress Cataloging-in-Publication Data
Rader, Karen A. (Karen Ann), 1967–
Making mice / Karen A. Rader.
p. cm.
Originally presented as the author's thesis (doctoral)—Indiana University, 1995.
Includes bibliographical references and index.
ISBN 0-691-01636-4 (cloth : alk. paper)
1. Mice as laboratory animals—History. 2. Jackson Laboratory (Bar Harbor, Me.)
I. Title.

QL737.R666R334 2004
616'.027333—dc21 2003054715

British Library Cataloging-in-Publication Data is available

This book has been composed in Sabon with Kabel Display Family
Printed on acid-free paper. ∞
pup.princeton.edu
Printed in the United States of America
10 9 8 7 6 5 4 3 2 1

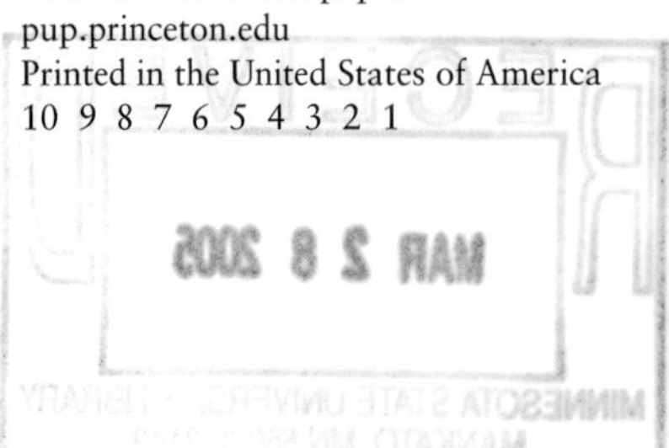

FOR JOHN

CONTENTS

ILLUSTRATIONS

ACKNOWLEDGMENTS

My work on this project brought me into contact with many social worlds, ranging from libraries to laboratories, and I managed to incur significant debts in all of them. The seeds were first planted by my teachers at Loyola College in Maryland, where I studied with Stephen Hughes, whose enthusiasm for the history of science nurtured my own. Years later, as a graduate student in Indiana University's Department of History and Philosophy of Science, I made contact with an extraordinary group of faculty and graduate student colleagues—including James H. Capshew, Ann Carmichael, Patrick Catt, Martha Crouch, Frederick Churchill, Alice Dreger, Thomas Gieryn, Brad Hume, Stephen Kellert, Richard Sorrenson, Elizabeth Green Musselman, John C. Powers, Judith Johns Schloegel, Corinna Treitel, Julliane Tuttle, Wini Warren, and Eric Winsberg—each of whom lent support and provided a good sounding board in the early stages, and many of whom continue to do so. After "birthing" my dissertation, my advisor Jim Capshew repeatedly supplied me with interesting new leads to explore, while taking every opportunity to offer solid professional advice and (even more importantly, during what felt like endless years on the academic job market) to express his confidence in me and in my work. Also, Tom Gieryn has since become a good friend, as well as an invaluable source of encouragement, thoughtful criticism, and last-minute citations.

My initial research introduced me to the beauty of Maine, as well as to some of the state's most distinguished institutions of higher learning and research. At the University of Maine's Fogler Library, Muriel Sanford assisted me with the C. C. Little Papers, and Elaine Smith later helped with illustrations. At JAX librarian Doug MacBeth provided me with convenient access to the archive, and he and Ann Jordan tended graciously to my needs on my return visits. Then-director Ken Paigen endorsed my presence at JAX, sat for an interview, and, together with his wife Beverly, helped me initiate relationships with key historical scientists there. Many of these JAX "old-timers" sat for one or more interviews, including Earl and Margaret Green, Skippy Lane, Joan Staats, Elizabeth Russell, Tom Roderick, and George Snell. Later, at Oak Ridge National Laboratory, so did Bill and Liane Russell. I am grateful to have had the opportunity to pick their collective brains for historical details not found in their scientific journal articles. A historian's production timeline nearly always passes slower than a scientist's, but I am nevertheless saddened by the knowledge that many of these people did not live to see their contributions properly acknowledged.

My dissertation related a small piece of the laboratory mouse's history, primarily through the lens of C. C. Little's vision for the Jackson Lab and mouse genetics. For the book, I thought it was important to expand beyond that. I wanted to tell a story that acknowledged the historical role of the consumers as well as the producers of these animals—of those researchers who literally bought JAX mice, but also of those widely varying public and scientific actors who supported their production in less direct but no less important ways. In the course of developing these ideas, I profited greatly from the support and suggestions of many people and many institutions.

MIT's Program in Science, Technology, and Society—through a Mellon Postdoctoral Fellowship in the History of the Life Sciences—was a hospitable first home after graduate school. My fellow postdocs, Devora Kamrat-Lang and Nadine Weidman, shared friendship and provided a core of stability and sanity for developing an intellectual community. Lily Kay and Charlie Weiner taught me a thing or two about teaching the history of biotechnology, and they were equally generous with their scholarly feedback (especially Lily Kay, whose enthusiasm for the history of biology still inspires me). Deborah Fitzgerald and Amy Slaton introduced me to the historiography of technology and gave careful and thoughtful readings of various chapters and drafts. With the support of a National Science Foundation Scholar's Award (#95–21621), I stayed on in Cambridge another year, this time at Harvard's Department of the History of Science. There I continued to partake of the area's vibrant science studies scene, and (like many young historians of biology before me) I had the much appreciated benefit of Everett Mendelsohn's expansive intellect and good mentoring.

During the following year at Princeton University's Davis Center for Historical Studies, I was socialized into "animal studies" by a wonderful weekly seminar series and my accomplished cohort. Katherine Grier, Nigel Rothfels, Ed Steinhart, and William Jordan were particularly supportive and friendly colleagues, always willing to offer constructive criticism on preliminary thoughts and to share their wisdom on how best to live a scholar's life. Norton Wise and Mary Heninnger-Voss warmly welcomed me into the Program in the History of Science, which provided an unrivaled atmosphere for developing connections between the cultural and intellectual pieces of my story. I feel fortunate to have known Gerry Geison, though I will always regret that I never got to tell him why: his honesty, generosity, and graceful intelligence made me want to be a better teacher in these same ways for my students. Likewise, I am deeply grateful to Angela Creager for many conversations, both formal and informal, about this project, and especially (many years later!) for her insightful reading of the penultimate manuscript.

While at Princeton, I met some contemporary "mouse people"—a creative group of molecular biologists whose abiding passion for the historical study of their organism sustained my work. Lee Silver encouraged me to write for practicing mouse researchers and endured endless mundane questions about mouse boxes, mouse food, and the like. With characteristic grace and good humor, Thomas Vogt allowed me to become a participant-observer in his working mouse genetics laboratory, so I could experience firsthand what it was like to handle the creatures I have spent so much time writing about. He also introduced me to other historic mouse researchers, who sat for formal and informal interviews: Anne McLaren, W. K. Silvers, and Salome Waelsch. Shirley Tilghman cheered on all of this and now follows in the footsteps of C. C. Little, thus raising to two the number of people who occupy the vocational category "mouse-geneticist-turned-college-president."

Since then I have received various forms of support and feedback from many more scholars and librarians in the history of science, technology, and medicine. Archivists knowledgeable in these fields helped me navigate a vast array of manuscript materials. I am especially grateful to Janice Goldblum (National Academy of Sciences), Marjorie Ciarlante (National Archives), Thomas Rosenbaum (Rockefeller Archive Center), and Beth Carroll-Horroacks (American Philosophical Society). Ever since I ran into him while doing my dissertation research at the Jackson Lab, Dan Kevles has been a valued source of encouragement and criticism. I would also like to record my particular gratitude to the following colleagues: Garland Allen, Rachel Ankeny, John Beatty, Adele Clarke, Joan

Fujimura, Jean Paul Gaudilliere, Scott Gilbert, Pat Gossell, Anita Guerrini, Elizabeth Hanson, Victoria Hardin, Adrian Johns, Stephanie Kenen, Bettyann Kevles, Evelyn Fox Keller, Robert Kohler, Sally Kohlstedt, Susan Lederer, M. Susan Lindee, Ilana Löwy, Jane Maienschein, Gregg Mitman, Paolo Palladino, Jennifer Price, Harriet Ritvo, Margaret Rossiter, and Alison Winter.

My editor at Princeton University Press, Sam Elworthy, is also a historian of science, and that made it possible for me to rely on his critical insights, as well as his encouragement, during the revision process. Sarah Harrington and especially Hanne Winarsky gave heroic administrative assistance as the manuscript was in production.

My students and colleagues at Sarah Lawrence College endured working with me during the final phase, and I cannot thank them enough for shepherding me through this crucial time. The Advisory Committee on Appointments granted me release time to write, as well as funding for research and publication expenses. Dean Barbara Kaplan stalwartly supported my work and at appropriate moments also helped me to laugh. Members of the Social Science Faculty Group and the Science Faculty Group embraced my interpolation between them and made me feel at home whatever I wanted to teach or talk about. Working with Leah Olson, Marsha Hurst, and Charles Zerner broadened my horizons, intellectually and pedagogically, in important ways. Melissa Frazier, Mary Porter, and Lyde Sizer saw me through numerous personal, scholarly, and writing crises. Librarians Janet Alexander, Geoffrey Danisher, Judith Kicinski, and Jenni McSpadden were themselves invaluable resources for obtaining published resources. James Freedland, Alexis Turner, Jamie Thompson, and Rebecca Frick provided diligent bibliographic and research assistance, each at different stages.

My final, and most profound, thanks are personal: to Maria Dans and Claire Gregory, for the mellowing effects of their old friendships, and to Tobe and Ralph Sevush, for the energizing effects of their new one. To Richard and Holly Rader, and to Thomas and Katherine Willetts, for reminding me of the meaning of family, as well as for encouragement and endless faith in my endeavors; to Charles and Kathleen Powers, for giving me a "second family"; to Samuel Powers and Grace Powers, for being the jewels of my life; and lastly, to John Powers, for true collaboration and true love.

ABBREVIATIONS

Over ten years of researching, I examined many unpublished archive and manuscript sources. The materials on which this book draws most heavily are asterisked below. I provide annotations only in cases where the status, location, or content of a collection is unclear from published finding guides or has changed since I did my original research.

***CCL-UMO** Clarence Cook Little Papers, Raymond Fogler Library, University of Maine, Orono, ME

As of December 2002, Little's papers remain only partially processed. My convention is to note the box number in all cases, but to note the folder name only when the folder is not self-explanatory (i.e., folders often bear the same name as the documents they contain, and they are not numbered or ordered).

DOE-MD Department of Energy Archive, Department of Energy, Germantown, MD

DOE-Hum Department of Energy Archives, Human Radiation Experiments History Project, Washington, DC

The DOE-Hum archive was assembled for the purpose of writing a history of human radiation experimentation, mandated by President Bill Clinton and Secretary of Energy

Hazel O'Leary in 1993. These federal documents, now declassified, are now being relocated to the National Archives. They consist mainly of documents related to the Division of Biology and Medicine in the immediate postwar decades.

HJM-IU H.J. Muller Papers, Lilly Library, Indiana University, Bloomington, IN

LCD-APS Leslie Clarence Dunn Papers, American Philosophical Society, Library, American Philosophical Society, Philadelphia, PA

***JLA-BH** Jackson Laboratory Archive, Jackson Laboratory Library, Jackson Laboratory, Bar Harbor, ME

In the last several years, the Jackson Laboratory Archive has been more fully processed. More than half of what was merely in self-named folders on open shelves is now in numbered boxes, and a web-based archive guide has been developed for both the manuscript and photograph collections. For the material that was not in boxes when I did my research, I simply list complete titles and folders, through which one could trace the location to the new box numbers, using newly constructed finding guides.

JLOH-APS Jackson Laboratory Oral History Collection, American Philosophical Society Library, American Philosophical Society, Philadelphia, PA

JLOH-KR Jackson Laboratory Oral Histories (by author)

The original JAX oral histories, supervised by Judith Swazey in 1986, remain the definitive collection. Transcripts of the interviews I did from 1993 to 1995, including unique interviews with William and Liane Russell and Earl and Margaret Green, remain in my possession.

***NACC-MD** U.S. National Advisory Cancer Committee Papers and Transcripts, National Archives (Archives II), College Park, MD

***NAS-DC** National Academy of Sciences Archives, National Academy of Sciences Library, National Academy of Sciences, Washington, DC

***RAC-NY** Rockefeller Foundation Archive, Rockefeller Archive Center, Tarrytown, NY

RRC-UT Radiation Research Collection, University of Tennessee, Knoxville, TN

MAKING MICE

INTRODUCTION

WHY MICE?

On October 23, 1947, fourteen people and tens of thousands of laboratory mice perished when the sleepy resort community of Bar Harbor, Maine, burned to the ground. A forty-mile wind-borne fire front triggered early evacuation of most of the town's estimated 4,300 human residents. Some escaped by car or bus, and thousands more rushed to the docks to await rescue; the scene, a Coast Guard official told the *New York Times*, "was reminiscent of Dunkirk." Many loyal caretakers of the island's nearly three hundred palatial estates stayed behind to fight the flames "with nothing but brooms." Elizabeth Russell, a scientist at the Jackson Laboratory, Bar Harbor's nearly twenty-year-old institution for research in mammalian genetics and cancer, remembered seeing a small plume of smoke on October 14, while at a staff meeting at nearby Hamilton Station, and marveled at how the "tiny fire had continued to grow." She and the rest of the staff quickly escaped the premises and were spared injury, but their experimental organisms fared less well. The fire completely destroyed the original lab building, and two new "mouse houses"—the second of which was under construction at the time—were seriously damaged. Except for the few hundred mice readied for shipment to researchers in a corner isolation room, all ninety thousand resident rodents (housed primarily in wooden mouse boxes) died in the blaze. When the embers cooled, those who first arrived on the scene remember two things: the strange and unforgettable smell of burnt mice, and the comment that the lab's founder, geneticist Clar-

I.1. Jackson Memorial Laboratory, c. winter 1935 [Source Credit: Jackson Laboratory Archives].

I.2. National Guardsman in front of the Jackson Memorial Laboratory, the day after the October 1947 "great Bar Harbor fire." [Source Credit: Jackson Laboratory Archives].

I.3. View of mouse rooms burned by fire, Jackson Memorial Laboratory, October 1947 [Source Credit: Jackson Laboratory Archives].

I.4. C. C. Little meets the press and surveys the damage, October 24, 1947 [Source Credit: Jackson Laboratory Archives].

ence Cook Little, made upon surveying the damage: "Now we can see the water"[1] (fig I.1–I.4).

The next day, as Maine's governor scrambled for federal disaster relief money to rebuild America's "Vacationland,"[2] Little received multiple unsolicited offers of aid to re-establish the "JAX" mice (as they had come to be known, from an abbreviation of the lab's cable address). The Rockefeller Institute for Medical Research and the Carnegie Institute both pledged facilities for maintaining the surviving mice, and before they knew the full extent of the damages, the boards of the American Cancer Society and the National Institute of Health (NIH) held special meetings where they decided to offer Little a replacement building for the continued production of mice in Bar Harbor. But perhaps most remarkably, individual geneticists and medical researchers who had previously received stocks of JAX mice began sending back breeding pairs of those same stocks to Bar Harbor. Little told the Rockefeller Foundation's Warren Weaver that there was "hardly a genetics or cancer research institute east of the Mississippi" that didn't respond to his lab's crisis. He analogized the animals' return to a biblical miracle: "The bread which we cast upon the waters several years ago, is now returning to us."[3] By contrast, Little claims to have received only one angry letter from a local "anti-vivisec-

[1] On the Bar Harbor evacuation, see Frank L. Kluckhohn, "18 Dead, Damage $25,000,000, as Forest Fires Sweep on in Wide New England Area," *New York Times*, 25 October 1947, A1+, and "Much of Bar Harbor Razed as 4,300 Flee Forest Fire; Whole Maine Towns Gone," *New York Times*, 24 October 1947, A1+. For a comprehensive list of fire-related media articles, see Jackson Laboratory Association folder, Box 735, and Fire folder (with sample articles), Box 730, both CCL-UMO. On the lab's losses, see 19th Annual Report of the Jackson Laboratory, 1947–48 (JLA-BH). On postfire memories, see interviews with George Snell (June 1995) and Joan Staats (June 1993), both JLOH-KR. Quotes from Elizabeth Russell are from her published recollection, "Mouse Phoenix Rose from Ashes," in *Perspectives on Genetics Anecdotal, Historical, and Critical Commentaries, 1987-1998*, ed. James F. Crow and William F. Dove (Madison: University of Wisconsin Press, 2000), pp. 29–30 (originally published in *Genetics*, October 1987). Russell remembered Little's postfire statement slightly differently: "Now we can see the sea."

[2] See George Lewis, "The Maine That Never Was: The Construction of Popular Myth in Regional Culture," *Journal of American Culture* 16, 2 (Summer 1993): 91–99.

[3] Emergency telegrams are contained in Box 735, CCL-UMO. For some sample scientific responses to the JAX Lab fire, see WW to K. Compton, 3 October 1947; WW to RBF, 7 November 1947; both RF Archives, RG 1.2, 200A, Box 134, Folder 1191, RAC-NY; "RBJ Lab," RF Trustees Bulletin, November 1947. See also CCL to WW, 21 and 28 November 1947, all RF Archives, RG 1.1, 200D, Box 144, Folder 1777, RAC-NY. For a specific example of a "mouse return," see CCL to Muller, 3 November 1947; Muller to CCL, 7 November 1947, CCL to Muller, 17 November 1947: all in the H. J. Muller Ms., Manuscripts Department, Lilly Library, Indiana University-Bloomington. For biblical rhetoric, see 1948 JAX Annual Report (public version) and a 1952 film produced by JAX Lab on the subject, *R_x Mouse*, c. 1950–52, both in JLA-BH.

tionist women's club," which expressed regret "that Dr. Little and his fellow scientists had not been burned up in the blaze instead of the mice."[4]

By 1949, national fundraising drives combined with additional governmental support to ensure that Jackson Laboratory would rise from its ashes. That year in a foundation endorsement letter, the lab's Board of Trustees noted that the institution had been completely rebuilt and had reclaimed its status as the "Bureau of Biological and Medical Standards." During the fall of 1953, Little felt so confident about the lab's future that he contacted his lawyer about his ultimate wish: to link the success of the JAX mouse to another popular mouse who had also weathered the Depression era. He wrote:

> I was very much interested in the article in *Life* on the 25th Anniversary of Mickey Mouse. . . . 1954 is also the 25th Anniversary of the Jackson Laboratory, that in a somewhat similar, but less sensational, way has done for the mouse in science what Disney has done for it in amusement. The possibility of arousing Disney's interest in doing something of a philanthropic nature along the line of a factual, or partly factual film, to tell the story of the mouse (which might easily be a brother or other relative of Mickey) has been in [our] minds . . . for some years.

Little did eventually correspond with Disney, but apparently nothing ever came of his idea. He later told a friend the moral he drew from this interaction, as well as from public responses to the 1947 fire: "In these days, when support of basic research by the American public is its chief and constantly growing hope, efforts of this kind, which might seem through Victorian eyes to be undignified, are not really as shallow and superficial as they may seem."[5]

[4] On the Bar Harbor antivivisectionists, see Arthur Bartlett, "The Big Mouse Man of Cancer Research," *Coronet* 26 (August 1949): 161–62. The letter's author also objected to the well-publicized JAX experiments on rabbits, which aimed at trying to create "good" and "bad-tempered" strains through genetic inbreeding. But Little refused even to respond to this argument of individual animal integrity and replied: "Dear Madame: The members of your club seem much more bad-tempered than the rabbit." Bartlett himself concluded that Little's mouse work represented the only noble social ethic: "scientific progress in service to humanity." For an overview of antivivisection in the United States circa 1900 and beyond, see Mary L. Westermann-Cicio, "Of Mice and Medical Men: The Medical Profession's Response to the Vivisection Controversy in Turn of the Century America," Ph.D. dissertation, State University of New York, 2001; Susan E. Lederer, *Subjected to Science: Human Experimentation in American Before the Second World War* (Baltimore: Johns Hopkins University Press, 1995).

[5] Cf. Richard W. Jackson (Roscoe's son) to Warren Weaver, 7 October 1949, plus enclosed endorsement, Rockefeller Foundation Archive, RG 1.1, Series 200D, Box 144, Folder 1778, RAC-NY. On Disney, see CCL to Roy Larsen, 5 November 1953, Box 12, Folder "L," JLA-BH. After several frustrated communications with Disney's associates, Little finally had

More than fifty years later, the "Great Bar Harbor fire" represents a relatively minor event in American history, and yet it raises many compelling questions for historians of science and culture. Why were such a large number of mice gathered in a little-known, nonprofit cancer research laboratory in Maine, and why did their animal deaths warrant national media attention alongside the direct effects of the fire on the human inhabitants of Mount Desert Island? Why did the NIH, as well as so many researchers and foundations, desperately want to reassemble JAX Lab and its mouse colony, and why did so few persons concerned with animal welfare or animal rights object to this project? Why was there more national financial support available for quickly rebuilding JAX's mouse houses than there was for rebuilding Bar Harbor's own natural resources (and thereby its local tourism industry)? And finally, why might Little have thought Mickey Mouse would prove a powerful tool for doing so—even while Walt Disney himself found this idea problematic?

This book seeks to answer these questions by examining the contingent process through which American biological and medical researchers developed the mouse into a standardized laboratory organism during the period from 1900 to 1955. Like the science it reconstructs, this book is based in large part on scientists' own accounts of their work—research articles, correspondence, and other bureaucratic paper trails of their administrative interactions—but it also mines the historical record for traces of this same science's more public culture: congressional testimony, publicity films, popular magazine feature writing, and so forth. By crafting a conversation between these rich bodies of primary and archival source material, it strives to explore the nature of laboratory mouse standardization from the perspectives of the animal's developers as well as its various users: mouse genetics experimenters in labs at JAX, medical researchers who paid to have JAX mice sent to their own labs, science policymakers who located a program for coordinating bench-top research in murine bodies, and the American public, who at once consumed laboratory mice as cultural icons of biological research and supported mouse experiments and production with their tax dollars. I situate my account at the locus of mass production that historians of technology have deemed "the consumption junction,"[6] but the engine driving my account is a concern for

lunch with Disney himself, "who seemed to be interested in incorporating the Laboratory's program and opportunities in connection with a television program which he is planning" (CCL to Benjamin Sherman, 5 April 1954, Box 12, Folder "S," JLA-BH). I have found no Disney program that meets this description and no further archival evidence that this collaboration progressed. See also C. J. LaRoche to Bea Little, 22 January 1954 Canning to LaRoche, 25 January 1954 and 4 February 1954, all Box 12, Folder "D (for Disney)," JLA-BH.

[6] Feminist historians first advocated this approach to highlight women's technological agency, as a corrective to accounts of early twentieth-century technologies made primarily

the complex interplay between science and society, so "users" is a theoretical category I employ very idiosyncratically. Taking cues, respectively, from the work of historians T. J. Jackson Lears and Phillip Pauly, I define "consumption" as a process of "individual choice and consciousness, wants and desires . . . in the context of social relations, structures, institutions, systems"—mainly because I am interested in perpetuating a definition of culture that emphasizes its historical and etymological roots at "the intersection of the biological and the technological" in America (especially during the early decades of the twentieth century).[7] Ultimately, then, this book describes the means by which scientists developed JAX mice into standard mammalian research organisms not just through the eyes of researchers doing experiments in laboratories, but through their encounters with the politicians and policymakers of the fledgling national system of biomedical research emerging in this period. At the same time, by considering how inbred mice became iconic symbols of the value of standardization within our culture's changing understandings of animals and science in the twentieth century, I am also suggesting that the public audience for this work must be considered another kind of scientific user. To understand how broader cultural imperatives shaped the practical nature of standardization in research, and vice versa, is to understand the social and scientific meaning of biology in twentieth-century American life.

Focusing primarily on the inbred mice produced by one institution—the Jackson Lab—my story chronicles both the specific evolution of one animal species (*mus musculus*, the common mouse) through its journey into the laboratory, as well as a key period of disciplinary and methodological reorganization in biology. Inbred strains were first developed and promoted for philanthropically funded cancer genetics research at the Jackson Lab, but financial deficits brought about by the Depression provoked director C. C. Little to circulate these animals more widely, as "pure" biological reagents for more diverse lines of medical research. After World War II, as the genetic etiology of cancer began to wane in experimental cancer work,

---

from the perspectives of their mostly male creators and producers. Cf. Ruth Schwarz Cowan, "The Consumption Junction: A Proposal for Research Strategies in the Sociology of Technology," in *The Social Construction of Technological Systems: New Directions in the Sociology and History of Technology*, ed. Wiebe E. Bijker, Thomas P. Hughes, and Trevor J. Pinch (Cambridge: MIT Press, 1987), pp. 261–80, and the excellent and updated historiographic discussion in Nina E. Lerman, Arwen Palmer Mohun, and Ruth Oldenziel, "Versatile Tools: Gender Analysis and the History of Technology," and "The Shoulders We Stand on and the View from Here: Historiography and Directions for Research," in *Technology and Culture* 38 (1997): 1–32.

[7] Richard Fox and T. J. Jackson Lears, *The Culture of Consumption* (New York: Pantheon, 1983); Phillip J. Pauly, *Biologists and the Promise of American Life* (Princeton: Princeton University Press, 2000), p. 8.

the social and scientific need for good mammalian models of radiation damage gave the inbred mouse a new mission. Along with these changes in scientific agenda, however, came shifts in the patronage of science and the commercialization of its infrastructure (now including standardized lab animals). These developments nearly rendered the coexistence of research and mouse production at Jackson Lab unsustainable.

In the early years, JAX scientists constantly fought back the tide of what they came to know as "operation bootstrap"—the piggy-backing of mouse research onto the development of the production colony—but in retrospect, their persistence paid off. In the 1950s, although JAX was widely acknowledged as (in the words of one trustee) "the bureau of mouse standards," C. C. Little could barely convince either medical genetics researchers or granting agencies that mammalian genetics was worth much investment. Today sales of JAX inbred mice to outside researchers exceed two million organisms annually.[8] Furthermore, since its inception in 1959, JAX's frozen mouse embryo repository has accumulated more than 2,400 strains of mouse mutants. These animals, instead of being bred, are stored more cost effectively as embryos in vats of liquid nitrogen. Kenneth Paigen, Jackson Lab director from 1989 to 2000, claims that "more than 95 percent of all mouse models used in the world come from the Jackson Laboratory." As the 2001 JAX Annual Report concluded: "Researchers around the world agree that JAX© Mice are the 'gold standard' of genetic purity in mouse models," citing a 2000 report from Michael Festing and Elizabeth Fisher that "at least seventeen Nobel prizes . . . have flowed from the Jackson Laboratory."[9] One of these Nobel Prizes was awarded in 1980 to a JAX researcher, George Snell. Snell's congenic strains, which he began developing in the 1940s and completed in 1957, enabled him to identify and characterize the key genetic locus of histocompatibility in mice. This work (along with that of Baruj Benacerraf and Jean Dausset on the analogous phenomenon in human tissue transplant) was honored by the Nobel Committee as "laying the foundation for our knowledge of 'self' from 'non-self.' "[10]

[8] Personal communication, JAX Public Information Office, June 1992. See also Jackson Laboratory Annual Report, 1991.

[9] Lee Silver, "Suppliers of Mice," appendix A to *Mouse Genetics* (New York: Oxford University Press, 1995), p. 285. Paigen quoted in Diane Harrison, "Jax Lab Moves into the Future," *Ellsworth American*, 29 June 2000. Jackson Lab Annual Report, 2001, p. 26; cf. Michael Festing and Elizabeth Fisher, "Mighty Mice," *Nature*, 404, 6780 (20 April 2000): 815.

[10] Cf. introduction to George Snell, J. Dausset, and S. Nathenson, *Histocompatibility* (New York: Academic Press, 1976). When asked in 1996 about the medical significance of his work, however, Snell demurred: "Everybody hopes that what they do will turn out to be useful." Similarly, until his death in 1996, he continued to regale visitors with stories about how his Bar Harbor neighbors mistook news of "George winning the prize" as an

The Jackson Lab's research successes since the 1950s have not been limited to Snell's work. In the late 1950s and 1960s, for example, staff scientist Leroy Stevens was doing tumor transplantation work on Strain 129 mice, and he made a leap that would "profoundly affect stem cell technology a decade later." When Stevens noticed that the primordial germ cells that gave rise to teratomas looked a lot like the cells of considerably earlier embryos, he decided to transplant cells from various stages of early Strain 129 mouse embryos, including inner cell mass cells, into testes of adult mice. Some of these early embryo cells gave rise to teratomas, which, when transplanted into mouse bellies, displayed the ability to generate an impressive range of tissue types. Stevens called these cells that could support differentiation "pluripotent embryonic stem cells"—the origin of the term "stem cells."[11]

By far, however, one of JAX's proudest accomplishments is that the National Cancer Institute has renewed the lab's designation as a "Cancer Center" for genetic research every five years since it initially bestowed on JAX this honor in 1983. "That designation," Paigen wrote in his 2001 Annual Report Director's Message, "is vital to the Jackson Laboratory because basic cancer research is a thread woven into the fabric of our very institution."[12] For the twenty-five years between 1955 and 1980, that thread was not always acknowledged by science policy-making bodies, but it is one of the arguments of this book that it was there all along, ready to be rewoven (by new techniques of mammalian genetic manipulation) into the tapestry that is modern biomedical research. In fact, this book's pre-1955 focus highlights how problems of genetics once considered unanswerable in mammals were later transformed into cutting-edge research fields. Thus the Rockefeller Foundation program officer who in 1951 wrote that "the most valuable export of the Jackson Memorial Laboratory is in terms of boxes of mice rather than scientific publications" failed to appreciate the important, but often unpredictable, connections between the two. *Mus musculus* and its many mutants were well poised to colonize the laboratories of the new organismal molecular biologists of the 1970s, and work with mice has ranked especially significant in recent cancer research, as well as in the emergence of other biomedical fields such as molecular immunology and genetic epidemiology.[13]

---

accolade for his gardening prowess, not his research accomplishment (interview with George Snell, May 1995, JLOH-KR).

[11] Quotes from Ricki Lewis, "A Stem Cell Legacy: Leroy Stevens," *The Scientist* 14 (5–6 March 2000): 19. Cf. Leroy Stevens, "The Development of Transplantable Teratocarcinomas from the Intratesticular Grafts of Pre and Post-implantation Mouse Embryos," *Developmental Biology* 21, 3 (March 1970): 364–82.

[12] Kenneth Paigen, "Director's Message," JAX Annual Report 2001, pp. 5–6.

[13] Scott Podolsky and Alfred Tauber, *The Generation of Diversity: Clonal Selection Theory and the Rise of Molecular Immunology* (Cambridge: Harvard University Press, 1977);

Ironically, even mammalian genetics—the field scientists and policymakers labeled too slow and laborious to invest in during the early twentieth century—has undergone what can only be described as an explosion in the last decade. The first mammalian gene ever cloned and sequenced was from a mouse.[14] Further, although mouse mutants have been the object of animal fanciers' fascination for centuries, the decoding of the mouse genome achieved in 2002 was possible because advances in mammalian gene manipulation technology (first recombinant DNA, then the gene "knock-out" technique[15]) combined with significant material investments, dating all the way back to the beginning of the twentieth century, to preserve genetically known strains of this animal created by and used in cancer research and radiation genetics. Mouse work has even begun to revolutionize basic Mendelian assumptions, especially the notion that a gene's expression is independent of the parental origin of the chromosome.[16]

Individual lives, however, are what connect larger structural shifts in the intellectual organization of science and the local modus operandi of research, and so it should not be surprising that I sustain my account of twentieth-century biology not through claims to institutional or organismic "greatness" but rather through more intimate knowledge of scientific biography. Thus I begin with and repeatedly emphasize the passion and drive of C. C. Little in the project of developing the inbred laboratory mouse.[17] During his testimony before the 1965 congressional hearings on cigarette labeling, Little asked lawmakers if they comprehended why he was focusing so much on the animal that was the basis of his scientific

cf. Albert Cambrosio and Peter Keating, "The New Genetics and Cancer: Contributions of Clinical Medicine in an Era of Biomedicine," *Journal of the History of Medicine and Allied Sciences*, 56, 4 (October 2001): 321–52.

[14] D. A. Konkel, S. M. Tilghman, and P. Leder, "The Sequence of the Chromosomal Mouse Beta GLobin Major Gene: Homologies in Capping, Splicing and PolyA Sites," *Cell* 15 (1978): 1125–32.

[15] On recombinant DNA's development and regulation, Susan Wright, *Molecular Politics: Developing American and British Regulatory Policy for Genetic Engineering, 1972–1982* (Chicago: University of Chicago Press, 1994). Cf. Mario Capecchi, "Altering the Genome by Homologous Recombination," *Science* 244 (1988): 1288–92, and "The New Mouse Genetics: Altering the Genome by Gene Targeting," *Trends in Genetics* 5 (1989): 70–76.

[16] Shirley Tilghman, "The Sins of the Fathers and Mothers: Genomic Imprinting in Mammalian Development," *Cell* 96 (1999): 185–93.

[17] For a biographical approach to model organisms, see Judith Johns Schloegel, "Intimate Biology: Herbert Spencer Jennings, Tracy M. Sonneborn, and the Career of American Protozoan Genetics," Ph.D. diss., Indiana University, forthcoming, and "Life Imitating Art, Art Imaging Life: Intimate Knowledge, Agency and the Organism as Aesthetic Object," *Videnskabsforskning* (Danish Newsletter for the Network of History and Philosophy of Science) 20 (1998): 2–18.

claims about smoking and cancer in humans: "I have spoken of mice as the servant of man. Why is this true? What made its truth evident? In other words, why mice?" Little was then nearing the end of a long career dedicated to "building a better mouse" for research, and his final project was controversial: as head of the Tobacco Institute Research Committee (precursor to the contemporary Council for Tobacco Research), he advanced the hypothesis that certain cancers developed in animals only if they possessed a preexisting genetic susceptibility. Indeed, there was perhaps no one for whom these queries held more personal meaning or urgency. On his eightieth birthday, Little penned a cartoon that summed up his views of the mouse's scientific and institutional achievements: it showed a likeness of Little himself dwarfed by a statue of the "JAX mouse, 1929–1968." The mouse carried a sack of money—presumably that which JAX made through the sales of mice to researchers—and addressed its scientist-muse: "You've had *80* years! Look what *my* family has done in *39* years!"[18] (fig I.5).

Little's question, "why mice?" did not merely reflect his own inner journey. Taken as a broader reflection, this query interrogates the central role of particular animals in the process of biological and medical knowledge-making. Exactly how and why are certain animals chosen for certain kinds of experimental research, while other creatures and other compelling research questions ignored? Most scientists who work with laboratory mice respond to the material part of this question by citing a "laundry list" of their creatures' many research-friendly biological properties. For example, they are small and relatively tame animals, which makes them easy to handle, house, and feed. They breed readily and often (several times per year), and three weeks after the females have mated, good-sized litters of pups are born, which allows for a quick yield of research results, whether in terms of providing a large sample or observing generational patterns. Finally, mice are mammals with a 99 percent genetic homology to humans, and they happen to get many of the same diseases as us (cancer, heart disease, etc.), which (by extrapolation) makes it possible to track and experiment on many human health conditions in situ.[19]

[18] C. C. Little, typescript, 1965 Testimony on the Cigarette Labeling Hearings (n.d.), Box 732, CCL-UMO. On Little's involvement with the TIRC, see Robert N. Proctor, *Cancer Wars* (New York: Basic Books, 1995), p. 107; cf. Little's eightieth birthday cartoon in Jackson Laboratory Photo Archive.

[19] See, for example, Patricia Lauber, *Of Man and Mouse: How House Mice Became Laboratory Mice* (New York: Viking Press, 1971), p. 49; Gina Kolata, "A Star Is Born: Even a Lab Mouse Needs an Agent," *New York Times*, 26 January 1997, p. E5; cf. Lee Silver, "Mice as Experimental Organisms," *Encyclopedia of the Life Sciences* (Nature Publishing Group, 2001), available at www.els.net (accessed 8 June 2003). The figure of 99 percent genetic homology comes from recent Mouse Genome Project data—e.g., Mark S. Boguski, "The Mouse That Roared," *Nature* 420 (5 December 2002): 515–16.

I.5. Cartoon drawn by C. C. Little to thank staff for their celebration of his eightieth birthday [Source Credit: Jackson Laboratory Archives].

From a scientist's perspective, enumerating such variables provides a concise statement of the important pragmatic qualities of a successful experimental animal in biomedicine. But from a historian's perspective, these seemingly universal measures of scientific success work to decontextualize the mice themselves from the places and the circumstances under which they were developed and used as experimental animals. By presuming that the contemporary understanding of what constitutes "good" research is timeless, they "black box" the values informing the research process and render invisible the very nexus of politics and practices that defined what counted as laboratory "success" (and therefore, which intrinsic qualities of the mouse were "useful") in the first place. Mice that entered scientific laboratories before 1900 were far more likely to be stray creatures looking for food or shelter. By 1960 mice had become laboratory fixtures in cancer studies and mammalian genetics (especially radiation genetics) embedded with multiple, co-existent meanings of their "usefulness." The former were animals trying to further their own basic survival. The latter were animals whose bodies and representations were re-engineered by humans, to further the local goals of particular research communities as well as the social aims of those people and institutions that surrounded and supported this work—including other scientists, foundations, and members of the American public. In short, what remained of their animal agency in the human world was far more complex than simply searching for scraps of food or warm shelter.

Spurred on by Robert Kohler's 1994 book *Lords of the Fly*, as well as by recent science studies work in the material cultures of experiment, both biologists and historians of biology are now paying attention to the role of model organisms. Some of this work follows the model of "great men" histories of science but substitutes "great organisms." But the bulk of it has yielded important new insights with regard to the social life of biologists in their laboratories, as well as the process of making biological knowledge. Developing model organisms—from flies, to corn, to bacteria, and eventually to viruses—was (and still is) one of the most resource-intensive aspects of being a geneticist in the early twentieth century. But seemingly mundane investments in these tools of the trade have yielded many rewards for both the individual and the collective enterprise: faster research results and greater consensus over their meaning, to name just two.[20]

[20] Robert E. Kohler, *Lords of the Fly: Drosophila and the Experimental Life* (Chicago: University of Chicago Press, 1994); Rachel Allyson Ankeny, "The Conqueror Worm: An Historical and Philosophical Examination of the Use of the Nematode *Caenorhabditis Elegans* as a Model Organism," Ph.D. diss., University of Pittsburgh, 1997; Ilana Löwy and Jean-Paul Gaudillére, "Disciplining Cancer: Mice and the Practice of Genetic Purity," in *The Invisible Industrialist* (New York: Macmillan, 1998), 209–49. Cf. Adele E. Clarke and Joan H. Fujimura's collection, *The Right Tools for the Job: Materials, Instruments, Tech-*

Model organism studies have also provocatively explored the relationship between human and material agency in the triumph of experimental biology over earlier natural history methodologies. Kohler, for example, argues that *Drosophila* "colonized" genetics laboratories by virtue of its natural fecundity. The fly's ability to generate new mutant forms of itself catalyzed T. H. Morgan and his colleagues to standardize and domesticate a variety of strains for chromosome mapping, and the subsequent deluge of experimental material displaced rival neo-Mendelian studies and organisms from biology's center stage. Angela Creager's analysis of the history of TMV (Tobacco Mosaic Virus) proposes another means of understanding the discipline's transformation—namely, through attending to the "everyday practice of finding and identifying workable precedents for innovative experiment." Wendell Stanley and other viral researchers, she argues, transformed biology by developing TMV into "a cluster of possible models and templates"—from the conceptual (viruses as genes) to the technical (viruses as crystallized proteins)—which themselves became a "set of resources for the creative borrowing and elaboration of previously unseen analogies" across diverse and unconnected fields, such as cell biology, cancer research, and bacterial genetics.[21]

Much of this work, however, still locates the value of standardized model organisms in universalized norms of scientific practice, rather than in particular means through which these creatures were first cultivated. As Adele Clarke notes in an insightful early essay: "In order to observe or produce the phenomena they study, all working scientists must obtain and manage research materials."[22] But standardization is often presumed to be an obvious next step in this process, undertaken to manage the

---

niques and Work Organization in Twentieth Century Life Sciences (Princeton: Princeton University Press, 1992), especially the essays on corn by Barbara Kimmelman and on Planaria by Gregg Mitman and Anne Fausto-Sterling; Andrew Pickering, ed. *Science as Practice and Culture* (Chicago: University of Chicago Press, 1992); Pickering, *The Mangle of Practice: Time, Agency, and Science* (Chicago: University of Chicago Press, 1995). For an excellent overview of recent work on model organisms, with careful attention to issues of agency and the material culture of experimental more broadly, see chapter 8 of Angela N. H. Creager, *The Life of a Virus: TMV as an Experimental Model, 1930–1965* (Chicago: University of Chicago Press, 2001). For a scientific perspective on model organisms, see the "Biology's Models" special issue of *New Scientist*, vol. 17, sup. 1 (5) (2 June 2003).

[21] Kohler, *Lords of the Fly*; Creager, *Life of a Virus*, pp. 6, 328–29.

[22] Adele E. Clarke, "Research Materials and Reproductive Science in the United States, 1910–1940," in *Physiology in the American Context, 1850–1940*, ed. Gerald L. Geison (Bethesda: American Physiological Society, 1987) p. 323. Howard Gest makes a similar point about the term "model" in science: see his "Arabidopsis to Zebrafish: A Commentary on the 'Rosetta Stone' Model Systems in the Biological Sciences," *Perspectives in Biology and Medicine* 39 (Fall 1995): 77–85.

natural "complexity and diversity of living organisms" and thereby simultaneously make the experimental systems in which biologists use them more "productive" while tending to the "practicalities . . . of scientific careers."[23] These assumptions are even made within historical accounts that (unlike scientists' "laundry lists") explicitly acknowledge the importance of local decision making in the development, dissemination, and adaptation of model organisms. Existing narratives beg larger questions about the underlying values motivating the process of adopting standardized animals and other model systems at the bench-top: "Complex" compared to what? "Manageable" and "practical" for whom, and why? "Productive" to what ultimate end?[24]

Recent case studies of standardization in the history and sociology of science stress how—for everything from techniques and instruments to classification and building schemes, and even human organ donation—achieving standards requires intense negotiation over what material, organizational, and conceptual categories can and should be deliberately controlled and therefore taken for granted.[25] Standardized organisms, therefore, need to be reconceived within a broader sociology of technoscientific work. These animals are the result, rather than the cause, of consensus among early twentieth-century experimental biologists, and a key goal of

[23] Creager, *Life of a Virus*, p. 319; Kohler, *Lords of the Fly*, p. 206.

[24] To counteract what he sees as a "hegemony of theory" in the social studies of science, Hans-Jorg Rheinberger has developed an epistemology of experimentation that treats research as a process simply for producing "epistemic things." See *Towards a History of Epistemic Things: Synthesizing Proteins in a Test Tube* (Stanford: Stanford University Press, 1997). My account counteracts what I see as another hegemony, namely, that of presumed universal scientific values.

[25] With regard to biology and standardization, see Kathleen Jordan and Michael Lynch, "The Sociology of a Genetic Engineering Technique: Ritual and Rationality in the Performance of the 'Plasmid Prep' "; Patricia Peck Gossel, "The Need for Standard Methods: The Case of American Bacteriology"; and Peter Keating et al., "The Tools of the Discipline: Standards, Models, and Measures in the Affinity/Avidity Controversy in Immunology," all in *Right Tools*, ed. Clarke and Fujimura, pp. 77–114, 287–311, 312–56, respectively. On the broader philosophical and sociological implications of standardization for science and technology, see Joan H. Fujimura, "Crafting Science: Standardized Packages, Boundary Objects and 'Translation,' " in *Science as Practice and Culture*, ed. Pickering, pp. 168–211; Theodore Porter, "Objectivity as Standardization: The Rhetoric of Impersonality in Measurement, Statistics and Cost-Benefit Analysis," in *Rethinking Objectivity*, ed. Allan Megill (Durham: Duke University Press, 1994), pp. 197–237; Linda Hogle, "Standardization Across Non-Standard Domains: The Case of Organ Procurement," *Science, Technology, and Human Values* 20, 4 (1995): 482–501; Geoffrey Bowker and Susan Leigh Star, *Sorting Things Out: Classification and Its Consequences* (Cambridge: MIT Press, 1999); Amy E. Slaton, *Reinforced Concrete and the Modernization of American Building, 1900–1930* (Baltimore: Johns Hopkins University Press, 2001).

this book is to map the historical credibility of the notion of "standardization" through the story of the mouse's development in the laboratory. What have been the values of those who pursued standardization of laboratory animals, and to what end? How did they convince others of the rightness of their methods and goals?[26]

As Kohler notes, one common-sense understanding of "standard" is simply "the things that everybody uses."[27] This definition plays on the primary meaning of the word, dating from fifteenth-century debates over weights and measures: a standard is an exemplar, an object or quality that serves as the authorized basis or principle to which others conform or by which they are judged.[28] In experimental biology, the material and practical aspects of "standardization" are synchronic, regardless of which is primary. For example, widespread research use of a species (one important practical concern) correlates with the extent to which that organism is first available or capable of being produced in large numbers (two key material achievements). In the case of genetically standardized mice, the number of animals in circulation started rising in the 1930s, and the creatures now represent (along with genetically standardized rats) at least 70 percent of all animals used in research.[29]

Still, in retrospect, the early success of the inbred mouse was underdetermined at the level of research practice; that is, its initial users did not necessarily commit to the genetic framework of experimentation in order to utilize this animal as a meaningful research tool. As Rachel Ankeny argues in her study of the nematode worm, *C. elegans*, model organisms

[26] The notion of mapping historical credibility I take from Simon Schaffer, "Accurate Measurement Is an English Science," in *The Values of Precision*, ed. M. Norton Wise (Princeton: Princeton University Press, 1995), p. 136 ("Undoubtedly a Victorian value, precision badly needs a cultural history which maps its historical credibility rather than assuming its methodological validity.") In this pursuit, my work is deeply indebted to earlier sociologists of standardization, in particular "social worlds" theorists. See introduction to Joan Fujimura, *Crafting Science: A Sociohistory of the Quest for the Genetics of Cancer* (Cambridge: Harvard University Press, 1996).

[27] Kohler, *Lords of the Fly*, p. 14.

[28] On the history of the word "standard," see the *Oxford English Dictionary*, which traces this usage to a 1429 parliamentary debate. For some interesting reflections on the history of standardization more broadly—especially regarding early French debates over state standardization of military production—see Ken Alder, *Engineering the Revolution* (Princeton: Princeton University Press, 1997).

[29] The numbers of animals are from the USDA/APHIS census of 1983 because the more recent census has not been released to the public. See U.S. Congress, Office of Technology Assessment, *Alternatives to Animal Use in Research, Testing, and Education* 5 (Washington, DC, 1986). On the standardization of the rat, which followed a parallel but very different scientific path from the mouse, see Bonnie Tocher Clause, "The Wistar Rat as a Right Choice: Establishing Mammalian Standards and the Ideal of a Standardized Mammal," *Journal of the History of Biology* 26 (Summer 1993): 329–49.

often serve as important vehicles of problem clarification—either for exploring new lines of work or for promoting a particular approach to biological work that is transferable *across* many different areas of research. Little clearly envisioned the inbred mouse as a vehicle for the latter—he sought to promote genetic approaches to *all* biomedical research—but both intellectual and practical constraints limited the straightforward achievement of this vision. The current meaning of gene (as functional piece of DNA) had not yet emerged in 1938, when Little gave a conference talk on "Some Contributions of the Laboratory Rodent to Our Understandings of Human Biology," and even when it did emerge, chromosomal manipulation in mammals proved technically impossible before recombinant DNA in the 1970s. One important historical question, then, is how did the standardization of the mouse at the locus of the gene become second nature (materially speaking) for biologists during a time when (practically speaking) precision control of mouse genes could not be experimentally achieved? In other words, how did the genetically standardized mouse initially succeed as a standard organism when mammalian genetics, the very science for which it was supposedly best designed, initially did not?[30]

My analysis attempts to resolve this paradox by resurrecting an even earlier meaning of "standard," originating in medieval warfare: a conspicuous object, such as a banner, carried at the top of a pole and used to mark a rallying point.[31] Of course, the ubiquity of these animals was what would literally make them conspicuous—as Little himself put it, mice had "served the avid maul of genetic researchers long and well" enough by 1938 to be granted a "titular partnership at a scientific meeting."[32] But while geneticists collected and developed more than fifty years' worth of genetically known mouse strains, without being able fully to exploit these materials for their own analyses, researchers found alternative uses for the animals. Some of these uses, such as tumor transplant studies, Little himself promoted, but others, such as the specific locus test, he did not and could not have imagined when he first began inbreeding the creatures. JAX mice,

[30] Steward Brand, *How Buildings Learn: What Happens to Them After They're Built* (New York: Penguin, 1995), p. 2.

[31] The 1989 OED notes that in 1138 the "standard" was so-named from " 'stand' because, it was there that valour took its stand to conquer or die." For science studies scholars, this rhetoric of warfare will invoke Bruno Latour's version of actor-network theory—see *The Pasteurization of France*, trans. Alan Sheridan and John Law (Cambridge: Harvard University Press, 1988)—although this is appropriate because this usage resonates with the rhetoric the historical actors themselves employ to describe the mouse's usefulness—e.g., for "the war on cancer." Cf. Wise, "Introduction," in *The Values of Precision*.

[32] C. C. Little, "Some Contributions of the Laboratory Rodent to Our Understanding of Human Biology," *American Naturalist* 73 (1939): 127–38.

then, functioned less as static research tools guaranteeing the dominance of a particular line of work than as ever-present totems of the genetic approach in American experimental biology. Because they were standardized at the locus of the gene, considerations of how their hereditary constitutions shaped results became a material part of all work in which they were used. But whether genetic considerations were made explicit by mouse researchers is itself a historically specific phenomenon that cannot be explained away at the level of what was technically possible or impossible. As genes have become increasingly valued entities in both American biology and American culture over the last hundred years, inbred laboratory mice have become increasingly valued for their ability to measure genetic effects.[33] My claim, then, is that the emergence of standard research organisms reflects changing social and disciplinary ecologies of knowledge. Genetically standardized mice were the standard-bearers for a genetic approach to biomedicine; their production represented, to paraphrase Karl Marx on technology, the power of genetic knowledge objectified.[34]

In making this argument, I do not mean to exaggerate the homogeneity of biological and medical research practices. In the early twentieth century, just as now, even biologists who openly embraced a genetic approach did not always agree on what Frederick Winslow Taylor called "the one best way."[35] Over the last hundred years researchers have argued frequently and vehemently over which kind of organisms are the right tools for research, the most famous sound bite from these debates being Jacques Monod and François Jacob's bold declaration: "anything found to be true of *E. coli* must also be true of elephants."[36] These controversies reflect genuine epistemological and methodological disagreement within the

[33] Economists and historians of mathematics often make this point about measurement more generally. See Ann Jennings, "The Social Construction of Measurement: Three Vignettes from Recent Events and Labor Economics," *Journal of Economic Issues* 35, 2 (2001): 365–71 (thanks for Marilyn Power for this reference): Theodore M. Porter, *Trust in Numbers: The Pursuit of Objectivity in Science and Public Life* (Princeton: Princeton University Press, 1995).

[34] "Nature builds no machines, no locomotives, railways, electric telegraphs, self-acting mules, etc. These are the products of human industry; natural material transformed into organs of the human will over nature, or of the human participation in nature. *They are organs of the human brain, created by the human hand*; the power of knowledge objectified." Karl Marx, *Outlines of the Critique of Political Economy*, 1857–61, trans. Martin Liclaus (New York: Penguin Books, 1973), p. 706. (Thanks to Shahnaz Rouse for pointing out this connection.)

[35] Cf. Robert Kanigel, *The One Best Way: Frederick Winslow Taylor and the Enigma of Efficiency* (New York: Viking, 1997).

[36] Jacques Monod and Francois Jaçob, "General Conclusions: Teleonomic Mechnisms in Cellular Metabolism and Growth," *CSH Symposium on Quantitative Biology* 26 (1961): 393.

community of practicing biologists. Should model organisms be simplified models of real life, as bacterial geneticists and C. *Elegans* workers often argue, or bits of real, complex life on which we can experiment, as mouse workers claim? Where individual biologists come down on these questions depends in large part on what jobs they think the discipline of biology itself should be doing. In 1975, for example, physiologist Hans Krebs championed the "unique characteristics" rationale articulated by August Krough in 1929 for choosing laboratory organisms: "for a large number of problems there will be some animal of choice . . . on which it can be most conveniently studied." But while Krough originally saw this as reason to pay attention to zoological diversity, Krebs assumed such diversity was "an exception, against a background of presumed generality."[37] Likewise, Gunther Stent has argued that in the 1960s a whole generation of young molecular biology researchers turned back to larger organisms and problems of development, largely because they believed that all the simple problems of the field had already been solved and it was time to move onward and upward (simultaneously in the chain of being and in their professional standing).[38]

Historically, however, model organism debates among biologists have overshadowed the equally important points of consensus that necessarily existed among biologists who worked on different organisms, as well as between the scientists who created and used standardized organisms and those supporting their work. Sometimes these concordances in values found voice, such as in government and foundation policy debates over what kinds of biological research, and therefore what kinds of organisms, to fund in postwar biomedical research. Nearly every scientist and policymaker in the late 1940s and 1950s—from those working on mice at the National Cancer Institute to those leading the Atomic Energy Commission's biological research projects in *Drosophila* and Paramecium—agreed that some sort of laboratory animal research was necessary to establish valid claims about everything from human cancers to genetic "fallout." The key question they disagreed on was: what kind would prove the most convincing? Scientific values of experimental proof had to be reconciled with political values of expediency and overall research coordination, but more often than not, even this was achieved without rancor.

[37] For a trenchant historical analysis of the Krough principle, see Cheryl A. Logan, "Before There Were Standards: The Role of Test Animals in the Production of Empirical Generality in Physiology," *Journal of the History of Biology* 35 (2002): 329–63, quote on 329–30.

[38] Soraya de Chadarevian, "Of Worms and Programmes: 'Caenorhabditis elegans' and the Study of Development," *Studies in the History and Philosophy of the Biological and Biomedical Science* 29 (1998): 81–105.

As Evelyn Fox Keller has argued, this scenario is not surprising: "by scientists' own internal ethic," results "must provide at least some predictive success to remain satisfying; and by the social and political ethic justifying their support, this predictive success must enable the production of at least some of the technological 'goods' the public thinks it is paying for." But more often than not, consensus in mouse research passed silently, either because the only historical actors to whom it would have seemed controversial literally had no voice—such as the mice who were being experimented on—or because the consensus among the human beings involved was so great that it seemed not to require comment. A final goal of this book, then, is to reinstate some of these "critical silences" in the consensual historical discourse of developing model organisms—in particular, those related to the scientific and social values informing animal research.[39]

For the historian, this approach presents a problem of sources as well as interpretation. One reason why scientific and technological controversies are so well-studied is that the day-to-day consensus assumptions that structure work in these fields are much more difficult to document. There are precious few "smoking guns" to be found that enable one to pinpoint the origin and development of agreement in areas where there was never any questioning of received values.[40] Nevertheless, agreed-upon values were important resources that biologists and policymakers used to articulate what jobs needed doing with laboratory mice and how they should do them. It is not just a coincidence, for example, that the initial funding for the development of the Jackson Lab's "mouse factories" came from Detroit car makers, who embraced the ethos of mass production. Nor is it immaterial that Little could speak, in a 1937 *Life* article, of mice as "Replac[ing] Men on the Cancer Battlefield" without fear of retribution from animal welfare groups.

In turn, I consider standardized laboratory mice not only as artifacts of particular knowledge-making activities but as what Ian Hacking has called *forms* of scientific knowledge—"what is held to be thinkable, or

[39] Evelyn Fox Keller, "Critical Silences in Scientific Discourses: Problems of Form and Re-Form," in *Secrets of Life, Secrets of Death* (New York: Routledge, 1992), pp. 91, 85. My historical approach to this problem neglects, but does not deny, the mice themselves as important "silent actors"; for a sociological analysis (and problematization) of agency in animals and other actors of the natural world, see Michel Callon, "Some Elements of a Sociology of Translation: Domestication of the Scallops and the Fisherman of St. Brieuc Bay," in *Power, Action, Belief: A New Sociology of Knowledge?*, ed. John Law (New York: Routledge and Kegan Paul, 1986), pp. 196–229.

[40] Pam Scott, Eveleen Richards, and Brian Martin, "Captives of Controversy: The Myth of the Neutral Social Researcher in Contemporary Scientific Controversies, " *Science, Technology, & Human Values* 15, 4 (Fall 1990): 474–94.

possible, at any given moment"—and my account strives to recover historical traces of what linguistic, material, and conceptual potentials were tapped while decisions about these animals' development and use were being made, both inside and outside the laboatory.[41] Ultimately, I argue that the usefulness of standardized laboratory mice was settled through consensus, not necessarily over the creatures' uses in specific experiments (although there was some of this), but over their ability to negotiate two key underlying tensions in life-science work of the early twentieth century.

The first tension is that between natural and technological systems in the realm of biological experiment. Vast controlled breeding populations of inbred mice, like many other living tools of basic biological research, would not exist save for the efforts of human scientists, so one key question facing those who used them involved how much of the knowledge obtained reflects aspects of mouse biology as it really is, and how much is an artifact of the experimental system. This question was raised repeatedly by the scientific actors themselves—sometimes by those with research programs that competed with Little's genetic vision (e.g., Maud Slye's pedigree approach to mouse cancer research during the 1920s), and other times by those who shared it (e.g., William Russell's specific locus test in radiation mutation studies during the 1950s). My narrative pays close attention to such liminal moments in this debate because they are the points at which scientific consensus was literally articulated. But it is also important for the larger argument of this book to note that the natural truths about the animal under discussion were framed by consequences beyond the laboratory. In the case of the Slye-Little debate, at stake was the institutional and intellectual framework for future cancer research funding: centrally coordinated, theoretically informed projects favored by Little and most experimental biologists, or locally centered, clinically framed case studies preferred by Slye and most medical researchers. In the case of Russell, it was the developing relationship between basic biological research and radiation policy-making in postwar America: would researchers themselves play a direct role in making policy, or would they merely present their data? These contexts, then, defined the intellectual, political, and social possibilities within which mouse users staked out their knowledge claims. The relatively peaceful coexistence of different mouse meanings within and between them suggests that persistent heterogeneity can stabilize (rather than undermine) projects to standardize scientific knowledge and the tools of its practice.[42]

[41] Ian Hacking, "Weapons Research and the Form of Scientific Knowledge," *Canadian Journal of Philosophy*, supp. vol 12 (1987): 237–60, quote on 243.

[42] A similar point with regard to laboratory techniques is made by Jordan and Lynch in "The Sociology of a Genetic Engineering Technique."

The second tension is that between human and animal subjects in the realm of American culture. This is an especially interesting theme that deserves its own history, especially in reference to scientific experimentation.[43] What animals are enough like us to make laboratory results obtained from them generalizable to humans, but not so much like us that we ethically prohibit their being the subjects of experiments? Social assumptions have shaped scientific considerations and uses of animals in this regard—particularly for dogs and other sentimentally valued creatures. Diane Paul details how the Rockefeller Foundation's Alan Gregg launched a postwar study of "Genetics and Social Behavior" in dogs because he sought to demonstrate to educators and doctors, the heritability of behavioral traits but recognized that existing work "had been demonstrated in organisms—such as fruit flies and rats—to which few persons could relate to emotionally." Also, Susan Lederer demonstrates that medical researchers at the Rockefeller Institute for Medical Research eliminated full-body photographs of laboratory dogs, as well as textual reference to their names, in published journal articles; in these ways, she argues, they achieved "the invisibility of the 'naturalistic' animal" and therefore shielded themselves from the criticisms of animal welfare advocates.[44] Occasionally, this tension was explicitly articulated or referenced by historical actors in the mouse's story—for example, in some of Little's many public appeals for mouse cancer research, or in debates by the Biological Effects of Atomic Radiation (BEAR) committee over radiation data in seven-locus mice versus Japanese atomic bomb survivors—and my analysis exploits these rare self-conscious reflections. But I also formulate this part of my argument comparatively, referencing existing case studies of other extrapolation debates in animals whenever possible, in lieu of referencing a broader cultural and conceptual history of the use of animal subjects in science that has yet to be written.[45]

Finally, a few words on my view of biologists who use mice and their work. Laboratory mice occupy a prominent place within recent biomedi-

[43] Susan Lederer's work on the history of human experimentation, *Subjected to Science*, is very attentive to this theme.

[44] Diane Paul, "The Rockefeller Foundation and the Origins of Behavioral Genetics," in *The Expansion of American Biology* (New Brunswick: Rutgers University Press, 1991), p. 273; Susan Lederer, "Political Animals: The Shaping of Biomedical Research Literature in Twentieth Century America," *Isis* 83 (1992): 61–79.

[45] Some more recent sociological monographs make important contributions to this goal: see, for example, Eileen Crist, *Images of Animals: Anthropomorphism and the Animal Mind* (Philadelphia: Temple University Press, 1999); Arnold Arluke and Clifford Sander, *Regarding Animals* (Philadelphia: Temple University Press, 1996). For a historical approach, see Daniel Todes's masterful account of Pavlov's dog experiments: *Pavlov's Physiology Factory:*

cal success stories: contemporary Americans need only open a major daily newspaper or listen to a national television news program to encounter the work now being done with these rodents—as animal models for the genetics of Alzheimer's disease or intelligence, to cite just a few recent reports. Perhaps in response to such developments, North American animal rights and environmental activists have begun to pay attention to mice in laboratories. In 1999 the Animal Legal Defense Fund called for a revision in the definition of "animal" covered by the 1972 version of the Animal Welfare Act, citing the "arbitrary" discretion exercised by the then-secretary of agriculture not to include the mice and rats in scientific laboratories among those animals covered when he first administered the act in 1966. Despite winning this suit, the U.S. Senate quietly approved a measure in February 2002 that would eliminate the federal funding necessary to change the regulations, setting the stage for a "tough battle" between university scientists and animal rights activists.[46] Likewise, in 2000 the Canadian Patent Office filed an appeal on the Canadian Federal Court of Appeal's decision to allow Harvard researchers a patent on Oncomouse in 1986. The commissioner of patents argued that the current Patent Act does not permit patenting higher life forms such as plants and animals, and in its most recent ruling on the case, the Supreme Court of Canada (by a 5–4 decision) agreed with him.[47]

My interest in documenting how standardized laboratory animals came to be is both academic and political, but not condemnatory. That current resistance to C. C. Little's "new deal for mice" took decades to materialize speaks to how mouse use in science has indeed become a "black box," in both science and society, but perhaps now is the perfect time to reopen that box.[48] How researchers and their constituencies determine what scientific things—objects, methods, theories—can be taken for granted reveals something very important about the nature of their work, as well as about received cultural values. My hope is that by returning to a time

---

Experiment, Interpretation, Laboratory Enterprise (Baltimore: Johns Hopkins University Press, 2002).

[46] See "Animal Welfare: A Petition for Rulemaking," *Federal Register* 64, 18 (28 January 1999): 4356–67. Cf. Ron Southwick, "Senate Votes to Block Expansion of Lab-Animal Regulations," *Chronicle of Higher Education*, 1 March 2002, p. 25; "Researchers Face More Federal Scrutiny on Animal Experimentation," *Chronicle of Higher Education*, 28 June 2002, p. 23.

[47] "Harvard College v. Canada (Commissioner of Patents)," 2002 Supreme Court of Canada 76, file no. 28155; 2002: May 21; 2002: December 5.

[48] As Geoffrey Bowker and Susan Leigh Star suggest in their work on standardization, "Black Boxes are necessary and not necessarily evil. The moral questions arise when the categories of the powerful become taken for granted." See Bowker and Star, *Sorting Things Out*, p. 320.

when the existence and use of these creatures first took shape, especially to Little's prescient vision of mouse use, we may learn more about how human agency shapes the course of science. In this way, we can better appreciate the scientific knowledge obtained from mice for what it is (as well as what it is not) and perhaps even begin envisioning new ways to make biomedical science a livable and workable space for all animals—human and nonhuman—to inhabit.

CHAPTER ONE

MICE, MEDICINE, AND GENETICS

From Pet Rodents to Research Materials (1900–21)

As a member of Harvard's class of 1910, Clarence Cook Little (1888–1971)—Paul Revere's great great grandson—busied himself doodling mice in the margins of his notebooks and preaching the value of science. "Pete" Little (as his friends called him) was a track star (fig. 1.1), but he was equally well known for being an enthusiastic student of the life sciences. He decided to study zoology with William Castle (1867–1962), one of the first American adherents of Mendelism, and immersed himself in its practical study. Castle would later recall later how "this handsome [track] team captain signed up for his genetics course and soon had persuaded most of the team to sign up with him." Little initially wanted to follow in his father's footsteps and breed dogs, now for genetic study, but Castle easily dissuaded him: in Little's description, his professor "skidded a live mouse at me across the benchtop and instructed me to learn everything about it" (fig. 1.2). In Little's sophomore year, Castle put him in charge of tending his colony of mouse mutants, and here, Little's interest in these animals took permanent hold.[1]

An earlier version of this chapter was published in *Studies in the History and Philosophy of Biology and the Biomedical Sciences* 30 (1999): 319–43.

[1] CCL's Zoology Notebook (c. 1910), pp. 80–81, Box 726, CCL-UMO. Castle incident quoted from Roberta Clark, "The Social Uses of Scientific Knowledge: Eugenics in the Career of Clarence Cook Little, 1919–1954," (M.A. thesis, University of Maine, Orono, 1986), p. 59, and is based on a letter she found in CCL-Mich from Little to colleague. I

1.1 C. C. Little (class of 1910) at a Harvard track meet [Source Credit: C. C. Little Papers, University of Maine].

But while Little was learning Mendel's laws and changing the ammonia-scented wood shavings of his rodent charges, "a great deal of stock-taking" was going on among biologists regarding the reform of their discipline. The year 1909 marked the one hundredth anniversary of the birth of Charles Darwin, as well as the end of an exciting decade of genetic research. "Genetics" as a distinct academic specialty was still very

have not yet been able to locate this letter in the Michigan files. See also George D. Snell, "Clarence Cook Little," *Biographical Memoirs, National Academy of Sciences* 46 (1975): 240–63; Cf. W. E. Castle 1954, pp. 598–99; George D. Snell, "Clarence Cook Little," *Dictionary of Scientific Biography*, vol. 7, ed. F. L. Holmes (New York: Scribner's, 1990), pp. 562–64. On C. C. Little's father and dog shows, see interview with Joan Staats, Bar Harbor, Maine, 21 June 1993, JLOH-KR. On Little raising animals, see Snell, "Clarence Cook Little," 1975, esp. p. 40. An excellent analysis of C. C. Little's childhood can be found in Clark, "Social Uses," chap. 2.

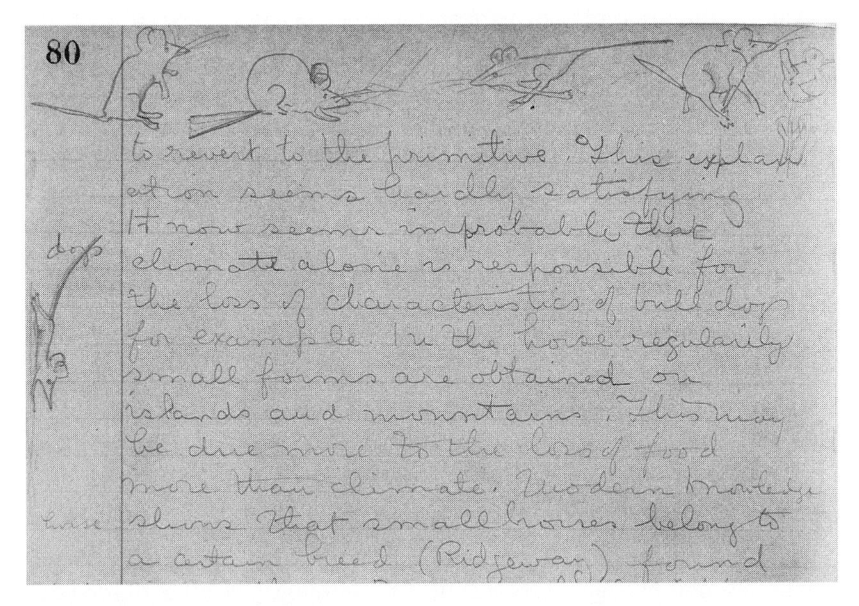
80

dogs

to revert to the primitive. This explanation seems hardly satisfying
It now seems improbable that
climate alone is responsible for
the loss of characteristics of bull dogs
for example. In the horse regularly
small forms are obtained on
islands and mountains. This may
be due more to the loss of food
more than climate. Modern knowledge

horse

shows that small horses belong to
a certain breed (Ridgeway) found

1.2. C. C. Little's doodlings of mice in his Harvard zoology notebook [Source Credit: C. C. Little Papers, University of Maine].

young—William Bateson coined the term only in 1906, and one could count on one hand the number of places offering graduate training in the field—but in its early years it had shown great intellectual and political potential. By 1910 biologists saw the validation of Mendel's laws in nearly every kind of animal and plant, and the idea of applying these principles to control human breeding caught on among educated and wealthy laypeople, some of whom founded the first American Eugenics Society that same year. Mendelian genetics was not the only face of biology in the period from 1880 to1920, but it featured prominently among the issues and approaches that defined the study of heredity as an academic discipline.[2]

[2] Quote from L. C. Dunn, "Testing Mendel's Rules," in *A Short History of Genetics* (New York: McGraw-Hill, 1965), p. 81. On what constituted "core" issues in early American biology, see Ronald Rainger, Keith Benson, and Jane Maienschein, "Introduction," in *The American Development of Biology* (Philadephia: University of Pennsylvania Press, 1988), pp. 3–14). By 1910, genetics was widely taught at American colleges and universities, and there were many American centers of graduate training in zoology that emphasized experimental methods, but only a few places had established a research base in Mendelism in this early period, and most of these were at the agricultural colleges. Cf. Barbara Kimmelman, "A Progressive Era Discipline: Genetics at American Agricultural College and Experiment Stations, 1900–1920," Ph.D. dissertation, University of Pennsylvania, 1987. By the end of the second decade of the twentieth century, Johns Hopkins (with H. S. Jennings and Raymond Pearl) and Columbia (with T. H. Morgan and E. B. Wilson) emerged as university

Around the same time, American humane societies renewed their attacks on another group of scientists who also worked with animals and were seeking new disciplinary legitimation: medical researchers. Earlier national efforts to regulate vivisection had met with little public support, but local opposition to animal experimentation was durable, and medical research work in cities like New York and Boston appeared genuinely threatened. In 1907–08 the American Medical Association established the Council for Medical Research, chaired by Harvard physiologist Walter B. Cannon. At the Rockerfeller Institute in New York, prominent bacteriologist Simon Flexner launched several media campaigns to prevent the passage of antivivisection bills (in 1910 and 1911) in the New York and New Jersey legislatures. Flexner himself questioned the "good faith of the entire attack"and sought to discredit antivivisectionist witnesses while at the same time arguing that "it is an imposition upon a lay public . . . to parade as wanton cruelty when applied to brute creation the very means which are in daily use to save life and limb in the best hospitals in this country."[3]

Such public criticism came at a moment when American medicine, like American biology, faced a series of internal reforms. In 1910 Abraham Flexner (Simon Flexner's brother) filed his landmark critique of medical education, in which he affirmed the importance of experimental research as a necessary component of training for physicians. Poorly equipped to sustain such programs—in terms of both personnel and facilities—U.S. medical schools experienced a crisis of confidence. Many closed (thirty between 1900 and 1910 alone) and those that remained scrambled to create new institutional structures (laboratories) and relationships (departments) that would sustain them. As Robert Kohler argues in his disciplinary history of biochemistry, "not since the reform of the German universities a half a century before had there been opportunities on such a scale for the creation or recreation of biomedical disciplines" in America.[4]

---

centers of genetics. On the complicated early history of Mendelism, see Jan Sapp, "The Struggle for Authority in the Field of Heredity, 1900–1932: New Perspectives on the Rise of Genetics," *Journal of the History of Biology* 16, 3 (Fall 1983): 311–42. Sapp argues that the genotype-phenotype distinction was instrumental in excluding non-Mendelian players from the emerging field of hereditary studies.

[3] On the persistence of the animal welfare movement in the early twentieth century, see Susan E. Lederer, "The Controversy over Animal Experimentation in America, 1880–1914," in *Vivisection in Historical Perspective*, ed. N. Rupke (New York: Routledge, 1987), 236–55). See also Lederer, "Political Animals," p. 62; George W. Corner, *A History of the Rockefeller Institute, 1901–1953: Origins and Growth* (New York: Rockerfeller Institute Press, 1964), pp. 86–87; "To Regulate Vivisection," *New York Times*, 3 February 1910, p. A1. Flexner quote from "Dr. Flexner Denies Cruel Vivisection," *New York Times*, 17 January 1910, p. A1.

[4] Paul Starr, *The Social Transformation of American Medicine* (New York: Basic Books, 1982) pp. 116–23. See also Daniel Fox, "Abraham Flexner's Unpublished Report: Founda-

In this shifting climate, Little found and pursued several opportunities to develop mice as genetically stable research organisms for both the biological and medical sciences. Within the conceptual and institutional frameworks provided by Harvard's Bussey Institution, he worked with mouse fanciers and their pet animals to establish and stabilize several inbred mouse strains for use in mammalian genetic research. His eugenic convictions combined with accidents of timing to gain him a position doing genetic research at the Harvard Medical School on mouse cancers, where in the subsequent development of inbred tumor mice Little and Ernest Everett Tyzzer created additional cancerous strains and made some of the earliest conceptual connections between biological and medical research.[5] But in 1919 Little returned to his Bussey roots and, together with E. C. MacDowell at the Cold Spring Harbor Station for Experimental Evolution, established large mutant strain colonies and communication networks among mouse geneticists, including the *Mouse Newsletter* (continued today as *Mammalian Genome*).

In all of this work, Little drew on reform ideologies—from biology, on the ethos of genetics as rigorous experimental method, and in medicine, from the drive for research-based medicine. Ultimately, however, his early success with mice as research organisms rested on some key scientific and social values remaining in tact. Little increased the material availability of the inbred mice and defined new scientific uses for them in genetics, which made the animals easier to obtain for research and gave them a scientific purpose, but he did so without violating existing cultural attitudes toward them as animals. In these ways, Little's actions transformed pet-fancier mice into laboratory mice without entering these creatures (or their scientist-users) into politically charged animal welfare debates.

MICE AT HARVARD'S BUSSEY INSTITUTION

American mouse genetics was born at Harvard's Bussey Institution.[6] Here, Little's own career first took shape in an unusual academic niche that provided important intellectual and practical resources. Here he established

---

tions and Medical Education, 1909–1928," *Bulletin of the History of Medicine* 54 (Winter 1980): 475–96; Robert Kohler, *From Medical Chemistry to Biochemistry: The Making of a Biomedical Discipline* (New York: Cambridge University Press, 1982), p. 121.

[5] On physiology's earlier disciplinary and institutional efforts to connect medicine and biological research, see the essays in Gerald Geison, ed., *Physiology in the American Context, 1850–1940* (Baltimore: American Physiological Society, distributed by Williams & Wilkins, 1987).

[6] Clyde Keeler, "How It Began," in Herbert S. Morse III, ed. *The Origins of Inbred Mice* (New York: Academic Press, 1980), pp. 179–93, quote on p. 185.

the laboratory mouse's first material identity by achieving the creation of stable inbred strains. But Little also imparted to these animals a scientific identity within a particular line of genetic work: at once, as organisms useful for investigating the basic Mendelian genetics of coat colors and the mammalian material achievement of the "pure lines" hypothesis.

From Castle's perspective, the timing of Little's interest in mice could not have been better, and he encouraged it for selfish, as well as pedagogical, reasons. Shortly after 1900, Castle's own research on mammalian sex ratios gave way to a broader consideration of the applicability of Mendelian laws in mammals. Castle bred and observed yellow mice in order to clarify Mendel's assumptions about purity of genetic gametes, while simultaneously beginning the first of a longer series of genetic breeding experiments for coat color in guinea pigs (1903–1907) and in rats (1907–1914).[7] He was searching for a young scientist-in-training to look after these new animal colonies, as it saved him the work (and perhaps even expense) of having to keep track of the nonscientifically trained technical assistants. Furthermore, like T. H. Morgan, Castle believed that experience with real organisms was the best way to learn genetics: "I jump at the chance to give them [the students] something concrete and first hand on the subject," he wrote in 1922.[8] So in 1907 Castle put Little in charge of the new mouse colonies, and this charismatic student quickly became known as Castle's mouse man.

Castle had planned to expand the scale and scope of his animal breeding efforts considerably when he moved in 1908 from Harvard's Cambridge campus to larger quarters at the Bussey Institute for Applied Biology, but this new space turned out to be advantageous in more ways than one. Formerly Harvard's undergraduate school of husbandry and agriculture, the Bussey had just reorganized as the "graduate school of applied biology related to agriculture and horticulture."[9] The main Bussey building was located in Forest Hills, Massachusetts, about ten miles from the main Cambridge campus, in the fields adjacent to Harvard's Arnold Arboretum. It had a neighboring farm with barns, outhouses, and even a make-

[7] W. E. Castle et al., "The Effects Inbreeding, Cross-Breeding, and Selection upon the Fertility and Variability of Drosophila," *Proceedings of the American Academy of Arts and Sciences* 41 (1906): 729–86; Garland Allen, "William Ernest Castle," in *Dictionary of Scientific Biography*, vol. 3, ed. C. Gillispie (New York: Scribner's, 1970), pp. 120–24; L. C. Dunn, "William Ernest Castle," *Biographical Memoirs, National Academy of Sciences* 38 (1962): 31–80, esp. 40–41.

[8] On Morgan as "unable to judge his colleague's experiments unless he knew from first hand experience how the experiments had been done," see Kohler, *Lords of the Fly*, p. 27. Castle to L. C. Dunn, 21 February 1922, LCD-APS. Castle spoke these words in the context of hearing that there would be a dog breeders' show in town during his genetics seminar class.

[9] W. E. Castle, *Mammalian Genetics* (Cambridge: Harvard University Press, 1940); cf. Karen A. Rader, " 'The Mouse People': Murine Genetics Work at the Bussey Institution of Harvard, 1910–1936," *Journal of the History of Biology* 31, 3 (Fall 1998): 327–54.

shift dormitory for the students. Moreover, the physical distance from Cambridge allowed the Bussey faculty to run de facto independent research groups, still under Harvard's fiscal umbrella but separate from the zoology department and therefore organized according to their own training methods and research priorities.[10] As geneticist L. C. Dunn later described Castle's new circumstances: "Here for the first time, Castle had the space and the freedom to develop an extensive program in mammalian genetics [and] none exceeded Castle in the vigor and pertinacity with which he pursued these questions." Shortly after arriving, Castle established a mouse room in what had been a greenhouse attached to the main Bussey building.[11] Both intellectually and practically, an animal breeding program of this magnitude would have been impossible within the confines of the department of zoology at the time—it demanded too much space, and genetics had strong links to agriculture while Harvard administrators were growing increasingly interested in using biology to develop strategic links between the college and the medical school.[12]

Years later Little told a correspondent that taking care of these mice transformed his own idea of the creature's usefulness for genetics: "I would say that it was the fact that Dr. Castle left me alone with a large number of mice and that the mice themselves provided a sufficient number of unknown factors to become interesting, which really focused and crystallized my enthusiasm."[13] But Little's enthusiasm for mice also drew strength from the hopes of fellow geneticists that mouse genetics promised greater utility for understanding human heredity. L. C. Dunn, a contemporary graduate student at the Bussey, would later reflect: "I don't think that human heredity was ever very far out of my mind while I was doing . . . things with the mouse."[14] Castle, the author of the first textbook on genetics and eugenics, would have encouraged such formulations.

[10] Castle's initial grant was $1,500: W. E. Castle to C. B. Davenport, 10 February 1908, CBD-APS. Cf. Karl Sax, "The Bussey Institution: Harvard's Graduate School of Applied Biology, 1908–1936," *Journal of Heredity* 57 (1966): 175–78.

[11] Dunn quote from George D. Snell and Sheldon Reed, "William Ernest Castle, Pioneer Mammalian Geneticist," *Genetics* 135 (April 1993): 751–53. Castle is not the scientist first credited with extending Mendelism to the animal kingdom: French biologist Lucien Cuenot began cross-breeding mice to analyze the heredity of pigmentation around 1900. About the same time, Englishman William Bateson conducted analogous research on guinea pigs and hens. On Cuenot's work, see Andree Tetry, "Lucien Cuenot," in *Dictionary of Scientific Biography*, vol. 3, ed. C. Gillispie (New York: Scribner's, 1970) pp. 492–94, and the entry for "Cuenot" in Jacques Ahrweiler, ed., *Écrits sur l'hérédité* (Paris: Editions Seghers, 1964). On Bateson, see Ernst Mayr, *The Growth of Biological Thought* (Cambridge: Belknap Press of Harvard University, 1982), pp. 544–45, 733.

[12] Kohler, *From Medical Chemistry to Biochemistry*, p. 316.

[13] CCL to George Potter, 29 September 1925, Box 3, Folder 6, CCL-Mich.

[14] Another reason for Dunn's choice was the chilly reception he received ("I was virtually ignored") on an undergraduate visit to Morgan's Columbia lab. L. C. Dunn Columbia Oral

The mice that Little cared for, however, were not just any mice: they were fancy mice, obtained through his mentor's connections with mouse-fancier organizations in Boston and beyond. The exact origins of mouse fancying are obscure, though collecting and breeding unique strains of mice in captivity dates as far back as seventeenth-century Japan (fig. 1-3), and geneticists avidly followed the history of this and other cultural practices associated with mice.[15] But the formation of many local and national mouse-fancier organizations in the early 1920s indicates that mouse fancying enjoyed increased popularity in Britain and the United States beginning in the early twentieth century. Mouse fanciers further domesticated *mus musculus* by selecting and preserving rare types with interesting coat colors or behavioral patterns. Judging from the records of an early breeder who sold mice to these murine hobbyists, fanciers prized unique strains and named them accordingly. For example, "cream sables" had milky brown coats, and "waltzer" mice were named for their characteristic "dancing" movement (later found to be the result of an inner ear defect).

Fanciers also selected for certain "standard" physical features and preserved the specimens that exhibited them. As described by a popular magazine article, fanciers thought "the perfect mouse should be seven to eight inches long from nose-tip to tail-tip, the tail being about the same length as the body and tapering to an end like a whiplash."[16] Fanciers most often kept these mice as pets and would travel with them to local or national "mouse shows," which awarded small cash awards to the owners of the most visually unusual and interesting specimens. Later, non-science-oriented mouse breeders emerged with more lucrative commercial interests in mind. In 1930's England, for example, mouse breeders could cash in on the demand for full-length women's coats made of mouse skins, which took four hundred skins and sold for $350.[17] These mouse breeders and fanciers essentially routinized the activity of mouse breeding in captivity well before scientists became interested in the mouse as an experimental

---

History, interview by Saul Benison (New York: Columbia University Oral History Project, 1960), p. 90–91, 41, 853.

[15] Mitosi Tokuda, "An Eighteenth Century Japanese Guide-Book on Mouse Breeding," *Journal of Heredity* 26 (1935): 481–84; cf. Clyde Keeler, "In Quest of Apollo's Sacred White Mice," *Scientific Monthly* 34 (January 1932): 48–53.

[16] See "The English Craze for Mice," *Reader's Digest* 30 (March 1937): 19.

[17] "Mice Beautiful," *Time* 30 (19 July 1937) 50, briefly discusses the American Mouse Fanciers Club, based in Stockport, NY, and its British counterparts; also, "Mouse Show," *Newsweek* 9 (23 January 1937): 40. On British mouse fancying and its associated trade, see "Big Results from a Small Hobby," *Fur and Feather* 69, 1708 (23 February 1923), which claims that one mouse breeder "makes no secret of the fact that last year he made over 100 pounds profit on his stud of mice" (quote from cover of issue). See Herbert S. Morse on Abbie Lathrop's breeding for the mouse fancy, in "Introduction," in *The Origins of Inbred Mice.*

1.3. Illustration of a boy holding an albino mouse from a seventeenth-century Japanese book, entitled *The Breeding of the Curious Varieties of the Mouse,* as described in the 1935 *Journal of Heredity* [Source Credit: *Journal of Heredity,* Oxford University Press].

organism. Fanciers thus provided scientists with both a unique mammalian material resource and a broader practical context in which mouse breeding was an accepted cultural activity.[18] Likewise, because mouse fanciers were the first group to take the systematic, domesticated breeding of this particular mammalian organism seriously, their experience and

[18] Other scientists also benefited from the knowledge and material supply of breeders and fanciers—most famously, Darwin from the pigeon fancy and early twentieth-century corn geneticists from the agricultural plant breeders. See James A. Secord, "Darwin and the Breed-

knowledge about doing so would be of great value to the emerging group of mouse researchers.

Although Castle took occasional foraging trips (for example, to South America) to look for new mammalian varieties, he obtained most of his mice from the Granby Farm in Granby, Massachusetts.[19] Granby Mouse Farm, as it came to be known, was owned and operated by former Illinois schoolteacher Abbie E. C. Lathrop. This institution was started around 1903 as an alternative to Lathrop's failing poultry business. Mice and rats provided an inherently quicker turnover because of their short lifespans and, more importantly, because the "mouse fancy" in the New England area constituted a built-in market.[20] Lathrop purportedly began her trade with "a single pair of waltzing mice," and her stocks gradually increased to 10,000 animals, including many exotic coat color strains. Lathrop's fancy mice, then, were Little's "raw" materials for the creation of laboratory mice.[21]

Access to these variants clearly shaped Little's initial genetic research on mice, which was directed at making Mendelian sense of mouse coat and eye color inheritance. Castle co-authored two papers with Little on mouse coat color genetics in the two years immediately after Little began caring for the animals. First was a 1909 study postulating that the paleness of coat color pigments in pink-eyed mice with black or brown fur was due to a different Mendelian factor from that which governed "diluteness" in other pale-colored mice. Second was a 1910 paper that demonstrated a unique Mendelian explanation of lethal genes in mammals. Castle and Little argued that, in observed crosses with yellow mice (always heterozygous), the missing expected homozygous yellow class was actually formed but simply failed to develop. Lethality, as both a prenatal and postnatal genetic phenomena, would soon be observed in other types of mouse inheritance, such as dominant spotting, which when inherited homozygously produces a form of anemia that so weakens the young that "they are not able to compete for their rightful share of mother's milk" (fig. 1.4).[22]

---

ers: A Social History," in *The Darwinian Heritage*, ed. David Krohn (Princeton: Princeton University Press, 1986), pp. 519–42; Kimmelman, "A Progressive Era Discipline."

[19] Letter from Castle to Dr. Michael Potter, 1958, reproduced in "Introduction" to Morse, *The Origins of Inbred Mice*.

[20] For a brief discussion of the American Mouse Fanciers Club based in Stockport, New York, as well as its British counterparts, see "Mice Beautiful," p. 50.

[21] See article on Lathrop in the *Springfield Sunday Republican*, 5 October 1913, p. 12.

[22] W. E. Castle and C. C. Little, "The Peculiar Inheritance of Pink-eyes among Colored Mice," *Science* 30 (1909): 313–15; "On a Modified Mendelian Ratio among Yellow Mice," *Science* 32 (1910): 868–70; C. C. Little, "A Note on the Fate of Individuals Homozygous for Certain Color Factors in Mice," *American Naturalist* 53 (1919): 185–87. Ultimately,

1.4. Illustration by Clyde Keeler from the 1932 Genetic Congress Program, showing the lethality in the inheritance of dominant-spotted mice [Source Credit: *Journal of Heredity,* Oxford University Press].

However, the fact that Little and Castle could collaborate with fanciers—a then-prominent group of animal producers and pet owners—also highlights the kind of cultural boundaries that separated mice from other species of pet animals. In early twentieth-century medical research, debates about the scientific uses of animals often pitted "innocent" dog and cat owners against "malicious and greedy" scientists, some of whom were purported to have scanned neighborhoods for pets on which to conduct their research. But the ethical yardstick by which antivivisectionists determined felines and canines to be the most threatened experimental animals

Haldane showed albino and pink-eye to be linked traits: see J.B.S. Haldane, A. D. Sprunt, and N. M. Haldane, "Reduplication in Mice," *Journal of Genetics* 5 (1915): 133–35. As Andree Tetry has noted, Lucien Cuenot crossbred yellow mice and first observed the phenomenon of lethality, but "he attributed the impossibility of obtaining pure yellow mice to selective fertilization" (Tetry, "Lucien Cuenot"). Years later, embryological evidence would confirm Castle and Little's hypothesis about the yellow mice: see Herman Ibsen and Emil Steigleder, "Evidence for the Death in Utero of the Homozygous Yellow Mouse," *American Naturalist* 53 (1919): 185–87. My thanks to Lee Silver for helping me sort through this early history.

was calibrated primarily on measures of an animal's social worth. Cats and dogs occupied an unambiguously positive place in American culture—as cherished companions, and "man's best friends." Mice, in turn, had been hangers-on to human culture for thousands of years, so their cultural identity as undesirable pests derived first and foremost from that relationship. Indeed, in the Linnean classification *Mus musculus*, the Latin *mus* derives from an ancient Sanskrit word, *musha*, meaning "thief," which points to the creatures' habit of scavenging from human food supplies. So while the cultural turn to mice-as-pets that the mouse fancy represents could potentially have resulted in increased emotional attachment to the species, in practice it did not.[23] By transforming mouse material into something of human utility, mouse fanciers instead appear to have lowered the practical and ethical thresholds for putting mice to other human uses. Thanks to the existence of fancy mice, mice did not need to be trapped from one's home to be obtained. Instead, mice could be ordered from a breeder, making contact with them in their "natural" state unnecessary. From an experimental biological perspective, this domestication averted trickier evolutionary questions about how their genetics worked in the "wild."[24]

Also, practical differences in genetic and medical research meant that Little's use of pet mice would not rouse the ire of those concerned with the plight of animals in research. An individual animal's perceived ability to feel pain was the key concern among American animal activists before 1920. Although in sheer numbers there were undoubtedly as many, if not more, mice in genetics laboratories as there were dogs in medical laboratories, the overt suffering and/or death of these animals was not built into the main genetic procedure of breeding. Usually, animal surgery was unnecessary—the mice that geneticists desired to mate were merely placed in the same cage at the proper times of day (i.e., at night, since mice copulate nocturnally)—and geneticists had a vested interest in keeping the animals alive to observe the outcomes of matings in subsequent genera-

[23] On the cultural meaning of certain animals as vermin, see Mary Fissell, "Imagining Vermin in Early Modern England," *History Workshop Journal* 47 (1999): 1–29. On the human construction of certain animals as pets, see Mary Midgley, *Animals and Why They Matter* (Athens: University of Georgia Press, 1983), chaps. 9 and 10. On the historical practice of pet keeping, see Harriet Ritvo, *The Animal Estate: The English and Other Creatures in the Victorian Age* (Cambridge: Harvard University Press, 1987), chaps. 2 and 3.

[24] C. J. Davies, *Fancy Mice*, 5th ed. (London: L. Upcott Gill, 1912). Harvard University Library holds only the fourth edition (1896) of this book, so perhaps earlier editions did not reach a broad audience. Mouse geneticists themselves were among the most thorough chroniclers of their creature's natural and social history. For the best overview, see chapters 1 and 2 of Clyde Keeler, *The Laboratory Mouse: Its Origin, Heredity and Culture* (Cambridge: Harvard University Press, 1931). See also Hans Gruneberg, *The Genetics of the Mouse* (Cambridge: Cambridge University Press, 1943), chaps. 1 and 2.

tions. For these reasons, mouse fanciers could rest assured that their favored animals endured no more suffering in the science laboratory than took place every day, in the confined spaces of rodent cages in their own homes or on their farms.[25]

By 1909, his junior year, Little had enough practical mouse breeding experience to begin striking out on his own, and he chose the scientific problem of inbreeding as one of his first independent efforts. Little's systematic inbreeding of mice started as a mundane, practical genetics project with two more immediate goals: accomplishing some independent research toward entering Harvard's doctoral program, and disproving what he thought was one of his advisor's overly conservative genetic hypotheses. Inbreeding was debated among animal and plant breeders as far back as Darwin's time, and it remained a controversial issue in genetics throughout the first decade of the twentieth century—especially at the Bussey Institute. Agricultural breeders argued over whether inbreeding that stabilized genetic variation had harmful side effects (including reduced fertility) that made certain stocks commercially unviable. Biologists, in turn, debated whether or not the practice of inbreeding produced truly "pure" lines—that is, strains of animals with identical genetic constitution at all, rather than merely a few, loci.

In 1909, while Little was still in the process of deciding whether to continue on with mouse work, Danish botanist Wilhelm Johannsen published a paper that seemed to settle this issue. Johannsen demonstrated that the result of continued self-fertilization in beans (*Phaseolus*) was a high degree of genetic homogeneity at all loci. Soon after, H. S. Jennings at Columbia and Raymond Pearl at Johns Hopkins published theoretical papers showing how mathematically Johannsen's "principle of pure lines" could apply to animals, even though they were not self-fertilized.[26] Little recalled later that he was immediately impressed by this new body of work on inbreeding and its potential applications for mouse work. He thought it suggested that "genetic homogeneity in other forms was also possible of attainment by continuous inbreeding. . . . It would merely take longer. I realized that once an inbred strain of mice was made genetically

[25] On the cultural legacy of the dog, see J. C. Russell and D. C. Secord, "Holy Dogs and the Laboratory," *Perspectives in Biology and Medicine* 28 (1985): 374–81. Susan Lederer, in "Political Animals," cites estimates that eight million Americans owned dogs in the 1930s.

[26] Wilhelm Johannsen, *Elements de Exakten Erblichkeitslehre* (Jena: G. Fischer, 1909); H. S. Jennings, "The Production of Pure Homozygotic Organisms from Heterozygotes by Self-Fertilization," *American Naturalist* 46 (1912): 487–91; Jennings, "Formulae for the Results of Inbreeding," *American Naturalist* 48 (1914): 693–96; Raymond Pearl, "A Contribution Towards an Analysis of the Problems of Inbreeding," *American Naturalist* 47 (1913): 577–614; Pearl, "On the Results of Inbreeding in a Mendelian Population," *American Naturalist* 48 (1914): 57–62.

homogeneous, one could keep it indefinitely and multiply its numbers to any degree necessary for experimental purposes over any desired span of years."[27]

But regardless of whether Little grasped the full ramifications of this work, his decision to take up mouse inbreeding represented a significant break from his mentor. Castle, who described himself as "experimental evolutionist," did not believe that inbreeding alone could ever produce stable genetic forms or truly pure strains. "The biologists' 'pure line,' " he wrote," is an imaginary thing," bearing "no more relation to animals and plants than a mathematical circle has to the circles described by the most accomplished draftsman." Furthermore, despite his own results from *Drosophila* inbreeding experiments showing that fertility was maintained even after fifty-nine generations, Castle doubted the viability of inbreeding as a stock-sustaining method.[28] Castle's beliefs on inbreeding were very influential, and in later papers he revealed his deep-seated doubt about human control over the production of genetic variability, so perhaps this attitude also had something to do with his rejection of the "pure line" hypothesis in 1909.[29] But viewed more broadly, Castle was well known for his recalcitrance toward new biological interpretations, and Little was still an idealistic undergraduate who had very little to lose by undertaking the inbred mouse project. Little recognized that it remained to be seen "whether continued inbreeding would weaken and eventually eliminate any [mouse] strain," and he had the practical breeding experience to know that mice naturally bred quickly and gave large litters—two advantages for overcoming an inbreeding depression. So in 1909, the same year that Johannsen's first results were published, Little began his first systematic effort to establish mammalian pure genetic lines by inbreeding several different pairs of brother and sister Bussey mice.[30]

As Castle predicted—and as mouse fanciers had themselves experienced—the material achievement of stable inbred mouse lines was not

[27] C. C. Little, "Your Question #2," CCL-UMO.

[28] W. E. Castle, "Pure Lines and Selection," *Journal of Heredity* 5 (1914): 93; see also Castle et al., "The Effects of Inbreeding."

[29] L. C. Strong, "A Baconian in Cancer Research: Autobiographical Essay," *Cancer Research* 36 (October 1976): 3545–53; W. E. Castle, "Piebald Rats and Selection: A Correction," *American Naturalist* 53 (1919): 370–75; cf. Dunn, "William Ernest Castle," pp. 42–45.

[30] Little, "Your Question #2." It was at about this same time that geneticist Helen Dean King began her long-term program of experimental inbreeding in rats at the Wistar Institute in Philadelphia. It is unclear whether Little and King were aware of one another's work at the time. King originally undertook the project to test whether inbreeding alters sex ratio, but she ultimately presented her results as proof that the effect of inbreeding may be to vigorously preserve certain traits. For a discussion of King's work, see Clause, "The Wistar Rat as Right Choice"; Cheryl Logan, "The Altered Rationale for the Choice of a Standard

easy. Most mice in the first generation died out quickly, and the inbred litters were smaller and sometimes not reproductively viable. But Little's efforts were not a total failure: one strain "prospered and showed no inherited weaknesses." This type of mouse was initially named "silver fawn" by the mouse fanciers from whom Castle had obtained the strain. Little synthesized his own breeding observations with other published studies by Bateson and Lucien Cuenot to argue that the strain should be renamed according to its distributive coat color factors or (in the parlance of the time) "allelomorphs." Thus Little argued that the strain contained recessive genes for "dilution" (its hairs and skin possessed a reduced number of pigment granules, compared with the "density" of pigmentation in common wild type mice); "brown" (compared with wild type "black"); and "nonagouti" (compared with wild-type "agouti," a pattern of black ticking that Castle had shown to be an independent unit character). He reported this "dilute brown" strain (in Mendelian shorthand: *dba*) in a note to *Science* in 1911, which was published only a year after he officially enrolled at the Bussey Institution as a doctoral student.[31]

Little's early success with the dilute brown strain appears to have inspired him to expand his horizons. As a graduate student starting in 1910, he continued caring for Castle's mouse colonies just as he had done as an undergraduate, but he also published more independent assessments of the Bussey mouse data. Because independent research was considered a valuable part of the Bussey learning experience, Little was encouraged by his teachers to extend his previous mouse work on a larger scale as his graduate studies progressed.[32] Thus, though he became interested for a brief period in the genetics of Castle's cat colony, he focused primarily on constructing and carrying out a mouse breeding scheme for further analyzing coat color phenomena.

Ultimately, Little recorded over 10,500 mouse young during the period from 1909 to 1912, and this experience increased his confidence in his own interpretations of mouse genetic phenomena. In 1912, for example, Little published a scathing critique of *Drosophila* prodigy Alfred Sturtevant's claim that there was a negative "association" (or linkage) between yellow and agouti factors in mice. Sturtevant argued that these two factors were genetically incompatible and never occurred in the same mouse. But Little chided him for basing this conclusion on too few observed mouse

Animal in Experimental Psychology: Henry H. Donaldson, Adolf Meyer and 'the' Albino Rat," *History of Psychology* 2 (1999): 3–34.

[31] C. C. Little, "The 'Dilute' Forms of Yellow Mice," *Science* 33 (1911): 896–97.

[32] For discussions of educational values at the Bussey Institution, see Rader, "The Mouse People"; William B. Provine, *Sewall Wright and Evolutionary Biology* (Chicago: University of Chicago Press, 1986), chap. 2.

matings and used coat color data from an obscure report by a little-known English mouse fancy breeder, "Miss Durham," combined with his own, to argue that yellow and agouti were often linked. By late 1912, Little had retracted this hypothesis—in part, he wrote in a note to *Science*, because he misread Durham's "ambiguous" unscientific mating tables. But also, by this time Little was working with fellow Harvard student John Phillips to expand his own studies; together, they published a 1913 report that detailed four independent pairs of Mendelian factors involved in mouse coat color. This latter study utilized a well-developed strain of "homozygous pink-eyed dilutes" (presumably the *dba*s) on which Little promised soon to report more.[33] Unfortunately, however, he got so wrapped up in this work, as well as his early family life—he married Katherine Andrews, the daughter of a prominent Boston architect, and they were expecting their first child—that he did not prepare adequately for his doctoral exams. His failure in 1913 was personally acute, but it also reflected the ongoing resentment directed at Bussey geneticists by Cambridge zoologists.[34]

Though Little would have to wait another academic year until he could officially be awarded his degree, the bulk of his dissertation on mouse coat color genetics was published as part of a Carnegie Institution monograph in 1913.[35] This document reflects Little's programmatic goals, while also testifying to the material transition still in progress from pet fancy mice to inbred laboratory mice. In it, Little debated with geneticists no

[33] Little, "A Preliminary Note on the Occurrence of a Sex-Limited Character in Cats," *Science* 35 (1912): 784–85; C. C. Little and John Phillips, "A Cross Involving Four Pairs of Mendelian Characters in Mice," *American Naturalist* 47 (1913): 760–62. "Miss Durham" is Florence M. Durham ("Further Experiments on the Inheritance of Coat Color in Mice," *Journal of Genetics* 1 (1911): 159–68), although little more is known about her. Little ultimately retracted his critique of Sturtevant's mouse coat color interpretation because he misinterpreted Durham's data reports: see C. C. Little, "Yellow and Agouti Factors in Mice," *Science* 38 (1913): 205. This was not the last time he would go after one of Morgan's Drosophilists for misunderstanding more complicated forms of animal heredity: cf. C. C. Little, "Alternative Explanations for Exception Color Classes in Doves and Canaries," *American Naturalist* 53 (1919): 186–87. But Little and Morgan appear to have remained friendly, as in 1916 they were still exchanging mice: see C. C. Little, "The Occurrence of Three Recognized Color Mutations in Mice," *American Naturalist* 50 (1916): 335–49.

[34] See Little's description of this event in his entries for the 1917 (third) and 1935 (twenty-fifth) anniversary Harvard College Class Reports, Harvard University Archives, Pusey Library, Harvard University, Cambridge, MA.

[35] Little's dissertation, supervised by Castle, East, and G. H. Parker, was entitled "Experimental Studies of the Inheritance of Coat Color in Mice and Their Bearing on Certain Allied Problems in Genetics" (ScD, Harvard University, 1914) but was published as C. C. Little, *Experimental Studies of the Inheritance of Coat Color in Mice*, Carnegie Institute of Washington Publication no. 179 (Washington, DC: Gibson Brothers Press, 1913).

less luminous than T. H. Morgan and William Bateson over the meaning of his new mouse data with regard to issues like dominance and epistasis. Also he was careful to report nomenclature of mouse strains based on their Mendelian factor composition, with the names given to them by fanciers in "scare quotes"—for example, "creams" or "sables."[36] But when Carnegie Institute published the work, Little included four pages of plates that were color-and-ink drawings of all the variations of coat color mice he used, and these invoke the imagery of fancier culture.

Little's mouse inbreeding efforts were important, then, because he approached them like a geneticist: he stabilized the hereditary constitution of murine material at the same time that he effectively connected this material to a well-understood set of research questions and approaches in the rapidly expanding field of Mendelian genetics. In this early period Little was not the only genetic researcher working on mice, and he was not the only scientist to see the methodological potential of homogeneous mammalian animals for genetics.[37] But because Little used inbred mice to demonstrate that inbreeding was an effective way to weed out variation and preserve a unique variant against a homogeneous genetic background, these mice were well positioned to move mammalian genetics intellectually and methodologically in the direction already proven fruitful by *Drosophila* and corn workers.[38] Selection of useful variants alone was not enough, for if it had been, fancier strains would have sufficed. Pet mice initially became laboratory mice because Little advanced their inbreeding while also rationalizing both their material understanding and their scientific uses.

USEFUL TUMORS: REDEFINING INBRED MICE FOR GENETICS AND FOR MEDICINE

When Little was at the Bussey, other researchers began exploring uses for inbred mice that were not governed by genetics—in particular, in experimental studies of cancer. Little came slightly later to this arena, but his effect on the scientific definition of the mouse there would be no less profound. In fact, Little's work on inbred tumor mice as the bearers of multifactorial genetic traits fashioned another unique dual scientific identity for laboratory mice: as organisms useful for simultaneously investigating basic genetic principles and basic cancer biology. But this identity was more

[36] Little, *Experimental Studies of the Inheritance of Coat Color in Mice*, pp. 39–41, 49–100.

[37] Cf. Clause, "The Wistar Rat as a Right Choice."

[38] Cf. Kohler, *Lords of the Fly*; Kimmelman, "Organisms as Ideology."

tenuous because (as Little's debates with researcher Maud Slye would make clear) the reformist ideology on which it was based threatened social, practical, and professional alignments between genetics and medicine.

Geneticists were not alone in their reliance on breeders for animal material: work on spontaneous cancerous tumors in mice also began through connections between medical researchers and some of the same mouse breeders who provided stocks to the Bussey. As early as 1908, Abbie Lathrop is reported to have observed skin lesions on some of her breeding stocks. She sent samples of the affected mice to the handful of medical researchers who regularly purchased animals from her, and she received a particularly interested reply from pathologist Leo Loeb of the University of Pennsylvania. That same year, Loeb had published work detailing his own successful transplantation of spontaneous tumors among strains of Japanese waltzing mice.[39] Loeb first confirmed the lesions as cancerous, and "by negotiations as yet unclarified," he and Lathrop embarked on a five-year program of joint research on the nature and transmission of the tumors in these mice. Surprisingly, although at this time women researchers were often denied credit for their work or treated like mere assistants, the Lathrop and Loeb collaboration appears to have been genuine, and Lathrop appeared as full co-author in all of the publications from the project.[40]

The ten articles Lathrop and Loeb wrote over the period from 1913 to 1919 represent the first work establishing the connection between certain strains of mice and the inheritance of cancer. In these papers, by contrast with Little's work, Mendelian terminology and methods are notably absent. Lathrop and Loeb used breeding experiments with the "silver fawn" and other Granby mice to show that the incidence of mammary tumors varied among different "families" of mice: for example, such tumors were high in the "English tan" and "sable" but low in "cream." Furthermore, they reported that ovariectomies reduced the frequency of mammary tumors, while pregnancies increased it. And finally, they observed that when members of high-tumor and low-tumor families were crossed, the incidence of tumors in the new generation emulated that of the high-tumor

[39] Leo Loeb, "Uber Enstanhug sines Sarcoms nachs Transplantation eines Adenocarcinoms einer japanischen Maus," *Zeitschrift auf Krebforschung* 7 (1908): 80. Cf. Leslie Brent, *History of Transplantation Immunology* (San Diego: Academic Press, 1997).

[40] There is some evidence that Lathrop herself envisioned using her mice and rats for science education purposes: cf. Michael Shimkin, "A.E.C. Lathrop: Mouse Woman of Granby (1868–1918)" *Cancer Research* 35 (June 1975): 1597–98. Herbert S. Morse, in *Origins of Inbred Mice*, like Shimkin, reports that these experiments "seemed to have been" designed solely by Loeb and then carried out by Lathrop, but there is no extant evidence for this belief. Cf. Marsha Richmond's argument that Bateson's women students from Newnham College assembled the initial evidence for Mendelism, while their "sisters" at Girton were rallying behind the biometricians ("Women in the Early History of Genetics: William Bateson and the Newnham College Mendelians, 1900–1910," *Isis* 92 [2001]: 55–90).

family.[41] Lathrop's scientific notebooks from this period also show that she had started methodically inbreeding her cancerous families of mice in approximately 1910, about the same time Little reported starting his inbreeding experiments with the *dba*.[42]

It is almost impossible to imagine that Little did not know of Lathrop and Loeb's work on spontaneous tumor mice. Though this work was published in medical rather than genetic journals, in the course of maintaining Castle's stocks Little probably would have had enough contact with Lathrop and her mice to encounter it.[43] Yet Little himself made no mention of any coat color/cancer correlation in his dissertation, and he would later claim that he independently observed the occurrence of high cancer (specifically, mammary cancer) in his own dilute brown nonagouti stocks around the same time.[44] No extant historical evidence remains that can definitively settle this issue. But regardless, from this point forward, Little's science changed course, and so, in turn, did the fate of the inbred mouse.

AT THE HARVARD CANCER COMMISSION: THE GENETICS OF MOUSE CANCERS

Contemporary with Lathrop and Loeb's work, another medical researcher also became interested in mice as systems for studying the cancer problem: a Harvard-trained doctor named E. E. Tyzzer.[45] After graduating in 1902, Tyzzer left Boston to travel worldwide and conduct research on the parasites associated with tropical diseases. When he returned in 1905, he was appointed by W. T. Councilman as director of the U.S. Public Health Service's Harvard Cancer Commission. This position afforded

[41] The complete list of Lathrop and Loeb's publications is in Shimkin, "A.E.C. Lathrop." At about the same time or slightly earlier, other scientists, such as Haaland, published observations that tumor transplants took more easily among mice of the same "race," but these scientists did not describe this "race effect" in genetic terms. I am grateful to Jean Paul Gaudilliere and Ilana Lowy for pointing out Haaland's work to me.

[42] Abbie Lathrop's Scientific Notebooks, 1912–1918, Box 9, JLA-BH: see esp. the entries for 11 April 1918 and the late 1915 breeding records. See Morse, *The Origins of Inbred Mice*, pp. 14–16, for his similar conclusions on the notebooks, as well as the calculation of when Lathrop began inbreeding.

[43] In a 1931 paper, Little refers to the work of "Miss Lathrop, a mouse fancier of more than ordinary care and scientific interest." C. C. Little, "The Role of Heredity in Determining the Incidence and Growth of Cancer," *American Journal of Cancer* 15 (1931): 2780–89.

[44] Little, "Your Question #2, CCL-UMO."

[45] One of Tyzzer's 1907 papers lists his full institutional affiliation as the Laboratory of the Caroline Brewer Croft Fund Cancer Commission of Harvard University, although his professorship was definitely in the medical school. The Harvard Cancer Commission, then, was a privately endowed cancer research laboratory on the Cambridge campus; archival materials related to its founding are now housed at the Countway Library of Medicine at Harvard.

Tyzzer increased status and clout within the medical school. As a result, he had the time and space to develop a significant mouse colony beginning in 1906.[46]

In 1907, in his first published work on the occurrence of spontaneous tumors in mice families, Tyzzer revealed his own appreciation of their scientific value. He argued that "since mice breed so rapidly and so frequently develop tumors," they were "especially adapted" to the "regulated" breeding experiments needed to provide medical researchers with more accurate conclusions about the heritability of cancer. But in a 1907 article, Tyzzer also praised "the demonstration of the inoculability of certain tumors in animals" and asserted that cancer work on transplantable tumors, especially with mice, had "opened up a whole new field of research in that an experimental basis is now furnished for the investigation . . . of problems related to tumor growth."[47]

In 1909 Tyzzer attempted to combine these two interests—in cancer's heritability and its etiology—when he published a study investigating a possible genetic basis for mouse resistance or acceptance of a tumor transplantation. Tyzzer reported that spontaneous tumors arising in the fancier strain Japanese waltzer could be successfully transplanted to all animals of that strain, but that common, "wild" mice were completely resistant to transplantation. Loeb himself had published the same finding a year earlier, but Tyzzer's innovation was to cross the waltzer mice with their noncancerous wild relatives in order to examine heredity's role in transplant variability. He noted that, whereas all of the first-generation hybrid mice were susceptible, none of the second-generation hybrids accepted the transplant. From these data, Tyzzer concluded that tumor susceptibility was not an inherited, Mendelian trait.[48]

Some people active in the field of cancer research at the time have noted that Tyzzer's hybridization studies were perceived as very exciting. L. C. Strong later wrote that many felt Tyzzer had shown that mice were "such

[46] Thomas H. Weller, "Ernest Edward Tyzzer," *Biographical Memoirs, National Academy of Sciences* 40 (1978): 353–73.

[47] E. E. Tyzzer, "The Inoculable Tumors of Mice," *Journal of Medical Research* 18 (1907): 137–53; Tyzzer, "A Study of Heredity in Relation to Development of Tumors in Mice," *Journal of Medical Research* 17 (1907): 199–211. The demonstration of the phenomenon of tumor inoculability was first made by A. Hanau, "Erfolgreiche experimentelle Ubertragung von Carcinom," *Forschung die Medizinisch* 7 (1889): 321. Cf. Henri Moreau, "Recherches experimentales sur la transmissibilite de certains neoplasm," *Archives de la Médecine: Experiementale et d'Anatomie Pathologique* 6 (1894): 677; Leo Loeb, "Development of a Sarcome and Carcinoma after the Inoculation of a Carcinomatous Tumor of the Submaxillary Gland of Japanese Mouse," *University of Pennsylvania Medical School Bulletin* 19 (July 1906): 113–16; Brent, *History*, pp. 383–84.

[48] Tyzzer, "A Study of Inheritance in Mice"; cf. Kenneth Paigen, "Seventy-five Years of Mouse Genetics: Some Perspectives and Lessons," ms., 1983.

a splendid tool available . . . that the great riddle of cancer might thus soon be resolved." There is, however, no direct evidence that Little knew of Tyzzer's work as early as 1909. If he did, perhaps an alternative explanation for why Little started inbreeding the *dba*s in 1909 might be that he recognized that another genetically uniform cancerous strain could be useful for pairings with Tyzzer's already highly inbred and cancerous Japanese mice.[49]

Nevertheless, by 1913, Little was clearly thinking about the relevance of Tyzzer's cancer work to his own inbred mice, as well as to larger issues in the field of mouse genetics. Specifically, Little began to argue that, although some patterns of mouse inheritance were too complicated to be explained by a simple factor analysis, a multifactorial interpretation could easily bring these cases back into the Mendelian fold. In 1914 he published a theoretical paper suggesting that "certain characters of an organism depend for their visible manifestation . . . upon the simultaneous presence of more than one mendelizing factor." In such cases, Little asserted, a Mendelian statistical analysis demonstrated that as the number of factors increased, the ratio of observed F_2 animals that do not show the trait to those who do would also increase rapidly. He later restated his hypothesis more succinctly: "Since they [genetic factors] segregate independently and recombine in the F_2 according to the laws of chance, it would naturally follow that the more factors there were, the *fewer* of the gametes in F_2 there would be which received the whole number necessary for susceptibility."[50] It was unlikely a pure coincidence that this example exactly mirrored the pattern that Tyzzer had observed in mouse cancer transplant susceptibility. The larger implication of Little's conclusion was clear, and it linked to his own interest in developing the mouse as a research organism: even more complicated patterns of inheritance such as those in higher animals could be probed through genetic experiments.[51]

When Little first made contact with Tyzzer in the fall of 1913, however, his timing was fortuitous. Since 1911 Little had been working part-time in the college adminstration as secretary to the Harvard Corporation, the university's governing body.[52] But as he was finishing up his doctoral work, he needed transitional employment that would not remove him from his family or the Boston Brahmin atmosphere of Cambridge that he had enjoyed since boyhood. Little later wrote that he only "reluctantly gave up the secretarial work" in the winter of 1913 and decided "to put

[49] Strong, "A Baconian in Cancer Research," p. 3545; Paigen, "Seventy-five Years."

[50] Little, "The Role of Heredity."

[51] C. C. Little, "A Possible Mendelian Explanation for a Type of Inheritance Apparently Non-Mendelian in Nature," *Science* 40 (1914): 904–906.

[52] "Clarence Cook Little," *Who Was Who in America*, vol. 5 (1969–73), pp. 435–36.

full time into research." Tyzzer was just then looking for a mammalian geneticist to assist him in making more refined Mendelian interpretations of his ongoing mouse cancer experiments.[53] Thus Little obtained an appointment under Tyzzer as research associate in genetics.[54]

During the next three years at Tyzzer's lab, Little moved to further "mendelize" the cancer problem and the inbred mouse materials needed to study it. First, Little initiated with Tyzzer a series of large-scale crosses between Tyzzer's cancerous Japanese mice and his own dilute browns. Notably, in one 1916 publication reporting their results, Little and Tyzzer emphasized that the Japanese mice were crossed with the "most homogeneous stock of common (house) mice that has been bred at the laboratory of the Bussey Institution. All present animals are direct descendants of a single pair of closely related dilute brown (fancier silver fawn) mice obtained in the spring of 1909. From the start the stock has been kept free from any outcross and has therefore an unbroken stretch of more than twenty-generations of inbreeding." These experiments with two highly inbred strains netted only 3 of 182 F_2 animals that were susceptible to the waltzer transplanted tumor. Nevertheless, Little and Tyzzer calculated backward from this number to argue that mouse tumor susceptibility could be interpreted as a Mendelian trait involving twelve to fifteen separate factors.[55]

Little noted that this interpretation of the Cancer Commission mouse data would be valid only if genetically homogeneous animals were used, but he concluded that it explained cases in which a character dominant in the F_1 generation appears to "almost completely disappear in the F_2." Little and Tyzzer's first collaborative work actually employed a fairly loose combination of genetic and medical research practices. The 1916 papers, for example, included tables of expected Mendelian types as well as pathological charts of individual mice and sketches of the progress of their tumors over the course of an experiment. But their conclusions ultimately confirmed the power of inbred mice as tools for studying multifactorial heredity. Because of the mathematical precision of Little's Mendelian-based theory predictions, the multifactorial hypothesis of cancer transmission could, by definition, be probed experimentally only with inbred animals.

[53] Undated manuscript (c. late 1950s), titled at the bottom: "The Genesis of the Laboratory," JLA-BH.

[54] Cf. Snell, "Clarence Cook Little," pp. 562–64.

[55] C. C. Little and E. E. Tyzzer: "Further Experimental Studies on the Inheritance of Susceptibility to a Transplantable Carcinoma (J.W.A) of the Japanese Waltzing Mouse," *Journal of Medical Research* 33 (1916): 393–427; also "Studies on the Inheritance of Susceptibility to a Transplantable Sarcoma (J.W.B.) of the Japanese Waltzing Mouse," *Journal of Cancer Research* 1 (1916): 387–89.

THE MEDICAL USES OF CANCEROUS INBRED MICE: A CLASH OF SCIENTIFIC PRACTICES

Cancer genetics remained at the intellectual boundaries of mainstream medical research, and very few workers were actually devoted to its investigation, so Little's rigorously Mendelian definition of the inbred laboratory mouse might at the time have gone completely unnoticed if not for Little's subsequent feud with cancer researcher Maud Slye of the University of Chicago. Beginning in 1915, while Little was still at Harvard, this well-publicized episode provides a window on the important practical and intellectual issues raised by Little's new interdisciplinary approach to research with inbred mice.

Maud Slye hardly followed a traditional path into cancer research, but her career does reflect important trends for women in science during the early twentieth century. She was a secondary school psychology teacher, trained at Brown University, who turned to the biological study of waltzing behavior in Japanese fancy mice on the suggestion of C. O. Whitman during a Woods Hole summer course. Whitman was so impressed that he invited Slye to Chicago as his research assistant. There she encountered pathologist H. Gideon Wells, who appointed her to Chicago's newly created Sprague Institute for Medical Research.[56]

From 1911 to 1913, Slye worked in relative isolation on what she framed as a study of spontaneous tumor occurrence in mouse "families." Methodologically, she employed a mixture of practices from genetic and medical research in this work. For example, she pen bred mice of well-recorded ancestry, keeping pedigrees as well as pathological charts for each descendant, and she autopsied each dead mouse individually to ensure that an accurate cause of death had been recorded. In 1914 Slye rose to prominence in cancer circles when she presented her initial findings to the American Society of Cancer Research meeting in Toronto. She subsequently published preliminary reports of her breeding studies on over five thousand mice in the *Journal of Medical Research*. In the first, Slye suggested that inheritance of cancer susceptibility was analogous to "albinism," and that it was transmitted as a single Mendelian recessive gene. This paper was well received by many medical researchers.[57]

[56] John Parascondola, "Maud Slye," in *Notable American Women: The Modern Period*, ed. B. Sicherman and C. H. Green (Cambridge: Belknap Press of Harvard University, 1980), pp. 651–52); "Maud Slye," *Current Biography* (1940), pp. 743–45. A more popular account of Slye's life is J. J. McCoy's *The Cancer Lady* (Nashville: Thomas Nelson Publishers, 1977). Cf. Margaret Rossiter, *Women Scientists in America*, vol. 1.

[57] Maud Slye, "The Incidence and Inheritability of Spontaneous Cancer in Mice," *Journal of Medical Research* 32 (1915): 159–72. Cf. McCoy, *The Cancer Lady* (1977).

Little reacted angrily and publicly to the scientific community's acceptance of Slye's conclusions. In a brief note to *Science*, he attacked Slye's experimental methodology, as well as her analysis. He noted that her preliminary discussion of the inheritance of mouse albinism was entirely at odds with the authoritative genetics work in this area, and he argued that because she presented no breeding data (i.e., total numbers of mice observed in cancerous and noncancerous breedings), she gave researchers no basis upon which to evaluate her "revolutionary" claims. He also objected to her randomly inbred mouse material and nonstatistical experimental design: "To those unfamiliar with the work of the geneticists," he wrote, "Slye's paper might be taken as presenting the well-known principles of Mendelian inheritance" in cancer when in fact no "type of inheritance she outlines" had "been observed in similar material" by genetic or medical investigators.[58] In short, for Little, the complex behavior of cancer inheritance in mice necessitated a geneticist's expertise: Slye was not trained in genetics and therefore could not properly breed her mice or interpret her data.

Slye's reply a few weeks later focused on both the theoretical and material points of Little's attack on her mouse work. First, she noted that Little's comment about albinism presupposed that her conclusions about mouse cancers were intended to support one particular theory of heredity, namely, Mendelism. In this regard, Slye contrasted her working assumptions explicitly with Little's: "Whether or not my strains of house mice have behaved in hybrid crosses in accordance with the established canon has no bearing whatever upon the behavior of cancer . . . it is an academic debate which lies in quite another field." Second, in reply to Little's insinuations about the irrelevance of her mouse strains, Slye wrote that she had chosen to employ pen inbreeding and the simple family pedigree analysis because she "deplored" the conclusions that "artificial" inbreeding encouraged. In strict brother-sister inbred strains, Slye claimed, "variation becomes wholly a matter of the environment."[59]

This accusation implied that Little's inbred mouse material was yielding artifactual—that is, nonnatural—data, and so he fired back a patronizing reply. Slye believed his work was "artificial," he suggested, only because she could not "understand the exact distinction between the gathering of valuable data and the interpretation of such data." Little then attempted to turn Slye's successes against her by citing several optimistic reviews of Slye's work in medical journals that praised her results for proving the inheritance of cancer, "given parents of pure breed." He argued that because Slye implied such definite conclusions about human

[58] C. C. Little, "Cancer and Heredity," *Science* 42 (1915): 1076–77.
[59] Maud Slye, "A Reply to Dr. Little," *Science* 42 (1915): 246–48.

cancer inheritance from such imprecise mouse materials, her work was misleading and worthy of an "extremely important postscript" to correct possible wrong impressions among clinical practitioners.[60]

Ultimately, the Slye-Little debates would continue for nearly thirty years, through attacks and counterattacks in forums ranging from specialist professional journals to popular magazines.[61] In its early stages, however, the fierce competition between Little and Slye to define inbred mice and their cancers illustrates how linking these laboratory creatures to a medical problem raised new social and professional stakes. Little now aimed to make inbred cancer strains a talisman of the intellectual and material power of Mendelian genetics. But even while medical doctors were not direct competitors toward that goal, within a research context, this goal had broader implications.[62] Who would get to wield the power of research as a tool for curing human health problems—geneticists or medical researchers? And how—through Mendelian research or traditional clinic-based or epidemiological studies? Finally, to what end—merely to obtain knowledge of cancer's biological causes or to effect public health interventions?[63] The Slye-Little debate made clear that such issues would now also need to be addressed in order to further stabilize the inbred mouse's scientific meaning and practical uses for cancer studies.

Ironically, just as Little's debates with Slye began to increase Little's reputation in medical circles, the pressure of doing mouse genetics in a medical setting began to lose its appeal. Other researchers began soliciting his mouse-breeding expertise: for example, Henry James of the Rockefeller Institute for Medical Research (RIMR) in New York, wrote to Castle in 1917 about the RIMR's difficulties in obtaining simple white mice. James asked Castle for any suggestions that Little in particular might have

[60] C. C. Little, "The Inheritance of Cancer," *Science* 42 (1915): 494–95. Cf. Phillip Pauly's insightful early argument that "the complex relations between biological and medical science formed a major part of the background to the tensions between naturalists and experimentalists that have interested historians," in his "The Appearance of American Biology in Late Nineteenth-Century America," *Journal of the History of Biology* 17, 3 (Fall 1984): 369–97.

[61] Little made the cover of *Time* on 29 March 1937; three weeks later (10 April), Slye made the cover of *Newsweek*.

[62] For an overview of the history of medical research on human cancers, see Charles Oberling, *The Riddle of Cancer* (New Haven: Yale University Press, 1944). For a contemporary critique of genetic research and its applications to medical problems, see Oscar Riddle, "Any Hereditary Character and the Kinds of Things We Need to Know about It," *American Naturalist* 58 (1924): 410–425.

[63] Slye herself would later invoke her mouse work to argue for the necessity of better record-keeping on human cancers, in the hopes that such information might eventually aid physicians in their clinical practices. See McCoy, *Cancer Lady*, pp. 88–90; see also Ludwig Hekoten (and other anonymous scientists) on Slye in Williams, *The Virus Hunters*, pp. 419–21.

about where to get mice or about things like cages and diet to help them start their own breeding stocks.[64] Also, in a letter to Harvard Medical School professor W. T. Bovie, Little admitted that he faced criticism from both geneticists and medical researchers. On the one hand, he was not supported by Castle, who "constantly" put him "on a defensive attitude" about his work in the applied context of the medical school. On the other hand, Little also noted: "unless my work led to good advertising material for [the Cancer Commission], I should have to walk the plank."[65] Shortly afterward, Little reported for Air Force duty during World War 1, and he left his mice behind for the first and only time in his life. While Little held various posts in Washington, DC, and Plattsburg, NY, Tyzzer maintained the *dba* strain.[66]

Before he left the medical school, Little articulated an ambitious vision for the proper scientific identity of inbred mice—namely, as a means of realizing a state of true cooperation between medical and genetic practitioners. In a 1916 essay on "The Relation of Heredity to Cancer in Man and Animals," Little concluded that it would be an "unfortunate" social consequence if neither medical researchers nor geneticists could recognize cancer as a problem "essentially biological in nature" and make it their priority to cooperate through research to solve it.[67] In private, Little also speculated that successful mouse genetics work on cancer and other "subjects of interest to medical men would lead to their conversion more rapidly" to the genetic approach to other human problems.

Upon his return from military service in 1919, Little quickly accepted a post as research associate at the Station for Experimental Evolution (SEE) in Cold Spring Harbor, New York. Perhaps Little's idealism about genetics moved him to escape the tense atmosphere of the medical school and revisit his Bussey roots. Both inbred laboratory mice and the man who created them had come full circle, back to mammalian genetics. Now, however, Little refocused his attention from research problems in

[64] Henry James to W. E. Castle, 8 November 1917, Rockefeller University Archives, Record Group 210.2, Box 1, Folder "Abderhalden-Councilman," RAC-NY.

[65] CCL to W. T. Bovie, 18 December 1918, Box 727, CCL-UMO. Cf. "It does not appear . . . that I could prove the value of genetics to a medical school as well under existing circumstances at Harvard." CCL to E. E. Tyzzer, 11 December 1918; also, E. E. Tyzzer to CCL, 9 December 1918, both in Box 739, CCL-UMO. As a result of some ongoing administrative work for Harvard College, Little had also been promoted to assistant dean of college and acting university marshal in 1916; see "Clarence Cook Little," in *Who Was Who in America*, vol. 5 (1974), pp. 435–36. Snell, "Clarence Cook Little," p. 243.

[66] In CCL to E. E. Tyzzer, 11 December 1918, Little refers to picking up some of the dilute browns from Tyzzer when he returns to the Boston area to get the rest of his affairs together. Box 727, CCL-UMO.

[67] C. C. Little, "The Relation of Heredity to Cancer in Man," *Scientific Monthly*, 3 (1916): 196–202.

this field to problems of material infrastructure: how could enough inbred mice colonies be sustained and made available to serve what Little would later call the "the avid maul of researchers . . . faithfully and well"?[68]

COLD SPRING HARBOR AND THE STATION FOR EXPERIMENTAL EVOLUTION: MOUSE GENETICS REVISITED

Founded by Castle's mentor, Charles Davenport, in 1904, the SEE was well funded by the Carnegie Institution, the same foundation that supported mammalian genetics work at the Bussey. It was a year-round independent genetics research center, but it also had a well-developed summer program for both graduate students and established geneticists who worked on all kinds of organisms. Castle, in particular, had extensive contact with the SEE.[69] When Little arrived, he was not the first mouse geneticist there. Davenport had tirelessly lobbied E. C. MacDowell, a fellow Bussey mammalian genetics student, to do research at the station right after his graduation in 1912. MacDowell at first declined Davenport's offer but eventually decided that Davenport's willingness to support mammalian projects represented a unique opportunity, and he soon made Cold Spring Harbor his full-time "scientific home."[70]

In 1919 MacDowell teamed up with Little and launched a full-scale effort to increase material investments in mouse genetics work. Drawing on their past experience at the Bussey, they focused their efforts not on setting a definitive research agenda, but on solving practical problems. Though both Little and MacDowell maintained other large animal colonies for genetic work during this period,[71] they worked methodically to increase both the availability of genetically identified inbred mouse material and the presence of a community of mouse workers at Cold Spring Harbor.

Seemingly the more powerful personality of the two, Little first installed his *dba*s and other coat color mutants into the existing SEE facilities and soon began scaling up their breeding. Soon both Little and MacDowell

[68] Little, "Some Contributions," p. 127.

[69] Castle was periodically appointed an in-residence research associate, and the full-time geneticists included G. Harrison Shull (from 1904 to 1915) and Oscar Riddle (from 1914 to 1945). See Garland Allen, "The Eugenics Record Office at Cold Spring Harbor: An Essay in Institutional History," *Osiris* 2 (1986): 225–64, esp. 230–31.

[70] CBD to E. C. MacDowell, 2 May 1912, plus MacDowell's [undated] reply; ECM to CBD, July 1914 and 24 September 1914; CBD to ECM, 17 April 1918, plus various letters from ECM to CBD during his wartime service in France; all CBD/CSH-APS.

[71] CCL to C. B. Davenport, 1918 [probably December], and MacDowell to Davenport, 2 September 1919, both mention plans for a dog-breeding colony: CBD/CSH-APS. Cf.

were encouraging other American mouse geneticists to visit the SEE and to make the institution the site of their summer research. Such workers included L.C. Dunn, William Gates, and George Snell. Just as T. H. Morgan would send his flies to Woods Hole every summer, Dunn initiated a regular practice of sending his entire mouse colony to Cold Spring Harbor for the warmer months.[72] Sometime in the early 1920s, Dunn appears to have completed a series of breeding experiments on mouse "mosaic" mutants at the SEE.[73] Anecdotal evidence suggests that these "researcher colonies" replicated the sparseness and cohabitation arrangements of the Bussey. Strong remembers setting up his "honeymoon residence" at Cold Spring Harbor in a tent, where he and his new bride kept the mice for his research under their bed, for fear of mouse typhoid contanimation. By 1919 Davenport himself noted that the SEE had accumulated all the makings of "a good little working group" in mouse genetics.[74]

In 1922, perhaps inspired by the successes that the Drosophilists experienced in their "summer colony" at Woods Hole, MacDowell requested a substantial budget increase to fund what he called his "extensive plans for the development of mouse breeding" at the SEE. Also, in case this strategy did not turn up enough additional mouse mutants for genetic analysis, MacDowell suggested that the SEE rely on a classic Bussey method: he wanted a special appropriation to visit mouse fanciers and purchase phenotypically deviant organisms from their stocks. As he told Davenport, he desperately wanted to preserve what he and Little had initially effected—the SEE's institutional role in developing the mouse genetics community: "I am thinking in terms of the mouse work as a whole and not alone of that in my immediate charge, for in connection with the mice has arisen a cooperative spirit that I wish to foster in every possible way." By the following summer, MacDowell himself was devoting nearly all his own research to mouse projects—ranging from the behavioral genetics of mouse alcohol intoxication in a strain of black mice desended from Abbie Lathrop's so-called Strain 57, to the role of genetic

C. C. Little, "A Note on the Origin of Piebald Spotting in Dogs," *Journal of Heredity* 11 (1920): 12–15.

[72] On Morgan and his flies, see *Lords of the Fly*, pp. 136–38 (which also notes that Cold Spring Harbor eventually became a rival center to Columbia for *Drosophila* production and exchange).

[73] LCD to CCL, 6 June 1929, LCD-APS; cf. Dunn, "A New Series of Allelomorphs in Mice," *Nature* 129 (1932): 130, and "Analysis of a Case of Mosaicism in the House Mouse," *Journal of Genetics* 29 (1934): 317–26. For a sample of Gates's work at the SEE, see Gates and Elizabeth Lord, "Shaker: A New Mutation in the House Mouse," *American Naturalist* 63 (1929): 435.

[74] L. C. Strong, "Inbred Mice in Science," in *Origins of Inbred Mice*, ed. H. Morse, 45–67. Cf. Davenport's "Memorandum for Dr. MacDowell," 16 December 1926, and CCL to CBD, 1918, and 26 May 1918: both from CBD-APS.

control in murine gestation phenomena. The number of mice in the SEE colonies had grown to nearly ten thousand.[75]

As a result, Davenport promoted Little in December 1920 to assistant director of the Carnegie Institution's genetics research facility at Cold Spring Harbor, and Little seized this moment to advance some of his own interests. Sometime between 1920 and 1922, Little invited cancer researcher Halsey Bagg to come to Cold Spring Harbor during the summer. Bagg, an M.D. who had a joint affiliation with New York's Memorial Hospital and the Cornell University Medical School, maintained a strain of albino mice that he had used for several years in his cancer experiments. The albino mutant was very common—it had been observed as far back as Aristotle's time[76]—but Bagg's albinos were more special than ordinary albinos, since they were already partially inbred. Given his own research at Harvard, Little might have gotten wind of Bagg's albinos through reading his many published papers in medical research journals. Whatever the case, by late 1922, "Bagg albinos—pen inbred" were listed among the mouse stocks available at Cold Spring Harbor, and in 1924 Bagg and Little published two papers reporting on the collaborative research project they undertook at the SEE. Presumably inspired by Muller's work in *Drosophila*, they attempted to induce cancerous mutations in Little's dilute browns with X-rays and then breed them to see if the mutations were heritable. The initial research was unsuccessful on its own terms, but nevertheless it encouraged more inbred mouse use because it produced one non-cancer-related mouse genetic mutation in the Bagg strain—"myelencephalic blebs," symbolized by *j*—that would become important in the genetic study of mouse head and jaw abnormalities. Bagg and Little also used the methods sections of their papers to promote inbred mice as the "pioneers among mammalian materials used for experimental breeding."[77]

[75] On the Little-Davenport clash that led to Little's departure, see the correspondence between them for March 1921 to May 1922. ECM to CBD, 31 August 1922 and 31 August 1923, where MacDowell describes himself as a "mouse geneticist"; cf. ECM's typescript Project Descriptions, 1923–24, as well as for 1926, "Phenomena of Gestation": all CBD/CSH-APS. On Abbie Lathrop's Strain 57, see Joe Wright, "The Mother of All Lab Mice," *All Things Considered* radio broadcast, 18 September 2002.

[76] Keeler, "In Quest of Apollo's Sacred White Mice."

[77] Obituary, Halsey Bagg, *New York Times*, 15 April 1947; C. C. Little, "Halsey Joseph Bagg," *Anatomical Record* 100 (1948): 397. On maintenance of the "Bagg albinos" at the SEE, see last page of an untitled Mouse Club Newsletter, c. 1922, Box 732, CCL-UMO. The Carnegie Institution 1921 Annual Report reports that Little and Bagg began their joint experiments in fall 1920: see pp. 112–13 of this report, Box 738, CCL-UMO; H. J. Bagg and C. C. Little, "The Occurrence of Four Inheritable Morphological Variations in Mice and Their Possible Relation to Treatment with X-rays," *Journal of Experimental Zoology* 41 (1924): 45–91; "Hereditary and Structural Defects in the Descendants of Mice Treated

1.5. "Mouse Club of America." In the middle of the top row (with a dark shirt and a tie) is E. C. MacDowell. Standing directly to his right are C. C. Little and (holding a cage, in a bow tie) L. C. Dunn. The remaining people are likely a combination of caretakers and women genetics students from eastern colleges. [Source Credit: C. C. Little Papers, University of Maine].

Even more importantly, however, Little formalized the exchange of material and practical information about mouse breeding by starting (around 1920) a group called the Mouse Club of America (fig. 1.5). Mouse geneticists affiliated with the SEE for summer research automatically gained membership: this roster included former Busseyites and some colleagues from their related medical research fields, such as Little, Dunn, and MacDowell, as well as Bagg and Rockefeller Institute geneticist and medical researcher John Gowen. Whenever mouse workers seasonally descended on the SEE, the Mouse Club would hold informal meetings and exchange information about mutant stocks and breeding experiments.[78]

with Roentgen-ray Irradiation," *American Journal of Anatomy* 33 (1924): 119–45. On work with myelencephalic bleb mice, see Gruneberg, *The Genetics of the Mouse*, pp. 229–31.

[78] Cf. CCL to LCD, 12 December 1928, LCD-APS. On the founding of the Mouse Club, see E. S. Russell, "Origins and History of Mouse Inbred Strains: Contributions of Clarence Cook Little," in *Origins of Inbred Mice*, ed. Morse, pp. 33–43. Russell says Joseph Murray, one of Little's students at the University of Maine, remembers Little starting this organization. Interestingly, Elizabeth Russell herself recalls this group as the "Mouse Men of

Also, starting in the early 1920s, the Mouse Club increasingly corresponded collectively—rather then through individually addressed letters—in a series of periodic mimeographs that emanated from what they called "Mouse Headquarters, The Animal House, Cold Spring Harbor, N.Y." In lieu of their official publications in genetics journals, which typically took many months to reach the scientific public, these newsletters were the means through which Little and MacDowell circulated basic information about mouse stocks, such as which strains and mutants were being bred at the SEE. They included lists of all mice available for shipping to a researcher's home institution, and of strains that were in the process of being made available from the expanding SEE colonies. Among them by the mid-1920s was a low cancer, black coated strain called *C57 Black*. Along with the lists, mouse newsletters also contained humorous stories about the trials and tribulations of day-to-day mouse work in specific institutional contexts. For example, in 1922 Dunn reported from "Storrs Local #15 Branch of the A.M.U. [Amalgamated Mousers of the Universe]" of his own problems with mouse maintenance. Infant mortality in the Storrs mouse colonies was so high, he wrote, that

> [My] generally pessimistic attitude in regard to the growth of population seems to have conquered [my] generally indefatigable scientific imagination.
>
> Each day we go through the ritual over each pen of extra skinny infants, saying earnestly and with conviction: "Every day in every way you grow fatter and fatter." In spite of the feeling we put into this formula, we have not yet pierced the consciousness of the mice.

These stories suggest that some of the practical hardships governing mouse genetics remained alive and well, despite Little's efforts. On the other hand, circulating these accounts—in "inside joke" form—transformed mouse people's individual experiences into an important source of community for their new working culture. Also, within such stories were often embedded useful preliminary results of mouse breeding experiments and reports on possible new mutants, as well as practical tips on mouse stock hygiene (e.g., whose colonies had recently endured a bout of "mouse pox," and speculations about what had caused it). In short, the Mouse Club newsletter was an informal yet powerful written of tool for sustaining mouse geneticists' communication network over the long distances and time periods that separated their contacts.[79]

America" or the "Founders Club": see also oral history interview by author with Elizabeth Russell, June 1993, Bar Harbor, ME; and oral history interview with Elizabeth Russell, Jackson Laboratory Oral History Collection, American Philosophical Society, p. 13; CCL to CBD, 14 January 1921, CBD/CSH-APS.

[79] See, e.g., Mouse Club newsletters from 1922, Box 739 and Box 732, CCL-UMO. In 1941, this form of communication became the official *Mouse Newsletter*—the forerunner of the contemporary *Mouse Genome* (now *Mammalian Genome*): see Rader, "The Mouse People."

Only a few months after his promotion, however, Little had a major clash of administrative authority with Davenport, and the younger man began entertaining a job offer from the trustees of the University of Maine to become their next president. Little officially accepted the job in March 1922, and that summer he began making plans to move his mice to their new laboratory home in the Science Building at Orono.[80] Little left behind breeding stock for all the standard inbred lines he had accumulated at the SEE. The Mouse Club of America and its newsletter continued beyond Little's tenure there as well; its members would still send each other stocks in between their working summers on the Long Island shore. But when Little left Cold Spring Harbor, he had succeeded more clearly in stabilizing the practical and professional aspects of inbred mouse work than in developing mouse or cancer genetics along the programmatic "lines of interest to medical men" to which he alluded when he arrived. Over Little's next six years as a university president, his new institutional circumstances would enable him to move these two scientific priorities into closer alignment.

CONCLUSION

The ties that Little forged among mice, medicine, and genetics in the first few decades of the twentieth century were strong enough to sustain his own career and to provide for the incorporation of these animals into some scientists' research repertoires. But even by Little's own formulation, successful inbred mouse work now required transforming not just the materials themselves, but the practices and professional alignments of both biological and medical research. As Little's debate with Maud Slye revealed, these fields had made some significant strides toward one another through work on transplantable tumors at Harvard and Cold Spring Harbor. But they were still separated by great divides: in terms of university departments and programs, geneticists had created many, but they were almost never integrated into medical schools, and in terms of social authority on issues of public health, medicine still wielded more despite inroads made by geneticists (for example, in immigration policy).[81]

Furthermore, although the arrangements Little and MacDowell made through Cold Spring Harbor provided some practical relief for geneticists,

[80] Little became assistant director when the SEE was reorganized into two sister institutions: the Department of Genetics and the Eugenics Record Office. See Clark, p. 77. The Davenport-Little clash is also described in detail in ibid., pp. 83–87; CCL to Davenport, 17 May 1922, and Davenport to CCL, 18 May 1922, both in CBD-APS.

[81] Diane Paul, *Controlling Human Heredity: 1865 to the Present* (New York: Prometheus Books, 1995).

inbred mouse creation, procurement, and use continued to be tied primarily to individual scientists, most of whom were geneticists. In the absence of better institutional infrastructures for maintaining inbred mouse materials, material scarcity of organisms continued to pose great problems for researchers. Former Bussey student William Gates, for example, wrote the *Mouse Newsletter* from his post at Louisiana State University in November 1922 to say that few of the strains he obtained from the summer spent at Cold Spring Harbor had survived the neglect that they experienced due to his heavy teaching schedule, and he declared his mouse work defunct.[82] His experience was not unique.[83]

These partial successes in terms of research, however, stand in sharp contrast to what appears to have been unanimous ideological agreement that inbred mice were appropriate for experimental use. The mouse's inferior ethical status combined with prevailing cultural stereotypes of both mice and antivivisectionists to provide a broader justification for scientific mouse work. According to historian Susan Lederer, challenges to animal experimentation in America began in the late nineteenth century over the issue of medical school vivisection, but these failed in large part because pro–medical research forces (primarily physicians) mounted well-organized public campaigns proclaiming the utilitarian benefits of animal experimentation. The motive force behind these clashes, in Lederer's view, was the "constant tension in the American debate between anti-vivisectionists, who attempted to identify themselves completely with animals undergoing experimentation, and medical researchers, who attempted to distance themselves from subjective judgments of animal distress." This subjective-objective tension was also reflected in the different political strategies the two groups took to make their ethical points. Antivivisectionists were far more likely to use visual appeals to display animal cruelty, while promedical forces avoided pictures of animals and relied on authoritative argumentation.[84]

Ultimately, the debate about the scientific uses of dogs in the 1930s appears to have defined the parameters of controversy over all animal use

[82] Gates, as quoted in the Mouse Club newsletter, November 1922, Box 739, CCL-UMO.

[83] Ilana Lowy and Jean-Paul Gaudilliere point out the ways in which Little's inbred mice might also be considered failures. They argue that genetically pure strains of mice have stabilized experimental cancer research practices, but not, as Little originally envisioned, the interplay between biological and clinical knowledge or the interplay of biologists and medical practitioners. See Lowy and Gaudilliere, "Disciplining Cancer."

[84] Lederer, "The Controversy over Animal Experimentation," esp. p. 237. For the connections between debates on animal antivivisection and human experimentation in the late nineteenth century, see Susan Lederer, "Human Experimentation and Anti-Vivisection in the Turn-of-the-Century America," Ph.D. diss., University of Wisconsin-Madison, 1987, and, more recently, Lederer, *Subjected to Science*.

for the same period. In sheer numbers there were undoubtedly far more mice in laboratories, but only if the entire ethical debate was recast according to quantitative criteria could the mouse have registered as more important to earlier twentieth-century animal activists.[85] Within the existing context, however, available evidence indicates that the cause of mice was not explicitly taken up by American antivivisectionists during this period.[86] During a time of heightened cultural awareness of issues of animal welfare, mouse experiments (by Little and others) appear to have excited no protests.[87] This silence, therefore, can be read as evidence of another, more profound success. Existing scientific and social values toward mice as animals remained intact, and these ensured that Little's inbred creatures could continue to cross the threshold into the laboratory in the decades to come.

[85] On the cultural legacy of the dog, see Russell and Secord, "Holy Dogs and the Laboratory." Cf. James Serpell, *In the Company of Animals: A Study of Human-Animal Relationships* (New York: Basil Blackwell, 1986), and more recently, Anthony Podberscek, Elizabeth Paul, and James Serpell, eds., *Companion Animals and Us: Exploring Relationships Between People and Pets* (New York: Cambridge University Press, 2000).

[86] As Lederer has noted: "Unlike medical researchers, the individuals who opposed animal experimentation did not, for the most part, deposit their correspondence and records in university archives." Therefore, historians have little material to go on in this area. My generalization is made based on a search of published records via the *Readers Guide* from 1920 to 1937, a search through the Rockefeller University Archives holdings on "Anti-Vivisection" (RG 600–1, Box 21, Folders 2 and 3, RAC-NY), and in consultation with Susan Lederer, who has done exhaustive research on the American Antivivisection Society's journals, *Journal of Zoophily*, *The Starry Cross*, and *The Anti-Vivisectionist*, in preparation for a book on the history of animal experimentation in the United States. For a broad overview, see Anita Guerrini, *Animal and Human Experimentation: An Introductory History* (Baltimore: Johns Hopkins University Press, 2002).

[87] The Cold Spring Harbor Laboratory Archive confirmed (personal correspondence, spring 2002) that there are no extant letters and/or new stories from local papers in the clipping files about any antivivisection activity at the lab during the 1910s–1930s.

CHAPTER TWO

EXPERIMENT AND CHANGE

Institutionalizing Inbred Mice (1922–30)

When Little moved from Cold Spring Harbor to Orono to assume the presidency of the University of Maine in June 1922, he lamented the disastrous transfer of his inbred mice: "It rained steadily for five days. This necessitated keeping the car curtains down and thus the atmosphere of the car was concentrated to a saturation point. In addition, the roads . . . were vile beyond description. . . . [S]ufficient jounces upset one or more of the mouse boxes occasionally, liberating all of its contents, animate and inanimate, all over the back of the car."[1] Still, the surviving *dba*, Bagg albino, and *C57* mice strains settled nicely into their new laboratory home, and Little remained as committed as ever to work with them. That fall he was inaugurated president of the University of Maine, and he described his planned educational experiments at the college to a Boston reporter, in words that echoed his previous arguments for the use of inbred mouse material in cancer research: "I am especially gratified at this appointment . . . because it will offer exceptionally favorable opportunities of testing certain theories and problems of education . . . [and] its [Maine's] population, composed largely of the best New England stock, is nearly as homogeneous as is possible to obtain at the present time." Little considered the chance to combine this work

[1] Mouse Club newsletters from 1922, Box 739, CCL-UMO.

2.1. C. C. Little as president of the University of Maine [Source Credit: C. C. Little Papers, University of Maine].

with his existing scientific research as the primary advantage of his new post (fig. 2.1)[2]

Likewise, three years later when he accepted the even more prestigious presidency of the University of Michigan in September 1925, one month shy of his thirty-seventh birthday, many of Little's friends and colleagues

[2] Downes, "Brookline Man," cover page; seventh (25th anniversary) Harvard College Class of 1910 Report, p. 450.

assumed that the increased administrative duties would finally mean the end of his scientific involvement with inbred mice. Raymond Pearl, a Johns Hopkins geneticist and Little's colleague on the Scientific Board of the American Birth Control League, wrote him a congratulatory letter in July 1925. He took credit for informing the Michigan trustees that Little was "the best man they could get for the place," though he conceded "it means the end of laboratory work for you, which is a pity, but no one can have his cake and eat it too." A newspaper reporter also noted the "apparent conflict between the pursuit of biology and the efficient running of a college," but Little himself rejected this dichotomy. His belief, he said, was that one should view "administration as the extension of pure experimentation" into the field: "the biological attitude towards human social problems has never been developed; it has been scarcely scratched." The reporter concluded that "above all else," Little remained "a biologist, who believes that experiment and change are the very breadth of life."[3]

What drew Little to college administration during this period was the same thing that drew him to inbred mouse research. These fields presented parallel opportunities to transform American science and culture into experimental systems that embodied his most cherished values: cooperation, efficiency, and social conscience. In practice, however, there was alternating synergy and competition between Little's endeavors as a college president and as a scientific researcher. Ultimately, through his institution-building efforts at Orono and Ann Arbor, inbred mice attained another scientific identity that incorporated these values into their production and use. Little's efforts succeeded in part because they coincided with broader shifts in the patronage of American scientific work—from small grants for individual scientists to programmatic funding for cooperative discipline-building—and because they resonated powerfully with what Eldon Eisenach has called the ethos of "managerial progressivism" among the industrial elite.[4] Little's entre-

[3] At the time of his appointment to Maine, Little was thirty-three years old and the youngest college president in the United States. He remained among the youngest when he was appointed to Michigan's presidency. See Frank Robbins, "The Administration of Clarence Cook Little," in *The University of Michigan: An Encyclopedic Survey*, vol. 1, ed. Wilfred B. Shaw (Ann Arbor: University of Michigan Press, 1942), pp. 88–98; Raymond Pearl to C. C. Little, 5 July 1925; Gardner Jackson, "Michigan Takes Live Wire from Maine in Clarence Cook Little," *Detroit News*, 3 July 1925; both in Box 742, CCL-UMO.

[4] Cancer researcher Michael Shimkin wryly commented on how alliance with business is a recurring pattern in Little's career. Little became the first director of the Tobacco Industry Research Council, Shimkin wrote in his 1977 history of cancer research, *Contrary to Nature: Being an Illustrated Commentary on Some Persons and Events of Historical Importance in the Development of Knowledge Concerning Cancer* (Bethesda: U.S. Dept. of

preneurship culminated in 1929 with the founding of the Jackson Memorial Laboratory in Bar Harbor, Maine. This institution stabilized his earlier research program in the genetics of mouse cancers and secured the material conditions of inbred mice as organisms for use in biological and medical research more broadly. It further embodied Little's own resistance to the organizational and intellectual particularism he thought had plagued both academic biomedical research and American culture in the immediate postwar period. Little's reformist impulses were essentially the driving forces behind the founding of the Jackson Lab and, in turn, the Jackson Lab "institutionalized" mice as research organisms that reified these ideologies.[5]

MOUSE WORK AT MAINE: OPPORTUNITIES AND CONSTRAINTS

In his first year at Orono, Little's desired career arrangement appeared to materialize exactly according to plan—in terms of both practical arrangements and research accomplished. Financially, Little brought with him from Cold Spring Harbor guaranteed support of $1,500 a year from the Carnegie Institution (the SEE's patron), and he negotiated with Maine's trustees for another $3,000 in research money as a condition of his ap-

Health, Education, and Welfare, Public Health Service, National Institutes of Health; Washington, DC: U.S. Govt. Printing Office, 1977), "probably because he preferred to receive research support from rich merchants rather than from the government" (p. VIII-C-2).

[5] Though "progressivism" is notoriously difficult to define (beginning with Daniel Rodger's landmark critique in 1982, a handful or more review essays written in American history during the last twenty years bear some version of the title "searching for progressivism"), Eisenach identifies "managerial progressivism" as one key strain of reform dominated by corporate liberals who sought change in the private sphere. See *The Lost Promise of Progressivism* (Lawrence: University of Kansas Press, 1994); Daniel Rodgers, *Atlantic Crossings: Social Politics in a Progressive Age* (Cambridge: Belknap Press of Harvard University, 1998). Little's reformist impulses were consistent with this ideology, in that they were nationally directed but not democratic; in fact, like other corporate and scientific attempts at reform, they were notably elitist in their heavy reliance on expert knowledge. See, e.g. Samuel P. Hays, *Conservation and the Gospel of Efficiency* (New York: Atheneum, 1959); and *Beauty, Health, and Permanence: Environmental Politics in the U.S.* (1987), as well as William Leach's magisterial history of the rise of consumer society, *Land of Desire: Merchants, Power, and the Rise of a New American Culture* (New York: Vintage, 1993). Regarding science, I presume Barbara Kimmelman's definition of "ideological" as simply involving "a politicized perspective," not external to the historical actors themselves but "arrived at through series of commitments over time" and "generated through their own perspective on their scientific work, which is always socially and institutionally situated." See Kimmelman, "Organisms and Interests: R. A. Emerson's Claims for the Unique Contributions of Agricultural Genetics," in *The Right Tools for the Job*, ed. Adele Clark and Joan Fujimura, pp. 198–232.

pointment. In the fall of 1922, Little informed his former SEE colleagues that his experimental organisms were "breeding in good shape" and "increasing rapidly" in their new home. Thus with the material conditions of his mice now secured, Little continued with the business of advancing his programmatic goals for mouse work. In 1923 he published a long paper in a pediatric journal that explained the results of mouse inheritance work on coat color and cancer, for a non–genetically trained medical audience. His stated goal was to increase the "recognition and appreciation" of mouse work among physicians in order to ensure "the adoption by medical men in general . . . of a biological point of view . . . which will aid immensely in reducing to an analyzable and predictable condition data that would otherwise remain unexplained." Again, in context, Little clearly equated "biological" with "genetic": embracing the genetic approach meant embracing the values of "analyzability" and "predictability" that he felt were important for all research.[6]

Little made some progress in the mouse work by enrolling more graduate students in his program of research. He brought with him to Maine one graduate student from Cold Spring Harbor: Beatrice Johnson, a former SEE assistant and Barnard College undergraduate. Johnson and Little had begun a collaborative project on the comparative genetics of normal and cancerous mouse tissue transplantation while they were at the SEE, and Johnson extended that work with Little's inbred strains to obtain a master's degree at Maine in 1925. Little also took on two more students once in Orono: the brothers Joseph and William Murray, former Maine undergraduates who began research on X-rays and mouse genetics.[7]

Shortly into 1923, however, trouble began to brew in Little's presidency, in part because of Little's continued commitment to his science. Maine's governor, Percival Baxter, demanded that the state's university

[6] CCL to Strickland (a Maine trustee), 7 March 1922, and CCL to Davenport, 21 March 1922; Davenport had strongly disapproved of Little going over his head to Carnegie's Merriam and requesting support for the Maine work, but Little did it anyway: cf. CCL to Davenport, 26 October 1921, and Davenport to CCL, 31 October 1921; all CBD-APS; fragment of a *Mouse Newsletter* by Little about transporting the mice to Maine, no date, Box 730, CCL-UMO; Little, "Congenital and Acquired Predisposition and Heredity," *Abt's Pediatrics* 1 (1923): 171–256, esp. 171.

[7] On Johnson, see *Who's Who of American Women*, vol. 1 (1958–59), p. 645; Little and Johnson, "The Inheritance of Susceptibility to Implants of Splenic Tissue in Mice. I. Japanese Waltzing Mice, Albinos and Their F1 Generation Hybrids," *Proceedings of the Society of Experimental Biology and Medicine* 19 (1922): 163–67. On the Murray brothers, see their entries in *American Men of Science*, 4th edition. Of the three students, Johnson most clearly internalized Little's programmatic vision: she later extended her graduate work in mouse breeding (a master's thesis on the "Effect of Ultra-Violet Light on Young Mice") to a project in the eugenics of birth control: see B. W. Johnson to Walter Taylor, 29 March 1929, Box 742, CCL-UMO.

become financially self-supporting, and he drastically limited the amount of funds available to the institution. Little strongly fought Baxter's proposal, using tactics that did not endear him to the governor; by 1925, Little's personal research stipend from the state had been cut to $300. It appears that Little made up for the cuts by relying on the personnel and facilities of the SEE in the summer months and on Maine's Agricultural Experiment Station, the other primary scientific unit of the Orono campus, during the academic year.[8]

Having graduate students contributed extra workers for mouse husbandry, but Little did not solve the practical problems of mouse work at Maine. He farmed out aspects of the mouse cancer problem in which he was still very interested but no longer had the time to complete himself. In 1924 he co-authored (with L. C. Strong) a new research study on the linkage between a mouse gene for cancer and one for coat color, and he published a new theoretical critique of the contemporary mouse transplantation studies done using Slye's pedigree methodology. That same summer he traveled to Bar Harbor to meet with several university trustees, and here he found someone unexpected who appreciated both the scientific and social aspects of this work.

MAINE'S SUMMER COLONY: THE DETROIT CONNECTION

Little met Roscoe B. Jackson (fig. 2.2) in Maine during the summer of 1924. Jackson and other prominent industrialists were among the wealthy seasonal residents of Bar Harbor during the early 1920s who were collectively known as "the summer colony." Year-round Bar Harbor residents tolerated the summer colonists, but a fairly rigid two-tier social system developed as a result of wide-ranging class distinctions. George B. Dorr, one of the few wealthy year-round Mt. Desert Island residents and a friend of Little's mother, owned large tracts of land in Bar Harbor, and in 1921 he donated a partition at the island's Salisbury Cove for the build-

[8] David C. Smith, *The First Century: A History of the University of Maine, 1865–1965* (Orono: University of Maine Press, 1979), pp. 120–22. Luckily, Little's troubles with Baxter did not set the tone for all of Little's future associations with Maine's state government: cf. Roswell Bates (Maine state representative) to CCL, 8 October 1954; H. J. Thorkelson, "Memorandum: University of Maine," General Education Board Collections, Record Group 2441, Box 680, Folder 7032, RAC-NY; Little and Strong, "Genetic Studies on the Transplantation of Two Adenocarcinomata," *Journal of Experimental Zoology* 41 (1924): 93–114; Little, "The Genetics of Tissue Transplantation in Mammals," *Journal of Cancer Research* 8 (1924): 75–95; cf. Little (with J. M. Murray and W. T. Bovie, "Influence of Ultra-violet Light on Nutrition in Poultry," *Maine Agricultural Experiment Station Bulletin* 320 (1924): 141–64.

2.2. Roscoe B. Jackson, official portrait, c. 1927 [Source Credit: C. C. Little Papers, University of Maine].

ing of what became Mount Desert Island Biological Laboratory (MDI-BL). Two summers after Little came to Orono, he brought his first group of zoology students to another site loaned by Dorr for a summer field study course. He most likely came to know Jackson and his family by circulating in Bar Harbor's small, insular, upper-class social circles during his own summer leisure time.[9]

Roscoe B. Jackson was, according to Little, a "small, well-built and handsome man"; by training, he was an engineer. Born in Ionia, Michigan, he attended the engineering college of the University of Michigan. After graduation in 1907, he designed cars for the Olds Motor Works, one of Lansing's most successful automobile manufacturers. In 1909, together with Roy Chapin, his college classmate and former Olds colleague, he established the Hudson Motor Car Company. This venture took its name from its primary financial backer, Detroit department store industrialist J. L. Hudson, who was also Jackson's wife's uncle. By introducing such innovations as the balanced crank-shaft and the first closed car, the Hudson Motor Car Company became a small but significant power in the automobile industry by the time Jackson assumed its presidency in 1923. Despite the high profile of the company, however, Jackson himself remained an enigma within Detroit social and business circles. He rarely appeared in public and was best known for shunning traditional executive cultural expectations.[10]

Perhaps because of his background as an engineer, Jackson was impressed with Little's administrative and scientific talents—so much so that he immediately began supporting his work. Jackson partially financed the biological field class at Bar Harbor during the summer of 1925.[11] His support for Little's Maine summer course increased modestly over the

[9] The MDI-BL always had a university connection: it was originally founded in 1898, under the name "Harpswell Laboratory," as a summer school of biology directed by J. S. Kingsley, zoology professor at Boston's Tufts University: see folder on "Weir Mitchell Station 1928" in Box 735, CCL-UMO. On Maine's "summer cottagers," see Charles E. Clark, *Maine: A Bicentennial History* (New York: W. W. Norton, 1977), pp. 153–56. In relation to the Jackson Lab, see Jean Holstein, *The First Fifty Years at the Jackson Laboratory, 1929–1979* (Bar Harbor: The Jackson Laboratory, 1979), pp. 4–7.

[10] "For years he [Jackson] had no desk in the main administration building of the plant, but maintained an office of his own out in the factory, close to the noise of the machines." (From Jackson's obituary in the *Detroit News*, 20 March 1929, front page.) For Little's description of Jackson,. see "Your question #10" [c. 1960s], Box 739, CCL-UMO; John B. Rae, *American Automobile Manufacturers: The First Forty Years* (New York: Chilton Company, 1959), esp. pp. 54–56, 154–55, 175–76.

[11] The first evidence of a financial exchange for the summer laboratory work is R. B. Jackson to C.C. Little, 13 July 1925, Box 742, CCL-UMO: Jackson gave Little $20. Little's resignation from the University of Maine was reported in *Time*, 13 July 1925, p. 20, and also *The New York Times*, 3 July 1925, p. 15. On the influence of the Hudson Motor Car Company in the 1920s, see Rae, *American Automobile Manufacturers*, pp. 56–58.

next few years, but like any businessman looking out for his investment, he soon began encouraging Little to enroll a wider audience, as the initial contributions allowed the biological work to expand in both scale and scope. In March 1926, for example, Jackson pledged $1,500 to Little for "expenses in connection with . . . your laboratory this coming summer," as well as a touring car for transporting students around the Maine island. He pragmatically directed Little to "spread" this latest contribution "among as large a number of individuals at Seal [Cove] as is practicable, not perhaps so much because of the matter of their financial assistance as to get as many of those possible interested in your activity." Apart from these general directives, however, Jackson apparently remained aloof from Little's actual scientific work.[12]

Little's ability to quickly forge a successful patronage relationship with Jackson shows that he had mastered one basic strategy for gaining resources within his contemporary system of scientific funding, namely, selling himself and his abilities in a one-on-one relationship between researcher and patron. To significantly scale up his mouse breeding and cancer research program as he desired, however, Little would require much larger sums and infrastructural support than this patronage relationship provided. Furthermore, the system of large-scale scientific patronage in the United States itself was in transition as a result of still-operative post–World War I concerns over scientific "manpower." As a matter of policy, individual grants-in-aid were less frequently given by large foundations such as Rockefeller and Carnegie than grants for departments or institutions that enabled the training and development of the scientific community as a whole. Little's ability to negotiate this transition would be tested in his later years at Michigan.[13]

"THE VALUE OF SCIENTIFIC RESEARCH": INBRED MICE WORK AT THE UNIVERSITY OF MICHIGAN

Little's Detroit summer colony connections clearly helped to place his name high on the University of Michigan's list of presidential candidates. Jackson Lab geneticist Elizabeth Shull Russell, who was the daughter of

[12] R.B. Jackson to CCL, 3 March 1926; CCL to RBJ, 10 March 1926; see also CCL to RBJ, 1 June 1926; RBJ to CCL, 8 June 1926; all Box 739, CCL-UMO.

[13] For early models of scientific patronage in the United States, see Howard Miller, *Dollars for Research* (Seattle: University of Washington Press, 1970) esp. chap. 6. On the individualistic character of the foundation patronage system in the first two decades of the twentieth century and the importance of a good patron-scientist relationship within this system, see Robert E. Kohler, *Partners in Science: Foundations and Natural Scientists, 1900–1945* (Chicago: University of Chicago Press, 1991), esp. pp. 71–72.

Michigan faculty member A. Franklin Shull and a Michigan undergraduate from 1925 to 1929, later explained: "Influential summer residents saw him on Mount Desert Island . . . [and] thought he would be good for the University."[14] But among the approximately forty-five suggested candidates laid before the committee, Little had a unique combination of university administration and laboratory experience.

The trustees and regents were openly looking for a proven leader in both these areas to ensure Michigan's continued transformation into a midwestern scientific powerhouse. Thus when Little's appointment as Michigan's next president was formally announced in July 1925, several of the committee members who chose him openly admitted that the prestige of his continued scientific work was a key factor in their decision. Two of the three members were from departments that had benefited enormously under the previous science-friendly administration of Marion L. Burton: Carl Huber of the medical school and Herbert Sadler of the department of marine and naval engineering. Sadler remarked glowingly that Little "had experience in a state university" and was aware of "the more direct service rendered to . . . the state by reason of its research laboratories." He continued: "That was President Burton's idea from the beginning. . . . I will not say that Dr. Little's reputation as a scientist and his investigations persuaded us to the choice, but I can say . . . the man who knows the value of scientific research through personal experience is likely to foster and encourage the truth-seeking urge in others. Such a man will understand the value of research to the university and to the community in general." Little himself visited the campus only two weeks prior and was reportedly just as "profoundly impressed" by Michigan's "dignity" and commitment to advancing knowledge through research, in contrast to the penny-pinching he had experienced at the University of Maine. Both the university and Little had high expectations of what each could do for the other's scientific reputation (fig. 2.3).[15]

Once again, Little's new employers went out of their way to insure that Little would carry on with his mouse work, but this required some financial and special finagling. Little was forceful about his needs: *The Detroit Saturday Night* reported that the only condition he attached "to the acceptance of his appointment was that he be provided with a private laboratory for research work, which he wishes to continue during his leisure

[14] Interview with Elizabeth Russell, 7 June 1986, JLOH-APS.

[15] Howard Peckham, "President Little Embattled," in *The Making of the University of Michigan, 1817–1967* (Ann Arbor: University of Michigan Press, 1967), p. 157. On the Burton administration's promotion of scientific research, see Robbins, "The Administration of Marion LeRoy Burton," in *The University of Michigan*, vol. 1, pp. 81–87; quote from Allen Shoenfield, ""U' Pins Faith in Dr. Little," *The Detroit News*, 7 July 1925, Box 742, CCL-UMO.

2.3. C. C. Little, University of Michigan presidential portrait [Source Credit: C. C. Little Papers, University of Maine].

from administrative work." Ultimately Little was granted $5,000 toward the maintenance of what was simply to be called "the President's Laboratory." This was an extraordinary sum considering both Little's paltry research allowance at Maine and the fact that Michigan had no general research fund for its regular faculty until Little himself wrote $3,000 into the budget to establish one during his first year there.[16]

Nevertheless, his fellow biologists appear to have welcomed him without resentment.[17] Shull wrote to Little that the biology department's "provision for mammal work is not very good" and recounted ten years of his own unsuccessful attempts to "attach" a vivarium (for animal breeding) to the construction budgets of various Michigan science buildings. Shull noted that basement space was available in the Natural Sciences building but perhaps not advisable, as "it would be a real thrilling sight" for those interested in Little's scientific work "to see the President of the University working in a basement!!"[18] Ultimately, Little decided to take the medical school space that anatomist Carl Huber offered and "hope[d] that somewhere over there it will be possible to find space for the mice." An official inventory of the President's Lab from early 1926 indicates that Little and his approximately 2,500 mice ended up occupying three large, nonadjoining rooms on the third floor of the new Medical Building.[19]

Laboratory budget ledgers for this year provide a revealing glimpse into the distribution of resources required for mouse work. During the

[16] Editorial: "New U. of M. President Seems an All-Around Man," *The Detroit Saturday Night*, 11 July 1925, Box 742, CCL-UMO; Holstein, *The First Fifty Years*, p. 10. While at Maine, Little received a total of $4,500 per year, but only $300 from the state of Maine to carry on research work: see H. J. Thorkelson, "Memorandum: University of Maine," 7 May 1925, GEB Collection, RG 2441, Box 680, Folder 7032, RAC-NY. By the 1927–28 budget, the Michigan Faculty Research Fund was granted a $30,000 appropriation; see Robbins, "The Administration of Clarence Cook Little," p. 93.

[17] G. Carl Huber (Medical Sciences) to CCL, 6 July 1925; CCL to Huber, 9 July 1925; A. Franklin Shull (Zoology) to CCL, 11 July 1925; CCL to Shull, 15 July 1925: all in Box 742, CCL-UMO; "G[otthelf] Carl Huber," *National Cyclopedia of American Biography*, vol. 42 (New York: James T. White, 1958), p. 693. See also Rollo E. McCotter, "Anatomy," in *The University of Michigan: An Encyclopedic Survey*, vol. 5: *The Medical School* (Ann Arbor: University of Michigan Press, 1951), pp. 818–19.

[18] Shull to CCL, 11 July 1925, Box 742, CCL-UMO. On Shull, see Garland Allen, "Aaron Franklin Shull," *Dictionary of Scientific Biography*, vol. 12, ed. C. Gillispie (New York: Scribners, 1970), pp. 416–18.

[19] The number of mice is based on an "Official Inventory of the President's Lab," dated 4 June 1926, which indicates Little had 384 small mouse boxes (which held approximately 2–3 mice each) and 175 large mouse boxes (approx. 8–10 mice each). Thus Little could have had anywhere from 2,096 to 2,794 mice; see "Appointments to Research Assistants" Folder, Box 742, CCL-UMO. Once again, I am indebted to Lee Silver for discussions on the history of mouse boxes, especially regarding their capacity and breeding uses. On the long history of problems obtaining space for mouse breeding on university campuses, see interview with Earl and Margaret Green, June 1993, JLOH-KR.

1926–27 fiscal year, the university appropriation for Little's laboratory remained $5,000, and a sizable portion of the money appropriated (56%) was used to equip the laboratory and to maintain the mice for scientific use. Nearly $3,000—an amount equal to what President Little requested for the total Michigan faculty research fund—went for the salaries of two full-time caretakers, mouse food (canary and hemp seeds), cotton batting, wood shavings, and other miscellaneous laboratory equipment.[20] Less than half of the laboratory's budget went to pay the salaries of two scientific workers, Joseph and William Murray. The Murray brothers had both studied zoology and done research on mice cancers with Little when he was at Maine: Joe as an undergraduate (B.A. 1925), and Bill as an undergraduate and a graduate student (B.A. 1921, M.A. 1925). Little's new arrangement at Michigan promised them steady employment as well as plenty of time for each to continue their research toward a doctorate, so they both followed him to Michigan in 1925. For Little at Michigan, just like Castle at the Bussey before him, it was advantageous to have such continuity among the workers that dealt with the mouse colonies. The Murray brothers had been around Little's laboratory for several years in Maine, and they already had valuable hands-on knowledge of both the mouse colonies and Little's desired work arrangements. Their presence provided added security that there would be scientific supervision of the mouse caretaking, even if Little's administrative duties made visiting the laboratory on a daily basis impossible.[21]

Little also tried to surround himself with other scientists who had practical experience with mice and an interest in his research program. Before he even arrived on campus and set up his laboratory, Little suggested to Shull that the zoology department consider hiring L. C. Strong, his former student from Cold Spring Harbor, to work part-time as an instructor, "with his salary increased from another source for research work" in the President's Laboratory. Despite an enthusiastic recommendation from T. H. Morgan, Shull wrote back that the department's hands were tied. Any new position that might soon open up, he said, would require much teaching, plus Strong wanted a salary twice as high as they could possibly offer. Notably, Shull took the opportunity to tell Little that he "would be very glad to see funds accredited to the department used for research."

[20] President's Laboratory, Department of Cancer Research, Budget 1926–29, Box 735, CCL-UMO.

[21] The only printed biographical sources on William S. Murray and Joseph M. Murray of which I have knowledge are their brief sketches in the 1938 edition of *American Men of Science*. On the shift of mouse workers from Maine to Michigan, see Elizabeth Russell, "Origins and History of Mouse Inbred Strains: Contributions of Clarence Cook Little," in *The Origins of Inbred Mice*, ed. Morse, p. 38.

Little was noncommittal on this larger request but conceded to waiting until finances were better to acquire Strong.[22]

Although Michigan provided Little sizable and regular support to enable his mouse work, he immediately began thinking beyond his initial arrangements. From his fifteen years of experience with mouse breeding, Little very likely was aware that it took more time and money simply to sustain a mouse colony than to work on simpler organisms also used for genetics research, such as *Drosophila*. To care for even one or two mouse strains to the point where they would be considered scientifically valuable for Little's genetics research (i.e., through eight generations of inbreeding) required financial resources more consistent than could be obtained by a state-run university.

In part, Little's urgency stemmed from the fact that the practical stakes were getting higher. Now if the lack of hygienic facilities and supervised caretakers for mouse breeding led to an epidemic—as it had once, just before he left Cold Spring Harbor—it could take many months to regenerate each inbred strain to the same level of genetic homogeneity, compared to several weeks with wiped-out *Drosophila* stocks. With his increased administrative duties, Little could not devote himself to such a program, nor could he afford to lose valuable research time rebuilding his organismic tools if he wanted to keep up with those working full-time on the cancer problem. Scaling up at this stage would provide insurance for his research program. To do so would involve the hard work of broadening the appeal of his scientific project, but it would have a large pay-off: it would effectively stabilize the material conditions of his resource-intensive research material.[23]

THINKING BIG: A NATIONAL BIOLOGICAL INSTITUTE

In July 1925, before arriving on campus to begin his presidency, Little wrote to Hugh Cabot, dean of the medical school, of his desire to make Michigan a paragon of cooperative scientific research in the Midwest: "I certainly am counting on you to do much towards helping us all realize some of the great things which everyone at Ann Arbor seems to expect. I

[22] Shull to Little, 11 July 1925; Little to Shull, 15 July 1925: both in Box 742, CCL-UMO.

[23] Cf. the material constraints and opportunities *Drosophila* work presented, as detailed in Kohler, *Lords of the Fly*, esp. part 1. It is possible, given Little's knowledge of genetics and Morgan's well-known experiences with *Drosophila* after scaling up the Fly Room, that Little anticipated that expansion of his own breeding efforts would also produce more mouse mutants for mammalian genetic study. In this early period, however, Little makes no explicit mention of such a rationale in any of his funding proposals.

was tremendously impressed by the opportunity existing there for medical work. There is apparently a real chance to work out the relationship between medicine, physics and biology on a scale which ought to exercise our ingenuity and appeal to the imagination." What Little did not tell Cabot, however, was that plans to realize this ambitious vision were already in the works. When Little returned to Bar Harbor in June 1925 to lead the biological field course, he and Harvard botanist E. M. East discussed the possibility of creating a national institute for biological research which would be similar in size and scope to the Rockefeller philanthropy's successful Institute of Medical Research.[24]

East's interests in such a project probably stemmed from his increasing dissatisfaction with trying to do what was perceived as "applied" agricultural genetics in the context of Harvard's academic biology environment. Ten years after Little's graduation, the Bussey remained a thriving place for genetics research in the United States, and East had many excellent students. Still, East confided in Little that he thought his research in agriculture and the genetics of plants was of much lower priority than Castle's mammalian work, and that he longed for a "larger sphere of activity in doing and directing research."[25] Little wrote back, echoing East's desire to make the new project comprehensive in a way that centralized all the involved parties at Michigan: "I should like to utilize all divergent interests in establishing a common graduate school which would be sufficiently large and comprehensive to appeal to the imagination of the Rockefeller." Right after the new year in 1926, Little initiated contact with the Rockefeller philanthropy by calling on agriculturist Dr. Whitney Shepardson of the General Education Board (GEB) to discuss the establishment of a national "Institute of Biological Research," headed by himself and East.[26]

Little's first meetings with the Rockefeller people passed uneventfully and without any promise of action on the big project, so two months later, Little called on H. J. Thorkelson, director of the GEB's division of colleges and universities, to pitch the idea another way: by stressing integrated research across departmental boundaries as a key form of science educators' social responsibility. By this time, Little had formulated

[24] CCL to Hugh Cabot, 15 July 1925, Box 742, CCL-UMO. On the creation of the Rockefeller's Institute for Medical Research, see Corner, *A History of the Rockefeller Institute.*

[25] William B. Provine, "Edward Murray East," *Dictionary of Scientific Biography,* 4: 270–72; CCL to East, 28 September 1925, in CCL Presidential Papers, as discussed in Clark, "The Social Uses of Scientific Knowledge," pp. 157–58; East to CCL, 15 October 1925, Box 4, File 23, CCL-UMich.

[26] CCL to East, 28 September 1925; East to CCL 15 October 1925 (in which East told Little: "I shall be glad to do anything you say."); CCL to East, 13 January 1926; all in Box 4, Folder 23, CCL-Mich.

a three-page outline proposal for the project, in which he requested a $5.5 million dollar preliminary endowment, with a significant amount of that money devoted to "secondary buildings" for animal breeding houses. He impressed upon Thorkelson that the University of Chicago and the University of Michigan were the only two midwest institutions that were distanced enough from their states' agricultural colleges to assure that the research would be truly independent, and despite his own problems finding room for his mice, he lauded the availability of campus space. He also noted his own administration's progressive attitude toward making new arrangements for collaborative research: "[they] wish to try *new* things . . . [and] . . . are proud of their tradition in this respect." Little suggested that the new biological institute would not cater to the immediate educational needs of the college by providing advanced degrees but rather would "fill two great human needs" by striving to maintain the "highest standards of research" in agriculture, physiology, and genetics; Little vaguely described the result as "preventative medicine" for American society. Not surprisingly, Thorkelson didn't bite: he indicated in his internal comments that he was "not impressed with many of the fundamental ideas of organization and financing" that Little had developed.[27]

Even after East's interest in the project had apparently waned, Little persisted. By February 1928, Little's original cooperative vision had mutated into something that blatantly served his own specific research interests. In one of his last interviews with Thorkelson, Little proposed that, in addition to a multiyear grant for his mammalian genetics research, the GEB fund a $250,000 one-story laboratory for experimental biology, which "will contain in the center general supply rooms for food and four wings each containing a series of bays" for holding mice. This wing would be specially designed to keep the groups of animals separate in case of an epidemic. Furthermore, Little had now added a eugenic research component to the proposal: a $100,000 addition to Michigan's new education building to be used for scientific studies on illegitimate infants from the state mental hospital "as material" for prekindergarten learning experiments.[28] Eventually Little attempted to attract the GEB's attention by the more traditional means of boosting the research and graduate training credentials of Michigan's biology program. In early 1928, for example,

[27] On Thorkelson, see Kohler, *Partners in Science*, pp. 204–207; H. J. Thorkelson, "Interview: C. C. Little, 5 March 1926," General Education Board Collection, Record Group 2239, Box 641, Folder 6712, RAC-NY.

[28] Thorkelson interview with C. C. Little, 17 February 1928, GEB, RG 2239, Box 641, Folder 6712, RAC-NY. On Little's interest in scientific education research to further a program of eugenical selection in schools, see Clark, "The Social Uses of Scientific Knowledge," chaps. 2 and 4.

Little sought to bring two prominent geneticists to the Ann Arbor campus. He lobbied to lure East away from his established niche at Harvard to head up the botany department at Michigan, and he tried to recruit mouse geneticist L. C. Dunn, then at the Storrs Agricultural Experiment Station in Connecticut. Ultimately, Little also tried using the personal connections of Roscoe Jackson to go over the heads of GEB officers and get the personal attention of John D. Rockefeller, Jr., for his project. But all of these strategies were too little, too late: the Rockefeller organization was growing visibly tired of Little's attempts to sell them on what Thorkelson sarcastically deemed "the general outline of a problem he has in mind for the advancement of human knowledge."

In March 1928, when Thorkelson paid a visit in Ann Arbor to formally deny the request, Little took him to his laboratory and showed him the inbred mice. Officially Little still clung to his reformist principles, but the practical argument for basic genetics research was what convinced the Rockefeller organization to contribute. According to Thorkelson's report, Little claimed resolutely that his "work in mammalian genetics cannot be carried on without the proposed new building."[29] Thorkelson had good reason to mistrust Little personally: his blatant self-promotion of the biological institute project almost destroyed the GEB's simultaneous negotiations to fund Michigan's physics department's path-breaking work in nuclear spectroscopy.[30] But under aggressive direct pressure from Little himself, Thorkelson finally acquiesced and recommended that the GEB finance half of the money needed to build Little's new laboratory.

Once he had Thorkelson's promise of funding in hand, Little pulled a bait and switch on all counts. He now estimated the laboratory building would cost only $150,000—a full $100,000 less than his previous estimate—and pitched the project to his Board of Trustees as having a

[29] CCL to W. H. Shepardson, 2 November 1925 and 8 December 1925; CCL to Wycliffe Rose, 29 September 1926 and 11 November 1926; Rose to CCL, 2 October 1926 and 15 November 1926; Thorkelson's memo, 5 March 1926; H. M. Randall to H. J. Thorkelson, 13 December 1926; all in GEB Collection, RG 2239, Box 641, Folder 6712. See also CCL to E. M. East, 29 September 1926, Box 739, CCL Papers-UMO; CCL to L. C. Dunn, 21 March 1928, LCD-APS. Dunn declined and went to Columbia instead (30 April 1928, Box 727, CCL-UMO), saying that he "appreciate[d] more than I can say the splendid opportunity you opened for me. I hope you have the sort of group you want at Ann Arbor." CCL to Roscoe B. Jackson, 18 January 1927; RBJ to CCL, 9 February 1927; both in Box 5, Folder 7, CCL-Mich. Thorkelson interview with CCL, 16 February 1928, GEB, RG 2239, Box 641, Folder 6712, RAC-NY.

[30] On Little, Randall, and Thorkelson, see Kohler, *Partners in Science*, pp. 225–30. Little selfishly refused to take Randall's request for matching funds to the state legislature and instead resubmitted to the Rockefeller organization a request for supplementing his own research on mice cancers as an implied condition for Michigan accepting any physics funding.

broader social service function: "[it] would serve as a means of educating the public to support research."[31] But it is important to note that Little failed to enlist the GEB on his own terms, largely because he misread the changing Rockefeller patronage system. With such specialized institutes as the one Little was proposing, the key question for the GEB was: might they develop into regional centers of research and training, or would they "remain one-man shows strictly of local interest"? In Little's case, it was unclear what was intended, other than to subvert the traditional department-centered funding system because it was inconvenient for his own purposes.[32] To the new Rockefeller leaders, such individualism clearly was not "cooperative" and thus must have appeared heavy handed. This strategy was doomed to failure in a system that now pivoted around the long-term goal of strengthening the whole scientific community through focused, local initiatives—and ultimately, it did not even meet Little's own criteria for expanding inbred mouse research in ways that would benefit all of American biology and society.

EXPANDING MOUSE WORK: THE CANCER FOCUS

Accordingly, after he obtained Rockefeller support to build the new building, Little redirected his energies toward gaining increased support from his Detroit patrons. More specifically, he sought to make his research program in mouse genetics accessible to a wider audience by linking it to a disease that many Americans perceived to be an increasing threat to public health, namely, what contemporary doctors referred to as "the cancer problem." In this way, Little made supporting inbred mouse research socially acceptable: his patrons were not simply funding efficient science, they were funding a project that would eventually cure an important disease.

Throughout the 1910s and 1920s, U.S. government statistics indicated that cancer mortalities were increasing. Just as importantly, as cultural historian James Patterson has shown, magazines, newspapers, and other

[31] Thorkelson interview with C. C. Little, 16 March 1928; see also Thorkelson's Memo, University of Michigan, 2 October 1928; both in GEB, RG 2239, Box 641, Folder 6712, RAC-NY.

[32] Kohler, *Partners in Science*, chap. 8, esp. pp. 221–2. Kohler notes that proposals to transform existing science departments into research institutes were also put forward by Arthur Compton (physics) at Chicago and Frank Lillie (genetics), but these efforts faced problems similar to those in the Little case; cf. the fate of Joseph Needham et al.'s 1935 proposal for an "Institute for Mathematico-Physico-Chemical Morphology," in Pnina Abir-Am, "The Discourse of Physical Power and Biological Knowledge in the 1930s: A Reappraisal of the Rockefeller Foundation's 'Policy' in Molecular Biology," *Social Studies of Science* 12 (1982): 341–82, esp. 362.

media from the period reveal that cancer captured the American public's imagination as a "dread" disease. The subsequent "conquest of cancer" challenge, which was constructed primarily by the American medical profession in response to public fear, began to attract interest among philanthropists looking for noble, popular causes. The Rockefeller philanthropy, for example, supported cancer research on a large scale, but primarily through its own Medical Institute and the individual initiatives of J. D. Rockefeller, Jr.[33]

At the University of Maine, a significant part of Little's scientific work on the mice he developed and maintained there had concerned cancer. Half of the dozen publications that he produced between 1922 and 1924 dealt with cancers in mice or the relationship between mouse and human cancers. Furthermore, while none of his Maine graduate students did doctoral or master's theses on mouse cancers, a bibliography of the Michigan lab between 1926 and 1929 indicates that many of these same workers were publishing in mouse cancer research as early as 1926–27. It is reasonable, then, to assume that Little's practical commitment to developing some lines of cancer research work originated early. Still, as of 1925, Little himself had not yet made any large-scale appeals to patrons based on his specific research interests in cancer.[34]

This is especially curious considering that Little had already gained notoriety among medical workers for his views on the cancer problem and on eugenics. As an outspoken advocate of birth control, Little became director of the scientific board of Margaret Sanger's American Birth Control League (ABCL) in 1925. Little probably alienated some physicians when he sided against Robert Dickinson, an eminent gynecologist and member of the American Medical Association, during debates over the licensing of birth control facilities that were sponsored by the ABCL's Maternity Research Council. Before this, however, Little's mouse work became known in medical circles through his highly publicized feud with Maud Slye of the University of Chicago (see chapter 1).[35] Little's unrelenting criticism of Slye won him many allies in the medical profession, which was then at the height of its own professional struggle against nonscientific "quackery." His subsequent work with M.D. and cancer researcher

[33] James T. Patterson, *The Dread Disease: Cancer and Modern American Culture* (Cambridge: Harvard University Press, 1987), chaps. 3 and 4, esp. pp. 76–77 and 97–98, on ASCC "Cancer Weeks" (an early 1920s' invention) and increasing media attention for cancer.

[34] See Little bibliography appended to Snell, "Clarence Cook Little"; "Bibliography of Papers Published 1926–29 Inclusive," Box 740, CCL-UMO.

[35] This controversy is discussed in detail in David M. Kennedy, *Birth Control in America: The Career of Margaret Sanger* (New Haven: Yale University Press, 1970), esp. chap. 7, pp. 198–99. See also Little's correspondence with Pearl on the matter in Box 6, File 26, CCL-Mich.

H. J. Bagg of New York's Memorial Hospital was also well received and published in respected medical journals. By the early 1920s, Little was even recruited to speak in more popular forums on the topic of cancer inheritance.[36]

But it was Jackson, not Little, who initially latched onto the public health aspects of the inbred mouse research as a way to get more funds. In March 1926 Jackson prompted Little to keep him informed of his plans pertaining to not only the summer field course, "but that part which you mentioned when you were last in Detroit of offering talks to the community of popular interest" on cancer research. If the research activities could be given a cancer slant, Jackson told Little, "it would make it easier to present the case to some of my friends." Little quickly responded by sending Jackson a statement concerning the laboratory's work on cancer-related problems. He also offered to "amplify" any of it to suit the context of Jackson's local letter-writing campaign for more funds.[37]

Although this proposal argued for the establishment of a special laboratory and a "Fund for Cancer Research" at Michigan, it is unclear how Little himself saw the full scope or mission of this new project. In a 1926 *Commonwealth Review* article, Little argued passionately for the urgency of a program to work out "the general principles of research on the cancer problem" with "some mammalian material which ages more rapidly than does man." But his subsequent letters to Jackson were more matter-of-fact. Little told him simply that funds would be necessary to hire new researchers, to purchase equipment, and to prepare more existing university spaces as animal rooms for his mice. Regardless, Jackson used Little's new emphasis on cancer to generate support for this work among his wealthy Detroit friends. He managed to convince his brother-in-law Richard Webber, head of Michigan's J. L. Hudson Department Stores, and Edsel Ford, son of automaker Henry Ford and head of Ford Motorworks, that funding Little's work on cancer would be a worthwhile investment. Webber, Ford, and their families summered in Maine with the Jacksons, so their paths probably crossed Little's enough that the scientist himself would have also been able to make a personal appeal on these grounds.

[36] On medicine versus quackery, see Richard Harrison Shryock, *The Development of Modern Medicine* (Philadelphia: University of Pennsylvania Press, 1936), pp. 241–42, 250–55, 303–42; Starr, *The Social Transformation of American Medicine*. For a broad view of professionalization movements (including medicine's) in early twentieth-century America, see Robert H. Wiebe, *The Search for Order, 1877–1920* (New York: Hill and Wang, 1967), esp. chap. 5; Little, "Relations Between Research in Human Heredity and Experimental Genetics, *Scientific Monthly* 14 (1922): 401–14; and Little, "Inheritance of a Predisposition to Cancer in Man," *Eugenics, Genetics and Family* 1 (1923): 186.

[37] R. B. Jackson to CCL, 3 March 1926, and CCL to R. B. Jackson, 10 March 1926; also, CCL to R. B. Jackson, 1 June 1926; R. B. Jackson to CCL, 8 June 1926; all Box 739,

Within one month, the three industrialists crafted a formal financial proposal for "the creation and financing . . . of a laboratory for scientific research and investigation of the biological phenomena of cancer" and presented it to the University of Michigan's governing body. Beginning in July 1927, they proposed, Little would receive $3,750 per month for his laboratory through October 1932—one-fourth provided by each Jackson and Webber and one-half provided by Ford. Little's patrons affirmed their trust that he would keep the venture's ultimate "end in view of discovering and promoting scientific knowledge of the causes, prevention and cure of cancer." Their only stipulation was that he be given full rein over the entire project.[38]

Michigan's regents quickly accepted the $45,000 proposal, and plans began to inaugurate what they called the "Laboratory of Mammalian Genetics, in Co-Operation with the Cancer Research Fund," which would succeed Little's President's Laboratory. The Murray brothers would stay on as senior assistants at slightly increased wages, and in February 1927 Little finally secured mouse geneticist L. C. Strong as senior research associate for the coming academic year. For reasons that are less clear, Little also initially hired Michigan physiologist Alvalyn Eunice Woodward as his other full-time research associate. Woodward did physiological research on *Orthoptera* as an associate professor of zoology at Maine, and though she was hired after Little's tenure as president, perhaps their paths had originally crossed in Orono.[39] But the bulk of the initial money from the five-year Detroit endowment went toward funding junior research assistants and nonskilled mouse caretakers. Little initially hired seven graduate students, including histologist Elizabeth Fekete, zoologists Charles Green and John J. Bittner, and medical school researcher Arthur Cloudman. Considering that Little also estimated that mouse maintenance expenses would more than double in the next academic year, he probably expected these workers to supervise mouse breeding and caring for the research animals.[40]

CCL-UMO. Although it is mentioned in these letters, I have found no extant copy of Little's description of his lab's cancer-related activities.

[38] CCL to Jackson, 18 January 1927 and 17 February 1927; Jackson to CCL, 19 February 1927; see "Preliminary proposal of R. H. Webber, E. B. Ford and R. B. Jackson . . . for the creation and investigation of the biological phenomena of cancer and other forms of growth"; all in Box 5, Folder 7, CCL-Mich; Little, "Genetic Investigations and the Cancer Problem," *Commonwealth Review* 8 (1926): 130–36; Holstein, *The First Fifty Years*, chap. 1.

[39] See "The Institutes" section of Walter A. Donnelly et al., eds. *The University of Michigan: An Encyclopedic Survey*, vol. 4 (Ann Arbor: University of Michigan Press, 1958), pp. 1538–39; I am indebted to Margaret Rossiter for this reference. A very brief biographical sketch of Woodward can be found in *American Men of Science*, 1938 edition.

[40] "Expenditures and Proposed Budget, July 1927–June 1929," Box 742, CCL-UMO. Cloudman obtained his bachelor's and his master's degrees from the University of Maine

Ultimately, however, this financial windfall was not as consequential as Little had anticipated. For example, it is not clear that mouse production was scaled up at the Michigan Cancer Research Lab along the broad lines that Little originally envisioned in the GEB Biological Institute proposal. There is no evidence that the university gave Little and his group significantly more room for mouse breeding. Furthermore, the survival of individual strains appears to have still been linked primarily to individual researchers and their particular animal needs. Each of the Murray brothers still bore responsibility for a different mouse strain—Bill Murray looked after the *dba* stock, and Joe, the *C57*s—and these husbandry assignments corresponded with the strains predominantly employed in their respective projects.[41]

Still, by shifting attention away from himself at times and focusing the scope of his reformist goals, Little had finally succeeded in obtaining the resources necessary to begin scaling up the inbred mouse work. Jackson and the other Detroit patrons did not appear to have increased their investment in Little's cancer work simply to gild their big business images with the patina of medical philanthropy. All three donors wanted their gift to remain anonymous and insisted that only the medically valuable results of laboratory work should be the target of media attention. Rather, as their proposal suggests, they increased their investment because Little struck the right balance between promoting his own scientific talents and agreeing to sustain a broad program of cancer research.

THINKING BIGGER: "A SMALL LABORATORY" IN MAINE

Over the next few years Little managed to remain in the good graces of medical practitioners, and recognition for his cancer work and eugenic ideas continued to grow. Throughout 1926 and 1927, Little spoke often to audiences of medical researchers and clinicians, and these speeches—not research papers on mice cancers from the Michigan lab—provided the bulk of his scientific publications for the same period. In February 1926 Little spoke to the American Congress of Internal Medicine twice: he gave a full paper on safe and socially responsible methods of steriliza-

sometime before 1926, when he became instructor of zoology at the University of Vermont. Thus Cloudman was probably Little's student at Maine in some capacity (either undergraduate or graduate), though I have not yet been able to locate his master's thesis for confirmation: for a biographical sketch of Cloudman, see CCL to Burton Simpson, 3 July 1936, Box 740, CCL-UMO.

[41] On the Murray brothers' husbandry tasks, see Russell, "Origins": cf. "Bibliography"; CCL to R. B. Jackson, 17 February 1927; CCL to RBJ, 19 February 1927; both Box 5, File 7, CCL-Mich.

tion and an informal address (as he described it: "from a layman") on needed reforms in medical education. In October 1927 he also gave the Fellowship Address at the American College of Surgeons on the dangers of overusing radioactive substances in medical procedures.[42]

During this time, Little also maintained his genetic interests in mouse material beyond cancer research. Like his mentor W. E. Castle, Little foraged the world for new, nondomesticated species of mice to be used in genetic comparison studies with his inbred strains. For example, he and his young family took a tour of Europe in the mid-1920s, and he planned it to end up with a trip to the Faroe Islands. The Faroes were a small Denmark territory, habited in part by members of the wild mouse species *Mus faeroensis*, and Little and his family trapped several dozen specimens to take home with them on the oceanliner *Aquitania*. Only a very few survived the trip, but this family activity resulted in the American introduction and eventual laboratory domestication of *Mus faeroensis* for linkage and developmental studies. Also, in 1927, Michigan acquired Horace Feldman, a former Bussey Institute graduate student, as instructor of zoology. Feldman and his colleague Lee Dice, curator of mammals in the Michigan Museum of Zoology, embarked on a decade of genetic research on variation within and among populations of wild and laboratory mice. Little sanctioned Lee and Dice's nonapplied mouse genetics work in early 1928 by giving them much-needed space for their large *Peromyscus* (deer mice) colonies in the newly completed Museum buildings. Clearly, noncancer-oriented genetics research on mice—and the material maintenance this required—also flourished at Michigan under Little's tenure.[43]

By contrast, Little's second college presidency progressed less smoothly. By 1927, the faculty gratefully acknowledged him for having increased their research fund from $3,000 to $30,000 in only two years, and students lauded his campus life improvements such as Freshman Week and

[42] C. C. Little, "A Discussion of Certain Phases of Sterility," *Annals of Clinical Medicine* 5 (1926): 1–4; "Agents Modifying the Germ Plasm," *Surgery, Gynecology and Obstetrics* 46 (1928): 155–58.

[43] This story is recalled by Little's son, Robert, in Holstein, *The First Fifty Years*, p. 76. Little's investments in mouse genetics at Michigan were fairly short-lived. Feldman ultimately became director of the newly christened "Laboratory of Vertebrate Genetics" upon Little's resignation. Green, Bittner, and Cloudman stayed at Michigan under his supervision and finished their Ph.D.s before following Little. But by 1934, interest in genetics at Michigan had waned and Dice took over the operation as director of the again renamed "Laboratory of Vertebrate Biology"; the original stocks of mice were largely discontinued. On Dice and Feldman, see Lee Dice, "The Institute of Human Biology," in *The University of Michigan: An Encyclopedic Survey*, vol. 4, pp. 1537–45; cf. budget and personnel listing in S. W. Smith to CCL, 31 May 1929, Box 8–7, JLA-BH. Still, work on *Peromyscus* mice at Michigan continued into the 1950s: see Emmet Thurman, Jr., *Dental Pattern in the Mice of the Genus Peromyscus* (Ann Arbor: University of Michigan Press, 1957).

women's dormitories. But Little's persistent struggles with the Michigan regents and trustees escalated when he proposed to use Michigan's undergraduate curriculum for a grand educational experiment. Little pushed to enroll all freshmen and sophomores in a separate University College under its own dean. This innovative system—which eventually became a model for most U.S. universities—would provide two years of common grounding in several basic subjects for all students. But even more importantly, Little argued, it would discern the genuinely promising students, who would be encouraged to go for a four-year degree, from students simply desiring a university stamp of approval by their name for job applications, who would be encouraged to leave. The Senate approved the plan, but the regents and the Literary College faculty provided stiff opposition. The regents noted that Little was not a man to back away from a "fighting attitude" on any issue on which he thought he was right.[44]

During these turbulent times in Little's administration, the Michigan Cancer Research Laboratory remained productive, and Little's relationship with his Detroit patrons grew stronger. The lab's scientists and scientific assistants worked year-round, taking almost no vacation for a year, and as a result the Michigan group published an impressive total of thirty-two articles on mouse research, on topics ranging from transplantable tumors in inbred strains to the effects of X-ray irradiation on mouse ovaries. In the summer of 1928, Jackson and Little were both elected to the board of the Mount Desert Island Biological Laboratory. Little finally got an official institutional foothold in Bar Harbor when the MDI-BL board voted to incorporate his annual field biology course into its own operations.[45]

Still, by early 1928, Little openly expressed discontent over both his scientific and administrative progress at Michigan. In March 1928 he wrote to L. C. Dunn:

> The research study plan has been making progress although not as rapid [*sic*] as I naturally would enjoy. . . . At the present time, the majority of

[44] On top of this controversy, in 1928 Little clashed with Michigan's legislature when he opposed a $10 million gift from a wealthy law school alumnus because it was earmarked by the donor for purposes Little felt were not in the best interests of the university. See "Little Forced Out of Michigan University," *New York Telegram*, 11 February 1929, in Box 743, CCL-UMO. Also "Dr. Little Explains Resignation," *New York Times*, 16 February 1929, p. 36: by banning automobiles on campus, he compromised his popularity with undergraduates. See Peckham, *The Making of the University of Michigan*, chap. 9; Ralph Stone to CCL, 19 July 1928, Folder 14, CCL-Mich.

[45] "Vacation Schedule" from the Michigan Lab, c. 1927–28 (based on workers listed), Box 730, CCL-UMO; "Bibliography"; Holstein, *The First Fifty Years*, p. 5.

> the faculties would, I believe, prefer that the initiation of educational reform here should be left entirely in their hands. With this I cannot agree because of their past inactivity and present illogical approach to the situation. As a result of this, it is possible that my stay here . . . will be decidedly more limited than I thought it would be a year ago.

In May of the same year, Little penned a letter to T. H. Morgan discussing possible options for his scientific future. Morgan replied enthusiastically to Little's apparent suggestion that he might leave university administration and head back to mouse research full time: "I need not tell you that it will be good news if you should decide to give all of your time to genetics and experimental biology. I have marveled that you have been able to keep up and active interest in these subjects and carry on the horrible duties of a President. . . . I can not but think that we [biologists] have a happier if less excitable existence."[46] Increasingly vicious debates with the regents provided the ultimate catalyst for Little's break with Michigan. In January 1929, he submitted a letter of resignation.

The national media followed this action with great interest: magazines and newspapers published several testimonials to Little's achievements in science and education, and writers engaged in wild speculation about his future plans. In a cover story on Little, *Time* reported that the two-time former college president was debating whether to "ally himself to the Rockefeller Foundation to continue his research on cancer." Little himself claimed: "I haven't a job in sight. It's the first time I've been out of work since my graduation from Harvard."[47] In private he took time to write to the RIMR's Simon Flexner and deny any role in devising the "unauthorized and fallacious" national media rumors that he sought cover for his

[46] CCL to L. C. Dunn, 21 March 1928, LCD-APS; T. H. Morgan to CCL, 29 May 1928, Box 739, CCL-UMO. Little requested a divorce from his wife Katherine in early 1929: see *New York Times*, 25 June 1929, p. 32: "Doubtless his divorce had something to do with his break from the university" (from Snell, "Little," *Dictionary of Scientific Biography*).

[47] Peckham, *The Making of the University of Michigan*, p. 169: cf. "Dr. Little Resigns as President of the University," *The Michigan Alumnus* 35 (26 January 1929): 327–30, in Box 8–7, JLA-BH. Clark ("The Social Uses of Scientific Knowledge," p. 179) shows that, after Little was appointed Michigan president in the summer of 1925, he received a significant number of congratulatory letters that expressed hope that he would one day be inaugurated "President Little of Harvard": cf. "Dr. Little Forced Out of Michigan University," 12 February 1929, *New York Telegram*, Box 743, CCL-UMO; "Jobless Little," *Time* 8 (4 February 1929): 36–38; and the letters in Box 737, CCL-UMO. The usually critical college paper, *The Michigan Daily*, was quick to praise Little's path-breaking leadership, as was the *Michigan Alumnus*. His self-cultivated status as the quintessential Boston Brahmin was once again resurrected in the form of rumors that he was to become president of Harvard: see also "Reports Dr. Little Will Quit Michigan," *New York Times*, 21 January 1929, p. 20, and 22 January 1929, p. 11.

mouse work under the Rockefeller medical research umbrella. That same week, he also acted magnanimously on behalf of Michigan's interests by contacting the Rockefeller Foundation's director for social science, Edmund Day, and emphasizing with regard to projects under consideration (one in biology) that he "still urge[d] the educational advantages of Michigan with the same enthusiasm." Perhaps it was public perceptions of this kind of gallantry that compelled pulp fiction writer Janet Hoyt to pen her 1929 novel *Wings of Wax* and to model her hero—the fictional University of Woban President Victor Marston—after Little and his Michigan experience.[48]

Privately, however, Little's world was unraveling and he was making plans to secure his own future as well as the future of his mouse research enterprise. His marriage to Katherine Andrews had almost completed deteriorated: he and his wife separated just before he announced his resignation. Four days after Little filed his letter with the regents, Jackson wrote to Little expressing "great regret. . . . [I]n spite of your natural leaning toward individual investigation, . . . I believe you have executive ability." More pragmatically, Jackson also wanted to know what Little's desired plans for the Cancer Research Fund were. He suggested a dinner engagement to discuss the matter. Little replied gratefully that, although the public had "imperfect information," he was more than willing to discuss things frankly with Jackson. It was at this meeting, on 1 February 1929, that Little probably first told Jackson what he saw to be the next step for himself and the mouse research. As Little recalled later: "Jackson asked me what I wanted to do. I told him that I wanted to establish a center of biological research on cancer and other constitution diseases as a continuation of the work which I had begun twenty or more years before. He asked me where I wanted to go and I told him to Maine—Bar Harbor—where George B. Dorr . . . had already offered land for a laboratory."[49]

[48] CCL to Simon Flexner, 21 January 1929, Box 739, CCL-UMO; Little himself described the rumors in a letter to Day: see E. E. Day to CCL, 22 January 1929, and CCL to E.E. Day, 26 January 1929; both in Box 743, CCL-UMO; Janet Hoyt, *Wings of Wax* (New York: J. H. Sears, 1929). For a brief reference to the connection between Little and Hoyt's book, see Smith, *The First Century*. A year later, Little himself published a more philosophical account of the trials and tribulations of university administration: see Little, *The Awakening College* (New York: W. W. Norton, 1930).

[49] Jackson to CCL, 23 January 1929, and CCL to Jackson, 25 January 1929; both in Box 8–7, JLA-BH; C. C. Little, "Your Question #5": one of a set of typescript answers to a reporter's questions, numbered #1–8, not dated but, from content, appear to be early 1960s, Box 737, CCL-UMO. Clark suggests that Little began discussing possible arrangements for a Bar Harbor facility with Jackson in 1927, but the earlier letters between Jackson and Little she cites as evidence were more likely in reference to negotiations with J. D. Rockefeller over the existing Biological Institute project. Little was committed to realizing this project

Where Little initially got the idea to move to Maine is not clear, but clearly he had begun to develop it before his resignation. Neither Joe Murray nor Bill Murray's widow recalled any mention from Little to his group about an independent research institute while they were at Michigan. Yet within ten days of their dinner meeting, Little forwarded to Jackson a detailed plan (possibly elaborated on from his original GEB Biological Institute proposal) that included a tentative operating budget for the institute's first five years in Bar Harbor, a proposed scheme of financing, and a very rough draft of a laboratory building floor plan, "along the lines of which we spoke." Around the same time, Little also contacted H. C. Bumpus, president of the Mount Desert Island Biological Laboratory, to convey his enthusiasm about the prospect of founding "a small laboratory" for his mouse work near what was then known as the Morrell Park Station of the MDI in Bar Harbor.

In retrospect, Little's plan was virtually devoid of more grandiose language like that used in his earlier philanthropic pleas, but it represented a reformist effort in that it comprehensively addressed the perennial practical issues that plagued the inbred mouse research. "It would be my plan," Little explained, "to live there myself all the year around and to have with me six or seven full time workers. The personnel is, I believe, available and I think I might be able to obtain the funds."[50] Little envisioned a place that combined the "biological resort" features and research material availability of Woods Hole Marine Biological Laboratory with the intellectual freedom, advanced facilities, and year-round productivity of the Rockefeller Institute for Medical Research.[51] Nothing like this existed in contemporary American biology or medicine: Little sought to create an institution that simultaneously promoted focus on particular research materials (mammals) and flexibility of research priorities.

Upon receiving Little's more detailed proposal, Jackson replied almost immediately with the Detroit group's tentative approval. Jackson confided to Little that he worried such an institutional arrangement might not be putting Little's talents as an administrator and program builder to

at Michigan, and his attempts get MDI-BL President H. D. Bumpus interested in the mouse work were clearly directed at motivating Rockefeller to fund it. Clark, "The Social Uses of Scientific Knowledge," p. 159, n. 20.

[50] CCL to Jackson, 11 February 1929; CCL to Bumpus, 4 February 1929; Box 8–7, JLA-BH.

[51] Joseph Murray and Francina Murray, interviews with Jean Holstein in Holstein, *The First Fifty Years*, p. 4; Sinclair Lewis, *Arrowsmith* (New York: Harcourt Brace, 1925), chap. 40. On Woods Hole, see Phillip J. Pauly, "Summer Resort and Scientific Discipline: Woods Hole and the Structure of American Biology, 1882–1925," in *The American Development of Biology*, ed. R. Rainger et al. (New Brunswick: Rutgers University Press, 1991), pp. 121–50 on the RIMR, see Corner, *A History of the Rockefeller Institute*.

best use: "I wouldn't think this decision would be of such moment if you were the usual type of scientist. . . . I can't help but feel very reluctant to have you sew yourself up on an isolated scientific adventure, although of course I can imagine possible happenings in this scientific undertaking offering compensation of far greater value to humanity at large than would be obtained by choosing the other line." But for Little, the decision had already been made.

As Little himself reminded public school students in a speech late that same year, he believed in taking leaps of faith if it could be demonstrated that the greater good would benefit: "opportunity," he noted, "means obligation."[52] In taking his Michigan resignation as a long-awaited chance to improve conditions for his research and for the long-term future of inbred mouse material, Little was living up to his own proverb. Two aspects of the new venture differ markedly from Little's past professional activities. First, its success was not predicated on working within the existing system of scientific institutions and patronage strategies. Second, never before had the maintenance of inbred mice been so critical for the continued prosperity of his own scientific work. Little had turned a corner, entrusting the fate of his entire career to inbred mice and the genetic research problems they were designed to solve.

FOUNDING THE JACKSON LABORATORY: THE BUILDING AND THE PROGRAM

In May 1929 the *New York Times* announced President Little's new venture to the public. The article noted simply that its institutional mission was understanding the causes of cancer, and that financially it was supported by "five wealthy Detroiters" who would act as trustees.[53] As Little later recalled, however, Ford, Webber, and Jackson were less concerned with the outcomes of the work and more concerned with the process: they were taking a well-considered investment risk based on shared managerial values that they understood could be applied to Little's research to make it more efficient and productive. "Jackson . . . being a good businessman," Little told a reporter, "saw the scientific value of the inbred mice and that they added accuracy and repeatability to the biological work."

Among Little's geneticist colleagues, word of mouth spread quickly of his plans. By early March 1929, fellow mouse geneticist F.A.E. Crew at

[52] C. C. Little, in "Ten Second Thoughts," Michigan's *Hamtramck Public School Bulletin* (April 1928), in Box 742, CCL-UMO.

[53] Jackson to Little, 13 February 1929, Box 8–7, JLA-BH; "Statement of Income and Expenditures, 27 May 1929 to 30 June 1930," Box 739, CCL-UMO.

the University of Edinburgh had heard of Little's plans to return to science full time. He wrote to wish Little good luck, and to register his hopes for the scientific success of the new institutional arrangement: "My dear Little, When [I heard] the news that you were quitting administration and heading back to active research, I thanked my Gods and Maud Slye, for I know that there is a job of work for you to do which will give you the peace and the joy that you must seek." The fact that Crew also took the opportunity to comment on the Slye-Little debate suggests that there was support among geneticists for Little's formulation of the cancer problem.[54]

Still, Little's lawyers and colleagues warned him not to get so caught up in the scientific planning of his new facility that he ran roughshod over the interests of the Bar Harbor community, whose practical support was critical for the lab's success. On the persistent urging of George Dorr, a local businessman, Little had decided to call the facility the Charles H. Dorr Memorial Laboratory, after Dorr's father and in deference to his gift of the thirteen acres of land on which it was to be built. Little designated his older brother, Boston architect J. Lovell Little, to design the laboratory building and solicit construction estimates from Maine firms. When they received the first bids in April, both Littles favored Bangor contractor Otto Nelson, who not only had put in the lowest bid but had promised the building would be ready for occupation by 15 September of that same year. The younger Little believed that the survival of his mice came first: to get the creatures into their new homes and resume their breeding before the colder weather set in was the best practical hedge against the inevitable loss of momentum due to the move itself. This approach would have won win high praises from the community of mouse researchers, but it created local tensions. Upon consulting with Dorr, Bill Murray urgently wired Little in Michigan to reconsider this choice because it would cause an "undercurrent" of "unfriendly local feeling" toward the whole enterprise.[55]

Before Little finalized these construction decisions, however, he got some bad news: Roscoe Jackson had died suddenly of influenza while vacationing in Mentone, in southern France. By this time the project had enough momentum that the loss of Little's greatest financial ally did not alter plans underway for lab construction. Jackson's widow, as well as

[54] "To Finance Dr. Little in Studies of Cancer," *New York Times*, 8 May 1929, p. 2; F.A.E Crew to CCL, 22 March 1929, Box 743, CCL-UMO; cf. Robert Cook, "Heredity and Cancer" [a journalist's summary of H. Gideon Wells, "The Nature and Etiology of Cancer," *American Journal of Cancer* 15 (July 1931) (supplement): 1919], in *Journal of Heredity* 23 (1932): 160.

[55] J. Lovell Little to CCL, 2 April 1929, 14 May 1929, and 27 May 1929; CCL to Bill Murray, 27 May 1929 and 31 May 1929; Murray telegram to Little and Little's wire reply, 27 May 1929; all in Box 739, CCL-UMO.

Edsel Ford and the Webbers, quickly reconfirmed their standing financial commitments to Little and his project.[56] The only thing Jackson's death appears to have immediately changed was the plan to name the new institution. Little's other Detroit patrons up to this point had categorically shunned any publicity related to their philanthropic activities, but upon Jackson's death, they approached Little with a request to name the new lab in memory of their colleague. Little presented this request to Mrs. Jackson and to Dorr, both of whom promptly agreed to the new plan.[57]

The Roscoe B. Jackson Memorial Laboratory was incorporated in Bar Harbor on 4 May 1929, with Little as president and Little, Bill Murray, and David Rodick, a Bar Harbor lawyer, as directors of the incorporation papers. In this way Little retained control of his new laboratory's business affairs, while generating philanthropic support from the Maine's upper-class "summer society." For example, the rest of the original members were Little's brother, George Dorr, and three prominent Bar Harbor and Bangor businessmen, and by October 1929 Little had expanded the corporation membership to include Army Colonel F. H. Strickland of Bangor, a former University of Maine trustee, H. S. Boardman, ex-president of the University of Maine, and three Boston businessmen. In enticing these prominent citizens to become members, Little emphasized the chance it would offer to partake of resort culture there. In what appears to have been a standard letter of appeal, he promised meetings would have "A good lobster feed and several other diversions. . . . No financial obligations are involved nor will there be any fees paid for attendance at the meeting other than giving you the best possible time we know how." This populist rhetoric conveyed the relaxed, informal atmosphere Little wanted to establish for the new facility, but its audience revealed his long-term interests—mouse work was serious science, but it could also provide appropriate entertainment for the local elite, who would in turn provide financial support for the lab.[58]

[56] Roscoe Jackson's obituary is featured on the front page of the *Detroit News*, 20 March 1929. For many years, it appears that Little incorrectly recalled Jackson's cause of death as a heart attack. See "Your Question #5"; CCL to Herbert Neal, 5 April 1929, Box 8–7, JLA-BH.

[57] Excerpt of letter from CCL to Dorr, 28 March 1929; CCL to Edsel Ford, 22 April 1929; both in Box 8–7, JLA-BH; CCL to George Dorr, 2 April 1929 and 22 April 1929; Box 739, CCL-UMO; "Will Be Memorial for R. B. Jackson," *Bar Harbor Times*, 8 May 1929, also Box 8–7, JLA-BH.

[58] Papers re: the incorporation of the Roscoe B. Jackson Memorial Laboratory, 4 May 1929; "Incorporators, Roscoe B. Jackson Memorial Laboratory (n.date)"; quote from CCL to Col. Strickland, 20 September 1929; all in Box 739, CCL-UMO; see also JLA-BH, Box 8–7, for a copy of the original incorporation papers. Little's attempts to promote the Jackson Lab resort culture were greeted skeptically at first within the scientific community: Little

Over the next several months, Little took advantage of the financial and institutional freedom he had to ensure that the new laboratory building itself was custom-made for large-scale, controlled mouse breeding. He originally estimated the cost of building construction to his patrons at approximately $50,000, and his negotiations with his architect brother show that keeping the cost of the laboratory within this figure remained a genuine priority. Landscaping costs were virtually nil, since the Dorr land on which the building was to be located had beautiful natural features. A contemporary local newspaper described the tract's no-cost aesthetic: "[It is] of great natural beauty. . . . Northward it looks across Frenchman's Bay and its island . . . south and west to the mountains of Acadia National Park, whose land it borders on two sides. There are no disfiguring surroundings. . . . [The site] is also supplied with water from the Tarn gravity supply from Eagle Lake. . . . Sieur de Monts Springs lies close at hand and the whole surrounding is one of exceptional landscape interest."[59] Mirroring these surroundings, Lovell Little's design for the building itself was simple and sparse. It was to be a 99 × 54 foot brick-over-wood building with a simple entry arch, a basement for storage, two main floors of labs and offices, and a large attic for mouse breeding.

Some costly modifications, however, Little deemed necessary to ensure that the institution would fulfill its practical mission of maintaining the genetic purity and health of the inbred mouse. For example, early on his brother informed him that dropping the height of the room-to-room partitions to seven feet would cut building materials and improve heating performance and costs. But Little decided at the last minute that "[this] item . . . would not be advisable. It would probably be better to have the partitions full height. This would prevent the interchange of vagrant animals from one room to another." Ultimately he spent an extra $600 per room to have the physical security measures that would insure the purity of his mouse material, as well to safeguard against the spread of disease among the stocks. Ironically, such modifications along with construction delays put the building two months behind and about 30 percent over its original budgeted figure—and made it more uncomfortable for its human occupants.[60]

---

expressed disgust at a rumor that the lab was really just a "scientific sex colony" in CCL to Joseph Murray, 23 December 1929, Box 739, CCL-UMO.

[59] "Cancer Research Laboratory Is to Be Established This Summer," *Bar Harbor Times*, 8 May 1929.

[60] For example, wood floors were left unstained, masonry walls unpainted, and room partitions in the library were changed from masonry to industrial-type steel: see Lovell Little to CCL, 2 April 1929 and 14 May 1929; A. G. Hopkins (associate of Lovell Little's) to CCL, 13 April 1929; CCL to Hopkins, 18 April 1929; Box 739, CCL-UMO. Holstein describes the original attic floor plan of the Jackson Lab as "divided by partitions that did not

But the extra time gave Little opportunity to wrap up his personal and scientific affairs at Michigan in preparation for the fall move to Bar Harbor. Chief among these was completing his divorce proceedings, and (presumably) planning his subsequent marriage to Beatrice Johnson. After she had graduated from the University of Maine, where Little supervised her master's thesis, Johnson went to work as laboratory researcher and executive secretary to Margaret Sanger's American Birth Control League, where Little was president of the Scientific Board. During this time, she collaborated once more with Little on mouse experiments designed to show that more living young result when births are spaced for recuperation of the mother. She appears to have followed Little to Michigan, where from 1926 to 1928 she was dean of women, and she may have also worked at the Cancer Research Lab.[61]

In a letter to Bill Murray late that summer, Little was cheerful about his new lease on life, and he summarized what he saw as the state of the research group in his last days there: "Joe [Murray] passed a very satisfactory Doctor's examination and is now through all of his worries and feeling much better. . . . The work at the laboratory goes well and when [Elizabeth] Fekete leaves on the 11th of June Joe will keep an eye on my mice. The numbers of mice stocks are much reduced preparatory to shipment."[62] Little surely recognized that moving to a new institution in Bar Harbor was a career risk even for established researchers like Strong, but especially for the graduate students who were in the process of finishing their doctorates. As a result, he decided that only the mice and a skeletal staff would move to the new facility during its first year and get things settled in preparation for the others. Little's secretary at Michigan, Mary Russell, took a "secretary-librarian" position at the Jack-

reach the ceiling," but perhaps this interpretation comes from an earlier floor plan: cf. Holstein, *The First Fifty Years*, p. 13. The laboratory building is listed as costing $66,489 in the "Statement of Income and Expenditures, Period May 27, 1929 to June 30, 1930," Box 739, CCL-UMO. Little would later remember the increased cost to be even greater ($69,000), but he emphasized that this was no fault of his brother, who donated virtually all of the architectural services: see "Your Question #5."

[61] Several pieces of correspondence from the Murray brothers at Michigan refer to "Johnny" (her nickname), but John J. Bittner also sometimes went by that name (though more typically the men referred to one another by last name). On the date of Little's divorce, see *New York Times*, 18 August 1929, p. 27, and various newspaper clippings detailing his second marriage to Beatrice Johnson on 3 September 1930, in Box 727, CCL-UMO. Little and Johnson's marriage was announced in the *New York Times*, 4 September 1930, p. 16. Beatrice Johnson's 1925 master's thesis was entitled "The Effect of Ultra-Violet Light on Young Mice" (see Johnson to W. P. Taylor, 29 March 1929, Box 742, CCL-UMO); cf. B. W. Johnson, "Experimental Breeding of Young Mammals," *Birth Control Review* 10, 9 (September 1926): 267–68. See also *Who's Who of American Women* (Chicago: A. N. Marquis Company, 1958–59 and 1968–69).

[62] CCL to George Dorr, 22 April 1929 and CCL to Bill Murray, 31 May 1929; both in Box 739.

son Lab starting on 3 September, and Bill Murray moved to Hampden, Maine, in May. For the next four months he commuted down to supervise the construction operation. Cancer researchers L. C. Strong, Joseph Murray, John Bittner, and Arthur Cloudman accepted positions as research associates for 1930–31, as did geneticist Charles Green, whose work dealt primarily with size inheritance in mice. Of the two women scientists at Michigan, only Elizabeth Fekete also signed on to go to Bar Harbor—as the lab's histologist.[63]

In retrospect, the first year of the Jackson Lab's existence passed uneventfully in both research and administrative terms. On 7 October 1929, the *New York Times* formally proclaimed the beginning of scientific work at the "Roscoe B. Jackson Memorial Laboratory of Cancer Work" and noted that soon "the structure will house 25,000 mice on which cancer experiments will be made" in "search for a possible cure."[64] But when the building was actually completed in November 1929, only Little, Bill Murray, and several small shipments of mice from Michigan occupied it.[65] Murray simply tended to the practical task of ordering the supplies and instruments necessary to equip the institution for further research.[66] To aid in establishing the first round of mice in the new quarters, in early October 1929 Little hired two local residents—Frank Clarke and Frank Oakes—to act as part-time manual laborers and mouse caretakers.[67]

Meanwhile, Little's reputation and visibility as a cancer researcher continued to grow, and a month after the lab opened he got another big break: the American Society for the Control for Cancer (ASCC) offered him a job as their managing director. The reasons that the ASCC recruited Little were manifold. The retiring director, sanitary engineer George Soper, advocated increased clinical facilities for cancer treatment, but Little's expertise on the experimental study of cancer better resonated with the ASCC's broadening consensus in support of medical research. Little was also known as a charismatic speaker, and the society was looking for

[63] On the hiring of caretakers, see "Interview with Frank Clark, Watson Robbins and Allen Salisbury," 16 May 1986, JLOH-APS. On the general hiring of scientific workers, see "Facts Concerning the Institution," in CCL to Alan Gregg, 24 May 1933, RF Archives, RG 1.1, Series 200D, Box 143, Folder 1773, RAC-NY. See also C. V. Green to CCL, 23 January 1930 and CCL to C. V. Green, 27 January 1930; J. J. Bittner to CCL, 23 January 1930; Cloudman to CCL, 23 January 1930; all in Box 735, CCL-UMO.

[64] "To Begin Cancer Research," *New York Times*, 7 October 1929, p. 10.

[65] On shipping the mice from Michigan to Bar Harbor, see Holstein, *The First Fifty Years*, pp. 76–77.

[66] Several of Murray's orders of supplies and instruments for the Bar Harbor facility (dated 7 November 1929 to 23 June 1930) are found in Box 739, CCL-UMO.

[67] See interview with F. Clark, W. Robbins, and A. Salisbury, May 1986, JLOH-APS. Later, Little would also commit lab funds to caring for the elderly Oakes and his wife, and he sought out a pension from the State of Maine for them: see Jackson Laboratory Annual Report, 1941, JLA-BH.

someone with such abilities to bring their message of cancer awareness and support for research to a wider national audience.[68]

He immediately accepted the ASCC job, and the final scientific product of the Michigan Cancer Research Laboratory group reflected his self-professed interest in consolidating philanthropic support for cancer under the umbrella of research, particularly mouse research: it was a devastating critique of Maud Slye's simple Mendelian-recessive interpretation of cancer. Since Slye never complied with Little's request that she publish all of her mouse data, Little used what existed of her published figures to offer a reinterpretation of her results. This reinterpretation, Little suggested, turned Slye's conclusions on their head. Mammary tumor inheritance was not recessive as she claimed, but instead it was dominant and sex-linked. Furthermore, although Little praised Slye for her "extensive" mouse breeding work, he also cast aspersions on almost every aspect of her experimental procedure. Notably, he accused her of inaccurate control of environmental factors (i.e., infectious mouse diseases), which might have caused deaths among her colony, and of using particular matings in more than one report to make her total data set look larger.[69]

Given Little's own plans, the timing of the attack against Slye was clearly self-serving. This latest diatribe against Slye's work came right on the heels of her successful appearance at the American College of Physicians' Annual Meeting. She had impressed many of the doctors there with her mouse research project, and the organization had finally voted to consider her plan to found a national bureau of human cancer statistics, which would function as the medical clearinghouse for the study of cancer inheritance in the United States. As a result of Little's critique, the Slye-Little debate over the nature of cancer heredity received renewed attention not only in the medical profession but also from the national media.[70] Little's next ten years at the Jackson Lab would be spent developing and propagating the methods of standardized mouse breeding he vehemently argued were necessary for "good" cancer research.

At this time, however, Little also sought a new base of support for his institution by casting his critique of Slye in terms of medical debates over eugenics. In a 1928 medical journal paper, for example, Little began by justifying his attack on Slye's work in terms of the ongoing battle of true scientific research against quackery in cancer studies:

[68] Donald Shaughnessy, "The Story of the American Cancer Society," Ph.D. diss. Columbia University, 1957, esp. chap. 8; Patterson, *The Dread Disease*, pp. 121–26; "Little," in *Who Was Who*. For more on the relations between Little, mice, and the ASCC anticancer alliance, see chapter 5.

[69] C.C. Little, "Evidence That Cancer Is Not a Simple Mendelian Recessive," *Journal of Cancer Research* 12 (1928): 30–46.

[70] On Slye and the effect of press coverage on the outcome of the Slye-Little debate, see McCoy, *The Cancer Lady*, chap. 11 and pp. 88–91.

> The *fact* that the tendency to certain forms of cancer is hereditary in mice has been established for some years. . . . For a time the medical profession was not very willing to accept this fact. Their experience had dealt with human beings, who had bred so slowly. . . . It was therefore entirely natural that the profession did not at once agree with the conclusions of those engaged in laboratory research. . . . [But] too many false hopes have been built up and later destroyed in many investigations on cancer to allow [Slye's] simple but incorrect genetic interpretation to go unchallenged.

He concluded by emphasizing that physicians' lack of trust in genetic methods for understanding complex problems was really the issue at stake:

> The complicated nature of cancer, its long and successful history of menace to the human race, and its social importance, together with the need of increasing the confidence of the medical profession in genetic methods of analysis, justify us in deciding definitely at this time that . . . continued study with more careful genetic analysis than that given by her [Slye] up to the present time.[71]

Little's rhetoric echoed the contemporary criticisms of medical doctors with regard to eugenics, and indeed, as plans for the opening of his new lab began to crystallize, Little became much more explicit in linking his critique of eugenics and his vision for the future of inbred mouse work. Most clinicians were predisposed to believe that if genetics played a role in human medical problems at all, it was much more complex than the hereditary mechanisms observed in laboratory organisms.[72] In 1926, Little had invoked what by now had become a predictable analogy to explain why the state of Maine would be a good place for doing genetics research. He described the geographical virtues of the place for sustaining humanity's pure Yankee stock in terms similar to the ones he used to describe his animal research organisms: "I happen to be working in Maine, where the proportion of the old New England stock is very, very high. . . . I don't want to see that particular element in the situation mixed up, or mauled up. I want to keep it in the way a chemist would prize a store of chemically pure substance that he wants to use for testing . . . for definite purposes when a certain element is needed."

A 1928 essay entitled "Opportunities for Research in Mammalian Genetics" was even more clear about his envisioned uses for the Bar Harbor

[71] Little, "Evidence," p. 45. Little makes this same point more generally in "Agents Modifying the Germ Plasm."

[72] A general discussion of the relations between medicine and eugenics in this period can be found in Kenneth M. Ludmerer, *Genetics and American Society: A Historical Appraisal* (Baltimore: Johns Hopkins University Press, 1972) pp. 63–72; see also Daniel J. Kevles, *In the Name of Eugenics* (Berkeley: University of California Press, 1985) esp. pp. 176–78, and Philip R. Reilly, *The Surgical Solution* (Baltimore: Johns Hopkins University Press, 1991).

mouse work. Experimentally based knowledge of mammalian genetics, Little argued, was needed to fill the "wide and somewhat awesome gap between the . . . structure of the germ cells in *Drosophila* and the wise direction of a program of improvements in human biology." Only when armed with such information, obtained from work with laboratory rodents, could "the laboratory geneticist" answer society's most pressing question to him: "Of what value to humanity are your findings?" And for Little, this was no less than the geneticist's social responsibility: "the problems to be investigated must be of interest to men."[73]

Finally, in a 1927 letter to John Kellogg, director of the Battle Creek Sanitarium and an influential member of the Race Betterment Foundation, Little described his lab's mouse cancer work as "concerning . . . the principles of human biology and eugenics." He then told Kellogg: "I thought you might be interested in seeing that we are making a real effort, although an admittedly inadequate one, to place a sound foundation of biology under the training of many people who now exercise a great influence in shaping the training of humans . . . it seems to me that this is the sort of work which [you] could well afford to endorse and support." Although Kellogg never offered Little money for his venture, he appears to have agreed with him at some level: Little was invited to participate in the Third Race Betterment Conference the following year. His talk there, provocatively titled "Shall We Live Longer and Should We?" made copious reference to the mouse cancer work and discussed the possible social fate of faulty "strains" of human beings.[74]

Over the next several years, Little published a series of articles and comments in the *Birth Control Review* that elaborated his eugenic views. Though not every piece mentioned the cancer work, they all had similar discretionary messages for medical doctors and the public alike: to the extent that eugenic social policy could be formulated from genetic science, it would need to be a science based on experimental evidence. For example, although he advocated birth control, Little urged physicians to "proceed with extreme caution" in the matter of eugenic sterilization, and he questioned whether it was a good method of "true" race improvement. Furthermore, in early 1928, the Committee on Formal Education of the American Eugenics Society, which Little chaired, published an extensive report on eugenics teaching in colleges and universities. Not surprisingly,

---

Cf. Debora Kamrat-Lang and Peter Weingart, "Healing Society: Medical Language in American Eugenics," *Science in Context* 8 (1995):175–96.

[73] C. C. Little, "Unnatural Selection and Its Resulting Obligations," *Birth Control Review* 10 (August 1926): 243–44; cf. Little, "Opportunities for Research in Mammalian Genetics," *Scientific Monthly* 26 (1928): 521–34; also *Sigma Xi Quarterly* 16 (1928): 16–35.

[74] CCL to Kellogg, 11 October 1927, Box 13, Folder 12, CCL-Mich; Little, "Shall We Live Longer and Should We?" *Proceedings of the Third Race Betterment Conference*, 2–6 January 1928 (Battle Creek: Race Betterment Foundation, 1928), pp. 5–14.

this report also criticized the tendency toward "sensational and pseudo-scientific" approaches to imparting eugenic knowledge, and it recommended that future eugenics classes be "intimately associated" with research work in genetics laboratories.[75]

Taken together, these activities reveal Little's strong programmatic stance on the relationship between the intellectual and practical aspects of biological research, and they suggest that the Jackson Laboratory represented the institutional incarnation of this very program. Little had supported eugenic research and social policies since his days at Cold Spring Harbor, and he had long drawn the analogy between mouse genetic research and human (physical and social) problems. It is thus not surprising that he still believed in some sort of eugenics and a role for mouse work in its realization. It is also not surprising that, at this juncture, Little might turn to a more ambitious version of the strategy that won him the funds and the power to set up the Jackson Lab in the first place. In context, however, it is certainly not trivial that the last work Little accomplished while at Michigan rearticulated his larger eugenic vision. The timing of these papers and talks coincides with Little's decision to establish the Jackson Lab, so they can also reasonably be taken as setting forth what he wanted to accomplish in his new institution—the founding of which he later described as "the fulfillment of a lifelong ambition."[76] Little explicitly set forth a research plan—breeding experiments with genetically homogeneous mice—and a social agenda—linking experimental genetics more closely with human eugenics—and saw both components as necessary to make biology both more scientifically rigorous and more socially applicable. The blueprint for the Jackson Laboratory was in many ways embedded in Little's mouse work from the beginning, but in 1929 he was taking pains to ensure that no one missed the ideological connection.

CONCLUSION

Perhaps more clearly than in any other period in his career, Little's professional activities in the twelve months preceding the opening of the Jackson Memorial Laboratory allow us to glimpse his view of the role that mouse genetics should play in science and modern society. By 1928 Little no

[75] C. C. Little, "A University President [on] Sterilization: A Symposium," *Birth Control Review* 12, 3 (March 1928): 89; Little, "Will Birth Control Promote Race Improvement?" *Birth Control Review* 13, 12 (December 1928): 343–44; "Report of the Committee on Formal Education of the American Eugenics Society Inc.," 1928, CCL-UMO. The other members of the committee were C. H. Danforth, H. D. Fish, H. R. Hunt, Ann Haven Morgan, E. L.Thorndike, and P. W. Whiting; "Propaganda versus Basic Progress," *Birth Control Review*, 14 (March 1930): 78–79.

[76] "To Begin Cancer Research," JLA-BH.

longer pandered to foundations and patrons for his livelihood. Instead, he spent his last months at Michigan speaking to a broad audience about his eugenic social vision—already expressed in the experimental ethos of his college presidencies and its connections with the mouse research plan. In founding the Jackson Laboratory, Little simultaneously sought to solve many of the practical problems of animal breeding and maintenance that had plagued his own research, as well as inbred mouse work more generally, for twenty years. From 1925 to 1929, he pursued every entrepreneurial strategy available to expand his mouse research enterprise to a level at which the material security of the organisms themselves would no longer be threatened. But ultimately, Little chose to construct the Jackson Lab's research program as the embodiment of an elite liberal Progressive alliance: an interdisciplinary exchange among physicians, medical researchers, geneticists, public health advocates, and social reformers. This ideological formulation of the relationship between mammalian genetics and human eugenics centered around the use of inbred mice for research. It effectively created constituencies for the knowledge obtained from Little's new lab and positioned his institution to serve all such interested parties.

But on 27 October 1929—two weeks before the Roscoe B. Jackson Memorial Laboratory officially opened its doors—the U.S. stock market crashed, and the nation plummeted into the most severe economic crisis in its history. Over the next ten years, the Depression simultaneously restricted the material wealth available for private funding of science and challenged Progressive reformers' faith in science as a positive agent of social change. Neither the Jackson Lab nor mouse work more generally remained unaffected by these events. In the difficult fiscal times ahead, Little returned to his original scientific ideology to justify patronage for the lab's research. But this strategy would bring only limited success for the institution's work in the years ahead.

CHAPTER THREE

MICE FOR SALE

Commodifying Research Animals (1930–33)

In November 1929, while many American workers and businesses struggled with the effects of the nation's most crippling economic depression to date, Little continued to preach the eugenic gospel. Now described by the press as a "cancer expert," Little drew headlines for condemning everyone from public health officials and the National Broadcast Company (NBC) for being unwilling to contribute to a genuine public discussion on this vital method of "true race improvement."[1] The following fall, in his official capacity as the American Society for the Control of Cancer's managing director, Little took his new wife, Beatrice Johnson Little, on a European tour that included nearly every hospital, cancer clinic, and cancer laboratory between Edinburgh and Warsaw. He returned from this trip with a profound respect for what he deemed the Europeans' more rational approach to the cancer problem. As he told the society's board:

> Europeans have toward the cancer problem a much more patient, long-time attitude than we do in America. Their main interest is research. They deliberately minimize the opportunities for cancer phobia and personal worries by preaching and practicing the impersonal, far-seeing attitude of

[1] "Birth Control Urged for Improving the Race," *New York Times*, 21 November 1929, p. 56.

> experimentation and research. As an incidental result . . . has come greatly increased public confidence and cooperation. This is true and permanent lay education.

Similarly, Little argued that the ASCC needed a well-organized group of human data-gatherers and statistical analysts to "establish constant and consistent types of records," in order to reveal demographically the true scope of America's cancer problem. Such an approach would mimic the hiring and training of data-collection workers by the Eugenics Record Office, run by Little's former employer Charles Davenport. Little believed that both of these rationalizing strategies needed implementation before the ASCC could even begin contemplating more personalized "lay publicity involving advice to the individual" on possible medical treatments.[2]

But in his subsequent public appearances, Little also emphasized the importance of research in connection with the mission of his own new institution. He told reporters that the Jackson Lab's commitment to researching cancer's *causes* was the first of its kind in America, and he claimed it "bids fair to equal, or perhaps outstrip, the work being carried on in Europe."[3] He also touted the Jackson Lab's valuable research materials, describing in detail for reporters the controlled, formalized mouse caretaking procedures. One impressionable *Portland Sunday Telegram* writer seems to have reproduced Little's entire narrative:

> The mice rooms [are] all kept at a temperature of 70 degrees Fahrenheit. . . . Every new-comer in the established pens has a number punched into its ear at birth, by which it is known and catalogued. In this way, the new cohorts are kept in line. . . . Cotton wadding and shavings constitute the pen. A quantity of hemp and canary food as well as broken dog biscuits are constantly before them. They drink water but small dishes of milk are given each box at noontime. . . . The glasses are washed and sterilized after each feeding.[4]

[2] Quotes from Little's "Report of the Managing Director to the Board of Directors," American Society for the Control of Cancer, December 1930, pp. 9 and 5, quoted in Shaughnessy, "The Story of the American Cancer Society," pp. 152–55. Later, in a published forum, Little complained that "European laboratories are full of experiments . . . on animal material which is genetically uncontrolled." See Little, "The Role of Heredity, 280–83. On the hiring of data collectors at the ERO, see Allen, "The Eugenics Record Office." Little himself made the explicit comparison between the ERO data-collection methods and what he had in mind for cancer research in "The Inheritance of a Predisposition to Cancer."

[3] Little's "welcome luncheon" speech before the ASCC is described in "Dr. Little Welcomed Here," *New York Times*, 11 October 1929, p. 30. Quote from "Preparing Mount Desert House for 30,000 Mice to Be Used in Cancer Research Interests: Science to Play Role of Pied Piper for Humanity's Sake," *Portland Sunday Telegram*, n.d., probably February 1930, JLA-BH. Little, "Education in Cancer," *American Journal of Cancer* 15 (1931): 280–83; see also "Urges More Cancer Aid," *New York Times*, 8 December 1929, p. 24.

[4] "Preparing Mount Desert House" (February 1930), JLA-BH.

Though most people still knew very little of what went on at the laboratory, Little's institution had already gained something of a reputation—especially among local townspeople—for being a "Mouse House" that embraced Tayloristic breeding principles.[5]

Getting these mice to their new home had been a more difficult matter than getting the human cancer researchers there. During June 1930, Joe Murray shipped most of the Michigan mice in their wire cages via rail, and because it was a temperate summer month these animals arrived in good shape. But L. C. Strong, the headstrong senior scientist of the group, chose to transport his mice separately, in an earlier rail shipment accompanied by a personal mouse caretaker. Ironically, these animals fared worse: the baggage car containing Strong's colonies got misrouted and traveled to Maine via the colder Canadian route. Strong lost some mice, but many survived the ordeal.[6]

In terms of plans to breed larger and better-managed mouse colonies, however, what Little planned to do with mice at Jackson Lab was not entirely unique. His institution's practices reflected recent shifts in how animals were typically obtained for American biomedical research. The shift of emphasis to experimental approaches in the life sciences—begun at the turn of the century in physiology and urged on by Mendelism in the 1910s and early 1920s—transformed practitioners' needs for animal materials: laboratory-based studies required inexpensive live (or at least fresh) materials that could be routinely and predictably accessed. On-site breeding colonies were the preferred solution to such problems, but not every institution could afford them; mouse geneticists experienced this lack most acutely. Furthermore, public anxiety over animal experimentation had increased such that not many institutions wanted to take on the challenge of developing and securing an in-house breeding facility for larger animals like mammals. In this context, biomedical supply houses emerged as key "infrastructural organizations." These enterprises existed with the sole purpose of maintaining animal resources over time for the research community, and their "standard" inventories represent the beginning of a separation of production of scientific organisms from their experimental users.[7]

Nevertheless, Little's decision to mass produce and sell the Jackson Lab's inbred mice to outside researchers was not a foregone conclusion

[5] Local Bar Harbor residents often referred to the Jackson Lab itself as "The Mouse House," but the name was unofficially used by some lab workers to describe the production facilities. See interview with George Snell and Joan Staats, June 1993, Bar Harbor, Maine, JLOH-KR.

[6] L. C. Strong, "Inbred Mice in Science," in *Origins of Inbred Mice*, ed. Morse, pp. 44–60.

[7] Adele Clarke, "Research Materials."

when the facility first opened its doors. Three years of plodding research performance combined with the Detroit patrons' near complete withdrawal of their funds to render the institution desperate for support after its initial funding agreement expired in 1932. When U.S. Public Health Service cancer researchers approached Little with a plan to have all their mice supplied by Jackson Lab, he was forced to approve it as a financial stopgap measure—a small "cottage industry" designed to keep the institution going. But the development of an inbred mouse sales venture, even on a small scale, necessitated a redistribution of the lab's resources: to transform animal production from a quasi-industrial practice to a genuinely industrial one required shifting the focus of the institution from in-house research to market research, and from controlled breeding and caretaking of animals to a system of standardized mass production. The development of inbred mice into laboratory organisms mass-produced and circulated by the Jackson Lab, then, was a process initiated and shaped not primarily by imperatives from within Little's original research program (as it had been in the case of *Drosophila* and genetic mapping) but by economic pressures that ran counter to it. Keeping the Jackson Lab financially viable meant selling mice. But selling mice meant adopting a consumer-centered ethos of production that, from its very beginning, threatened to undermine the lab's original goal of breeding and using inbred mice themselves to solve the genetic riddle of cancer.

SCIENTISTS AND MICE IN THEIR NEW HOMES

As director of the ASCC, Little quickly earned a reputation as a supremely self-confident, well-connected leader. Many of the most outstanding researchers and hospital administrators in the cancer field, like the ASCC itself, called Manhattan home, and Little seems to have visited them frequently while staying at the Harvard Club in New York—for example, Halsey Bagg and James Ewing, who worked at Memorial Hospital (the precursor to today's Sloan-Kettering Institute), and Francis Carter Wood, who was director of the Institute of Cancer Research at Columbia University.

Little also became known for his intolerance of bureaucratic and scientific inefficiency.[8] He was the society's most outspoken advocate of educating rank-and-file medical doctors about the benefits of experimental cancer research. Previously, the ASCC had directed its efforts toward educating the public about early cancer detection. But in Little's first "Report of the Managing Director," he harshly admonished his predecessors

[8] Shaughnessy, "The Story of the American Cancer Society," p. 160.

for this approach: "It would seem unnecessarily stupid to fail to include encouragement and practice of research as an integral part of the Society's program." As an experimental scientist and the head of a cancer research lab, Little himself clearly had a personal stake in this claim. But his "trickle-down" views on the social value of scientific research also originated from observing a successful research-intensive approach to cancer in action. The society's Board of Directors overwhelmingly sanctioned Little's philosophy by accepting virtually all of his recommendations.[9]

In turn, on Little's 'home front,' the first three years of the Jackson Lab's history also reflected an emphasis on rationalization and efficiency, and as a result, most of the successes this new institution could claim were more important internally than externally. While Little spent much of the initial year away from Bar Harbor on ASCC business,[10] Bill Murray appears to have devoted his to overseeing transitions in the mouse colony. Before the scientific researchers arrived, caretakers Frank Clark and Frank Oakes worked with Murray to develop strategies for improving caretaking procedures. Their first major project was building new living boxes for all the mice. Little requested simple rectangular wooden boxes, joined in identical pairs, since wood would provide added insulation from heat and cold and the pen arrangement would provide for efficient housing of large numbers of animals, as well as accuracy in identification. Murray was in charge of the construction project, and the subsequent interior design: bottoms of the boxes were to be covered in wood shavings, and small amounts of cotton batting were added for bedding.

The strains needed for the Michigan group's cancer research received breeding priority, although there was also some attention paid to maintaining unique variants for genetic research. For example, there were five thousand dilute (*dbr* and/or *dba*s) but only five hundred common wild house mice. The newspaper reporter also noted the presence of a "breed of pure bred blacks" (presumably *C57Black*s), inbred albinos, and several "nude" or hairless varieties, but he seemed more interested in some of the more exotic coat strains like "pale yellows" and a variety of black

[9] The only recommendation Little made that the society's board appears not to have heeded was the creation of a fellowship program to fund cancer research by medical and graduate students; the ASCC gave small denominations to researchers during Little's tenure, but it appears the Depression-induced lack of funds prevented a more organized grants program. See Stephen P. Strickland, *Politics, Science, and Dread Disease: A Short History of United States Medical Research* (Cambridge: Harvard University Press, 1972); Shaughnessy, "The Story of the American Cancer Society," pp. 155–56.

[10] An overview of the *New York Times* index for the period October 1929 to April 1931 reveals that Little traveled to New York to give speeches or attend ASCC fundraisers every two or three months.

mouse with "perfectly white faces." These were undoubtedly small colonies of mutants.[11]

The well-planned laboratory structure and new caretaking procedures worked wonders for increasing mouse supply. It is unclear precisely when the first shipment of mice arrived in Bar Harbor, or how many mice of each strain survived the trips, but in February 1930 a Maine newspaper reported that the lab housed ten thousand mice and was making plans for twenty thousand more. By April the mice had become so prolific that Murray was struggling to find a local cotton supplier who could sell the lab a cheap grade of standardized cotton batting in five-hundred-pound lots. And by May, mouse food orders for five hundred-pound bags of canary seed and ten hundred-pound bags of hemp, which Murray had formerly placed every three months, were now being placed once a month. This was a far cry from the small colonies and limited food stores kept at Michigan.[12]

Still, when Little took part in drafting a report for the National Research Council (NRC) Committee for Experimental Plants and Animals that same year, he compartmentalized his own institution's mouse breeding successes from his thinking on the broader problem of animal supply in biological research. Founded in 1928 by Bussey alum L. C. Dunn of Columbia University, this NRC group was composed primarily of geneticists interested in developing better arrangements for access to mutant materials. In his letters to committee members, Dunn frequently stated that their "eventual object" was to create a new "service or several services" similar to the American Type Culture Collection's centralized information and materials repository for bacteriologists. But Little had a different idea: he thought scientists should rely on already existing scientific institutions that bred mammalian materials: specifically, the untapped resource of the zoological parks.[13] Initially this debate was resolved by fiat—in 1929 William Mann abruptly nixed two years of efforts at collaboration with rodent geneticists and built a reptile house at the National

[11] "Preparing Mount Desert House" (February 1930), JLA-BH.

[12] Jackson Lab Secretary (Mary Russell) to George C. Frye Company of Portland, Maine, 18 April 1930; Ray Cotton Company to Jackson Lab, 7 June 1930; Bill Murray to Blamberg Brothers Inc. [food supplier], 27 December 1929; 5 March 1930; 15 May 1930; 17 June 1930; all in Box 739, CCL-UMO. A Michigan Lab supply inventory dated 4 June 1926 ("Appointments to Research Assistants," Box 742, CCL-UMO) shows that the researchers kept on hand only 75 pounds of cotton, 200 pounds of canary seed, and 30 pounds of hemp seed, which probably lasted them several months. Cf. Holstein, *The First Fifty Years*, p. 77. On Strong's general overprotective attitude toward his mouse colonies, see Strong, "Inbred Mice."

[13] LCD to Robert Hegner, 28 September 1928. In this early letter, Dunn mentions that Little was very interested in the zoo connection, but Dunn had already asked William Mann to become a member of the committee, so it is difficult to know who had the idea first.

Zoo instead—so the group decided their next step should be a "census" of the biological and medical research communities' problems with animal supply. Before the census could be made, however, Little and the remaining active members of the committee filed a report in December 1929. This document made one key recommendation: that any service established for the supply of materials "be retained by scientific or educational institutions and that commercial houses not be involved in such service." Little particularly wanted this committee objective made explicit to all researchers and institutions receiving the proposed survey, and eventually an explanatory paragraph was included in the census letter. Thus, although Little was interested in stabilizing mouse mutant strains for genetics, as well as making policy that would facilitate animal supply networks in the larger research community, he did not at this point envision the Jackson Lab as a means of solving these problems. Instead, he advocated cooperative, noncommercial arrangements with existing institutions.[14]

Little's preference reflected a narrow understanding of the material problems faced by mouse workers, especially mouse geneticists. For example, mouse geneticists ultimately came to represent more than one quarter of the total Bussey Institution alumni, a group of Harvard-trained scientists that Jack Weir argues "would have carried the United States [to] a position of leadership in genetics . . . even if there was no Columbia group."[15] But in the late 1920s, many "mouse people" who took academic positions upon graduation faced isolation from their community, lost research time, saw disruption of their animal breeding colonies, and experienced even more uncertain mouse maintenance arrangements (in terms of space and money allotted) than they had at the Bussey.

Some of these researchers landed at places with adequate resources. Dunn, for example, went to Storrs Agricultural Experiment Station in Connecticut even before he completed his 1920 dissertation.[16] But Dunn survived at Storrs only by displaying a characteristic Bussey flexibility about the kind of genetics work he was willing to do. In turn, he drasti-

[14] LCD to C. E. Allen, new chairman of the NRC's Division of Biology and Agriculture, 9 December 1929; CCL to LCD, 17 December 1929; both in LCD-APS. See also Karen Rader, "The Origins of Mouse Genetics: Beyond the Bussey Institution; II. Defining the Problem of Mouse Supply," in *Mammalian Genome* 12 (2001): 2–4.

[15] J. A. Weir, "A History of the Bussey Institution of Harvard University" (Grant Report, Penrose Fund), *American Philosophical Society Yearbook* (Philadelphia: APS, 1969), pp. 684–85. Weir is currently working on a full-length institutional history of the Bussey: see J. A. Weir, "Harvard, Agriculture, and the Bussey Institution," *Genetics* 136 (April 1994): 1227–31.

[16] Dunn Oral History, pp. 179 and 21, where he also notes there was a "well-beaten path from Harvard to Dartmouth."

cally scaled back his mouse linkage studies and devoted the better part of the next ten years developing inbred strains of chickens for studying economically relevant problems like the genetics of egg "hatchability." Dunn did not return to full-time murine work until 1928, when the Morgan group left for Caltech and Dunn moved to Columbia University. "I am now installed in my new-old quarters in Schermerhorn Hall and expanding the mouse business," he wrote to Castle in the fall of that year. "I shall have pretty good opportunities for work until someone objects to housing a large stock of mice in the center of the University."[17]

But far more of their classmates went to smaller institutions or teaching colleges, where they could not obtain material resources for their work at even the meager level that the Bussey had provided. Snell and Gates, for example, took jobs teaching zoology at Dartmouth College and Louisiana State University, respectively.[18] Gates wrote from Baton Rouge in November 1922: "I can not hope to do very much this winter, as my teaching schedule is heavier than I have had in years."[19] And by his own admission, Snell was extremely unhappy during his years at Dartmouth. In his first year there, he only had enough time and space to keep fifty pens of mice and one line of work going on the dwarf mutant he brought with him from the Bussey. This contrasts sharply with the wide-ranging, multiple mutant linkage studies of his dissertation research. By the middle of his second year at his new institution, Snell, too, was feeling the burden of teaching so heavily that he wrote to Dunn and asked him if he knew of any positions that would suit his research needs. Restless, he decided to take a graduate teaching job at Brown University for the coming academic year and was replaced by Sheldon Reed, another Bussey alumnus. But this new setting improved neither Snell's overall situation with regard to mouse material nor his research productivity. Like Dunn, Snell eventually broadened his research to other organisms in order to survive professionally.[20]

[17] Theodosius Dobzhansky, "Leslie Clarence Dunn," *Biographical Memoirs, National Academy of Sciences* 49 (1978): 79–104; Dunn to Castle, 6 October 1928, LCD-APS.

[18] "George Snell" in *Current Biography*, May 1986, pp. 48–50; W. E. Gates' entry in *American Men of Science* (1933). Keeler eventually also fell into this category: he stayed at Harvard until 1939, then went to the Wistar Institute for three years, after which he ambled through a series of one or two year teaching positions and eventually landed at the Georgia State College for Women in 1945. See *American Men of Science*, 8th edition (Lancaster, PA: Science Press, 1949).

[19] Gates, as quoted in "Mouse Club Newsletter," November 1922, Box 739,CCL-UMO.

[20] See Snell's C.V., provided to author by Snell; Oral History interview with George Snell, May 1986, in the Jackson Lab Oral History Collection, American Philosophical Society; Snell to Dunn, 29 December 1929, 7 January 1930, 14 March 1930, 27 May 1930 (where he expresses distress at his access to strains of mice needed for his linkage tests); all LCD-APS.

The diaspora of mouse genetics workers beyond the Bussey Institution also created two additional problems: a decentralized and therefore time-consuming network of communication and a general lack of stable colonies of mutant materials. If the extant correspondences of Dunn and Little from the 1920s are any indication, mouse geneticists spent much of their time tracking down the mouse mutants they needed, making sure that no one else was working on the particular problem they had in mind, or looking for their next place of employment. Dunn wrote to Castle in May 1925, for example, that even though L. C. Strong was "just at the point of reaping valuable results after long hard breeding work" on his well-respected mouse cancer genetics experiments, he was "out of a job for next year" at St. Stephens's (now Bard) College in New York, which had simply "decided that it can't carry his mouse experiments any longer, in view of the smallness of the place."

Mouse geneticists had developed several strategies for dealing with these problems. Dunn, for example, like Castle before him, worked hard to sustain an extensive network of mouse fanciers in the United States and abroad, and he frequently contacted or visited them in search of new variants. He had especially close contact with Carworth Farms, a commercial animal breeder in Rockland, New York, through which he obtained unique mouse mutant material.[21] Also, in addition to the ongoing *Mouse Newsletter*, by the late 1920s members of the so-called Mouse Club of America—the informal organization of mouse workers that made seasonal use of Cold Spring Harbor's mouse colonies—tried to replicate the collaborative atmosphere of that summer oasis by convening informal meetings at various academic year zoological conferences. There, presumably, they discussed the results of their experiments as well as how to procure or create the specific mouse materials they needed.[22]

But overshadowing such broader resource concerns and strategies, Little and his full staff from Michigan occupied themselves first and foremost with the task of settling into their own institutional routine at Bar Harbor. By the mid-1930s, new bureaucratic concerns arose concerning how to manage people, as well as mice. That July, at the first formal Meeting of Laboratory Workers, the staff agreed after a vigorous discussion to implement two blanket policies as part of their credo. The first dealt with making all research cooperative: "We should adopt a system of telling

[21] On Strong, see Dunn to Castle, 15 May 1925, LCD-APS. On English fanciers, see LCD to C. H. Danforth, 23 February 1928, LCD-APS. Carworth Farms was founded by Dunn's former Harvard colleague, Freddy Carnochan, in the 1920s: cf. Dunn Oral History, p. 678.

[22] Oral History interview by author with Elizabeth Russell, June 1993, Bar Harbor, ME; Oral History interview with Elizabeth Russell, Jackson Laboratory Oral History Collection, American Philosophical Society, p. 13.

one another everything we are trying to do." The second dealt with what should be routine procedure in caring for the mice: they noted that it was "important to change every box every week" and to label each researcher's mouse boxes with the proper initials so there would be no confusion regarding the origins of experimental organisms.[23]

Significant challenges to the meaning of Little's mouse transplantation genetic research program had also begun to emerge from medical circles. Cancer researcher James B. Murphy, for example, argued that since "the cancer cell was limited in its transplantability much as the normal cell is," transplantation could not reveal much about carcinogenesis. Instead, Murphy suggested that investigations of spontaneous tumors in laboratory animals might be more appropriate.[24] But Little and his group continued to assert that the genetics of tumor transplantation in inbred mice constituted a means for understanding the causes of malignancies. In 1929 Strong demonstrated that one of the genes involved in mammary tumor susceptibility was sex linked. Bittner subsequently pursued the genetic connection between mammary tumors and mouse coat color, in a series of transplant experiments. Strong used transplantation of multiple tumors from one individual inbred mouse to examine the genetic variability of tumors themselves. Cloudman extended this work in a study of eight transplantable tumors arising in the same inbred mouse line.[25]

Simultaneously, perhaps in response to the criticisms, Little and his group began to develop and use "naturalistic" but still genetically controlled models of cancer development. Ilana Löwy and Jean Paul Gaudillière have noted that Little's founding of the Jackson Laboratory roughly coincided with his group's increased production of a wide variety of spontaneous tumors in inbred mouse strains. This practice involved observing which strains of mice showed a predictable natural tendency to develop certain cancers, followed by brother-sister mating of individuals within those strains to stabilize that tendency. The immediate goal of this work,

[23] Minutes of Meeting of Jackson Laboratory Workers, 8 July 1930, Box 739, CCL-UMO.

[24] James Murphy, "Certain Etiological Factors in the Causation and Transmission of Malignant Tumors," *The American Naturalist* 60 (1926): 227–33; Maud Slye, "The Relation of Heredity to Cancer," *Journal of Cancer Research* 12 (1928): 83–133.

[25] L. C. Strong, "Transplantation Studies on Tumors Arising Spontaneously in Heterozygous Individuals," *Journal of Cancer Research* 13 (1929): 103–15; John J. Bittner, "A Genetic Study of the Transplantation of Tumors Arising in Hybrid Mice," *American Journal of Cancer* 15 (1931): 2202–47; "Linkage in Transplantable Tumors," *Journal of Genetics* 29 (1934): 17–27; Arthur Cloudman, "A Comparative Study of Transplantability of Mammary Tumor Arising in Inbred Mice," *American Journal of Cancer* 16 (1932): 568–630; cf. Bittner, "A Review of Genetic Studies in the Transplantation of Tumors," *Journal of Genetics* 31 (1935): 471–87 (mostly of work done at Jackson Lab).

then, was to combine two modes of mouse production: to make mice with the same genes (useful for mouse genetics work), while at the same time making mice with predictable tumor incidence (useful for cancer research more broadly).[26]

This work created several interesting leads, and its pursuit greatly increased the number of mice in Jackson Lab quarters. Little and his group used spontaneous tumor inbred mice in Mendelian breeding studies to further characterize the complicated genetic behavior of tumor development. For example, Bittner crossed two high mammary tumor inbred strains and observed that the tumor incidence in the F1 hybrids was slightly higher than in either parent stock; he attributed this to chromosomal heterosis.[27] Little himself got some surprising results from hybrid crosses between the high mammary cancer *dba*s and the sarcoma-prone strain *Y*—inbred by L. C. Strong since 1925—which suggested that certain cancers may have no parental or hereditary origin. Strong isolated a substance from the blood of these mice, the presence of which he correlated with both general growth and certain advanced stages of mouse cancers. Finally, Bittner and Cloudman started a large, joint project exploring the genetic variability involved in tumors arising from all the available strains of inbred mice. As spring approached, Little had enough extra materials that he could afford to ship gratis sizable numbers of *dba*s and *C57Black*s to Maine's Mount Desert Island Biological Lab summer students, who had requested them in order to do research at their home institutions.[28]

Despite the new emphasis on the spontaneous cancer projects, however, Little also made sure that his researchers did not neglect entirely traditional genetics work on the unique Jackson Lab mouse materials. In the summer of 1930, for example, he employed a local college student, R. A. Hicks, to undertake a large mouse serological study to determine the evolutionary relationships between several presumed-different species of mice. They reported in *Genetics* the following year their evidence that *mus musculus* and *mus faeroensis* were identical, as were *mus wagneri* and *mus bactrianus*. Meanwhile Charles Green continued his investigation of genetic factors for size in mice, using crosses between *dba*s and his own *mus bactrianus* stocks, and Little convinced Bittner to pursue genetic

[26] Löwy and Gaudilliere, "Disciplining Cancer," pp. 9–12.

[27] John J. Bittner, "Tumor Incidence in Reciprocal F1 Hybrid Mice—A X D High Tumor Stocks," *Proceedings of the Society for Experimental Biology and Medicine* 34 (1936): 42–48.

[28] "Cancer Prevented in Some Animals," *New York Times*, 28 June 1930, p. 18; Stanley Warner to CCL, 9 February 1931; CCL to Warner, 14 February 1931, Box 739, CCL-UMO; cf. Bittner, "A Genetic Study of the Transplantation of Tumors"; Cloudman, "A Comparative Study of Transplantability."

analyses of occasional coat color mouse mutants that appeared from his numerous cancer research crosses between *dba*s and Bagg albinos.[29]

When Little traveled to Cleveland in April 1931 to deliver his presidential address to the American Association of Cancer Researchers, the relationship between inbred mice and cancer genetics research was the topic at hand. Maud Slye, the Little group's long-time nemesis, had garnered media attention at the previous summer's AMA meeting in New York for claiming that cancer could be partly prevented through genetic planning.[30] Now speaking to an audience of medical workers, Little resurrected the rhetoric of his early career and praised the "growing interest of medicine in genetic research" and "the continued growth of that cooperative spirit." He went on to discuss the three prevalent approaches to understanding the role of heredity in determining the incidence and growth of cancer: spontaneous, induced, and inoculated tumors, all of which involved the use of inbred mice (since these were organisms in which these phenomena had been best studied).

Little briefly attacked Slye's "too simple" single-recessive view of mammary cancer—pointing out the recent change of heart expressed by her mentor, H. Gideon Wells, about her conclusions—but ultimately, he spent less time talking about the applied meaning of the results than he did emphasizing the organisms themselves: "genetically controlled" mice were necessary for obtaining rigorous experimental demonstrations of how cancer behaves. Previous tumor transplantation data, he argued, were full of "statistical artefacts due to the mixed genetic nature of the mice used." Use of inbred mice would serve to accomplish more quickly and carefully both geneticists' and physicians' research goals of understanding the problem they were studying. Little closed with a rejoinder to the medical profession's contemporary critique of the potential for eugenic applications for genetic research. "The medical profession," he wrote, "should give up the idea that genetic research seeks a practical method of applied eugenic control in the field of cancer and should realize that it is only another type of approach, another habit of thought, like serology, immunology or the use of tissue cultures." Here he held up the work of the Jackson Lab group—especially Murray, Strong, and Bittner on transplantable tumors—as a model of this proper genetic-but-not-eugenic attitude.[31]

[29] C. C. Little and R. A. Hicks, "The Blood Relationship of Four Strains of Mice," *Genetics* 16 (1931): 397–421; C. V. Green, "On the Nature of Size Factors in Mice," *American Naturalist* 65 (1931): 407–16; John J. Bittner, "A Color Mosaic in the Mouse," *Journal of Heredity* 23 (1932): 421–22.

[30] McCoy, *The Cancer Lady.*

[31] Little, "The Role of Heredity."

Little's retreat from eugenic claims regarding the applicability of work with inbred mouse material is curious but perhaps understandable given what was happening at his institution. The research results his scientists were obtaining showed that cancer genetics was an exceedingly complicated problem, and that same year he had already expressed fear about the possible "hysteria" that would result if lay people tried to read too much into the existing studies.[32] Just as importantly, Little's rhetoric strengthened the connection between the Jackson Lab's improved inbred mouse husbandry procedures and his scientists' research productivity. Simply put, Little was caught up in the practical and programmatic successes of his new venture, and his immediate concerns lay in putting the animals and the results they produced to the best use for his own workers.

FEELING THE DEPRESSION

Throughout this initial period of great activity and attention—perhaps because of it—Little's correspondence exuded confidence in the Jackson Lab's ability to remain unaffected by the contemporaneous financial crisis. Despite grim economic predictions for the banks, in May 1930 Richard Webber told Little he could expect an increase in the budget for the coming fiscal year. In January 1931 Little wrote enthusiastically to Mrs. Roscoe Jackson that everyone was "getting along splendidly here. . . . I feel that we are making very good progress." He requested—and obtained—more money from her for Strong to "visit a biochemical laboratory somewhere in Europe" and sharpen his serological skills. In a letter inviting Mrs. Jackson and the other patrons to visit, he enclosed a copy of the lab's summer schedule, which included (on the model of Woods Hole) Wednesday evening public lectures by lab personnel and prestigious visiting cancer researchers such as James B. Murphy.[33] That summer the lab inaugurated its research and teaching summer program, which would be the initial institutional mainstay of its interactions with the larger biological and medical research communities.

But the following fall marked the beginning of a prolonged, and seemingly unexpected, period of financial stringency for Little's institution. Little's original three-year financial agreement with the Detroit patrons was scheduled to expire in May 1932, and the effects of the Depression on the Detroit banks had grown progressively worse. Displaying his characteristic bravado, Little contacted the Fords, the Webbers, and Mrs.

[32] Little, "Education in Cancer."

[33] Richard Webber, 7 May 1930, Box 8–7, JLA-BH; CCL to Mrs. RBJ, 16 January 1931; CCL to Mrs. RBJ, 21 July 1931; both in Box 739, CCL-UMO.

Jackson shortly after Labor Day 1931 to request that they simply renew their original commitment to the lab for another three years. Mrs. Jackson was the first to reply: she would have to reduce substantially her contributions after 1 May 1932. Then in December 1931, Edsel Ford wrote Little that "the laboratory will no doubt come up again for consideration" in May, but he could make no guarantee of funding. The Webbers, who were Little's last hope, decided to support the Jackson Lab, but at levels far below the monthly financial stipend contained in the original agreement.[34]

Up to this point, Little had only minimally pursued other sources of funds. For example, he had encouraged Strong, his most senior researcher, to apply for a $7,000, two-year outside grant to the Josiah Macy, Jr., Foundation for developing his mouse cancer blood-work project into a viable serological test, with potential human clinical applications. Strong ultimately got this grant, which began paying him in the fall of 1931.[35] But because that money was earmarked for supplies and caretaking, to underwrite the mouse colonies necessary for Strong's project, Little needed to begin looking for more long-term institutional funding sources.

The stress of pending financial hardship had a demoralizing effect on Little, who began withdrawing from day-to-day life at the lab and neglecting his own mouse research. In October 1931, Bill Murray sent him a stern warning note: "Get ready to get mad, your [*sic*] on the carpet. It seems that the number of boxes has slowly but surely risen to 300 a week and your [*sic*] one of the chief offenders. It seems that you have 25% more boxes than would be necessary were your mice properly weaned etc. . . . As a matter of fact we are running way over budget and, like the national government, must do something." Murray softened the complaint by explaining that he simply didn't want the "young PhDs" to think that Little was doing bad science: "it gives them a crack at you and anything you publish." But mouse husbandry, because it represented such a large portion of the lab's operating budget, would increasingly become a direct target for cost-cutting measures in the months to come. Wooden boxes, in particular, were expensive items.[36]

[34] CCL to Mrs. Roscoe Jackson, 19 September 1931; Edsel Ford to CCL, 30 December 1931; there is no archival copy of any letter from the Webbers explicitly responding to a request from Little at this time, but there are letters of ad hoc contributions by the Webbers starting in July 1932; all contained in Box 739, CCL-UMO.

[35] Macy Foundation to CCL re: Strong's grant, 9 July 1932 and 1 October 1932, Box 739, CCL-UMO. The Macy Foundation, as others Little appealed to, falls in the tradition of Progressive philanthropy of the 1920s outlined by Judith Sealander in *Private Wealth and Public Life: Foundation Philathropy and the Reshaping of American Social Policy from the Progressive Era to the New Deal* (Baltimore: Johns Hopkins University Press, 1997).

[36] William Murray to CCL (handwritten), 28 October 1931, Box 739, CCL-UMO.

On New Year's Eve 1931, Little did some soul searching: he made a list of the "most outstanding features" of the Jackson Lab's work. Judging from this document, Little's vision of the lab's mission still clearly focused on research rather than mouse production. For example, first he mentioned "maintenance and investigation of non-cancer lines" of mice as significant, for it "provid[ed] other investigators with material." But above all, he emphasized, scientific results would be the feature that would "establish the reputation of the laboratory." In a gesture that can be read as equal parts personal egoism and faith in the lab's original program, he listed his own two experiments, in which he crossed high tumor strains *dba* and *A* with low tumor strains *C57Black* and *X*, to be the most important because they "provided information as to the type of inheritance involved" in cancer.[37]

It is notable that at this critical moment—with the lab's financial woes evident, with the NRC committee's discussion of animal supply still percolating in his thoughts, with tens of thousands of prolific inbred mice on hand—Little did not give voice to the possibility that the Jackson Lab could provide research materials to the scientific community in exchange for money. Perhaps having been scientifically socialized in the "free exchange" culture of mouse genetics precluded this thought. Or perhaps having not experienced the effects of the Depression first hand, Little was more deeply opposed to the vulgarity of commercialism in matters of science and education than some of his medical research counterparts.[38] In late 1928, for example, he had told one of his last audiences at the University of Michigan: "we are apt to forget that the production of a standardized product so highly successful as a business method is fatal in education." This belief initially extended to the production and sale of Jackson Lab inbred mice for outside researchers and persisted despite his willingness (like that of other cancer researchers at the time) to forge relationships with industrial patrons for purposes of research.[39]

A few weeks into 1932, Little reluctantly informed the lab's scientific Board of Directors of the impending financial crisis and tried to solve the problem by more traditional means: renewing his entrepreneurial activi-

[37] C.C. Little, typescript, "My personal work this year has been as follows," 31 December 1931, Box 730, CCL-UMO.

[38] Contrast Little's initial resistance to commercialism and industrialization with the strategy (starting in the late 1930s) of Memorial Hospital's Cornelius Rhoades in the development of the Sloan-Kettering bequest: see Robert Bud, "Strategies in American Cancer Research after World War Two: A Case Study," *Social Studies of Science* 8, 4. (November, 1978): 425–59.

[39] CCL handwritten mss. on University of Michigan stationery, n.d., Box 742, CCL-UMO. On the prehistory of postwar industrial funding for cancer research, see Bud, "Strategies in American Cancer Research."

ties with the foundations. "I am afraid," Little warned one board member, "that this is going to be a very bad year for scientific work that has not got immediate human appeal." He started by contacting John D. Rockefeller to ask for funds, but he received no reply.[40] Another board member, Herbert Neal, a Bussey Institute–trained geneticist then at Tufts University, encouraged Little to think in broader and more optimistic terms. Neal noted that because expenses over the past two years went largely toward setting up the lab, "we should be able to get on, even if our income is reduced fifty percent." The real problem, he emphasized, was a lack of permanent endowment, and along these lines he encouraged Little to "get strong enough letters in support of our Laboratory to make a very strong appeal to one of the great foundations."[41]

Little did not heed this advice, and in subsequent appeals he showed no signs of thinking about the lab's problems and approaches any differently. For example, he put together an application for outside funding from the [Anna] Fuller Fund of New Haven, Connecticut, to support Bill Murray's project on ovary secretions and spontaneous cancer occurrence in *dba* mice. He ultimately requested a three-year grant of $3,000 per year, but a close look at the budget shows that this money would have gone primarily for mouse-upkeep expenses associated with the larger colonies of this strain that Murray would require.[42] Meanwhile, he continued to press the Detroit patrons for a long-term commitment. In April 1932 he presented them with two possible budgets for Jackson Lab's fiscal year beginning in July, one of which cut expenses 17 percent and the other, 23 percent. To Little, the 23 percent cut really represented "very definite retrenchment" in terms of research: "It should be considered as a last line of defense . . . beyond which we cannot go without complete reorganization." But neither budget indicated a reduction in the money spent on mouse breeding and caretaking. In fact, while Little mentioned that salary cuts would also involve a reduction in "general operating expenses," he noted the total mouse maintenance budget of $900 a month "could not be reduced . . . without seriously hampering the work."

[40] For example, see CCL to Homer W. Smith at NYU Medical School, 11 January 1931, Box 739, CCL-UMO.

[41] CCL to Herbert Neal, 11 January 1932; Herbert Neal to CCL, 13 January 1932; Box 739, CCL-UMO.

[42] "Since large numbers of animals must be used in such work and they must be kept over a long period of time . . . it is hoped that the greatest liberality may be shown in not demanding finished results in too early a period. The mice must be kept from one to two years after the experiment is started in order to enable them to reach cancer age." See Application to the Fuller Fund, enclosed in CCL to E. M. East, 11 November 1932, Box 735, CCL-UMO.

The original donors' reactions to the heightened intensity of these appeals must have been positive, for Little wrote to the Rockefeller Foundation's Alan Gregg (a former classmate of his from Harvard) that he left the meeting feeling "greatly encouraged . . . that we could keep the laboratory going in these extremely upset times." Ultimately, the Fords, Webbers, and Mrs. Jackson committed to a budget of $30,725 for the fiscal year 1932–33—a nearly $25,000 cut from the previous year and more than half from the founding year. Nevertheless, with this promise in hand, Little appears to have backed off somewhat from soliciting other major philanthropic organizations.[43]

The lab's research program experienced some setbacks as a result of the immediate financial instability caused by the crash of 1929, but they were short-lived. Although the institution weathered 1931 without a significant loss in the numbers of mice available for research, Little lost one researcher. In the fall of 1932, Arthur Cloudman was awarded the Harvard Medical School Littenauer Fellowship in Pathology. Cloudman, who worked on the occurrence and diagnosis of multiple spontaneous cancers in individual strains of mice, told Little that he hoped the position might sharpen his nongenetic, medical research skills and techniques, but Little encouraged him to take it in the hopes of ensuring his financial and professional stability. While away, Cloudman retained friendly scientific relations with the lab; he corresponded with Little and occasionally diagnosed a difficult mouse tumor sent to him by one of his former co-workers. When Cloudman's supervisor, cancer pathologist Shields Warren, expressed interest in having him stay for a two-year academic/hospital rotation, Cloudman wrote to Little expressing regret that he had abandoned the Jackson Lab's mission. Little replied that he shouldn't worry: "I think it will be a wonderful chance for you to get training which ought to be of great value to you and to us. If the laboratory funds continue and the place runs at all you know, I am sure, that I am counting on you coming back. We must both take this for granted always." Cloudman rejoined the Jackson Lab the following year, and in his absence his name was mentioned in the grant applications that Little made.[44]

[43] Richard Webber to CCL, 17 February 1932, Box 8–7, JLA-BH; CCL to Mrs. Jackson, 4 April 1932, Box 739, CCL-UMO. See also Little's handwritten draft of these budgets in Box 735, CCL-UMO. For the budget 1932–33, see CCL to Alan Gregg, 24 May 1933, RF Archive, RG 1.1, Series 200D, Box 143, Folder 1773, RAC-NY.

[44] Cloudman's fellowship reported in CCL to Mrs. Roscoe B. Jackson, 4 April 1932; Cloudman to CCL, 28 November 1932; CCL to Cloudman, 30 November 1932; see also letter from Mary Russell (Jackson Lab Secretary) to *The Maine Alumnus*, 2 December 1932. Cloudman spent the summer of 1932 at Bar Harbor, but this potential research time was interrupted by the unexpected death of his wife: M. Russell to Dr. D.S.B. Young, 19 July 1932; all in Box 739, CCL-UMO.

Even with the Detroit funding renewed, however, Little and his staff were now facing massive budget cuts, and it was this practical realization that forced them to rethink Jackson Lab's institutional priorities. At the time of Cloudman's departure, Little approached the remaining staff scientists and requested their feedback on the latest budget-cutting woes. Minutes of this meeting note that the group unanimously agreed to use the small amount of cash on hand to secure the future of the inbred mice: they bought enough food for existing animals to ensure their continuation (under the same well-managed caretaking conditions) for a year. In providing for themselves they were much less generous: the researchers voted to cut their own salaries back to $100 a month—a reduction of more than half their original level. To further reduce living expenses, and thus create some spare funds for research, they also decided to combine households whenever possible and to grow a large community garden.[45] Despite these Draconian measures, it took outside encouragement to prompt them to move beyond a concern for preserving the mice for their own research production—in turn, to make them believe that what they thought of as mere animal caretaking work was itself producing something of broader value to the community of biologists.

MAKING THE DECISION TO SELL

Cancer researcher Howard Andervont spent the summer of 1932 in Bar Harbor as one of Jackson Laboratory's first official visiting summer investigators. Andervont (1898–1981) did his doctoral work on the relationship between cowpox and vaccinia viruses under a Carnegie Institution Fellowship at the Johns Hopkins University Medical School (Sc.D., 1926). His first job was as an instructor in Harvard's Department of Preventative Medicine, under Dr. Milton J. Rosenau. But in September 1930, upon Rosenau's death, he was appointed as a Public Health Service (PHS) officer and the first professional staff member of Harvard's Office of Cancer Investigations, under fellow pathologist J. W. Schereschewsky. Schereschewsky had risen in the PHS ranks during the late 1920s and early 1930s by becoming one of the surgeon general's most outspoken advocates of cancer as a public health problem. "Sherry" (as Schereschewsky was fondly nicknamed) recruited Andervont but refused to assign him a

[45] On salary cuts, see budgets enclosed in CCL to Mrs. RBJ, 4 April 1932, Box 739, CCL-UMO. For summer 1932, see also Holstein, *The First Fifty Years*. For a sense of the hardship such living conditions entailed for members of the lab community, see CCL to R. H. Webber, 22 December 1932, Box 739, CCL-UMO.

specific experiment: "he was a firm believer in freedom of research."[46] Andervont chose to work on transplantable tumors in mice, a problem of great interest to him from the perspective of virology, even while it was very different from his original experimental work.

In May 1932 Schereschewsky wrote to Little about a new line of experiments that he and Andervont conceived, "the object of which is to modify the production of spontaneous tumors in mice susceptible to their development." Schereschewsky and Andervont had been using two strains of spontaneous tumor mice in their experiments—one that was an agouti strain, obtained from L. C. Strong several years earlier. But despite the concerted efforts at selective breeding, mice in both of the strains had ceased producing spontaneous tumors. For their new set of experiments, Schereschewsky told Little, "the material at hand is inadequate . . . [but] you have colonies in your laboratory in which the spontaneous tumor rate is stabilized." He arranged for Andervont to come to Jackson Lab begin these experiments with the inbred mice materials there.[47]

Little was away from Bar Harbor soliciting donors when Schereschewsky's letter arrived, but his initial response upon hearing of Andervont's plans suggests that he had begun to envision the possibility of separating Jackson Lab's research materials from the genetic science for which they were created. Little replied immediately: even though the Harvard group's working hypothesis directly contradicted his own genetic view, he would be "very glad to have Andervont come here for the summer and work on [our] material." Furthermore, it is clear that Little was less concerned with the prospects of a research collaboration with the PHS scientists than he was with the potential economic gain of a materials-exchange relationship. Little offered Andervont all the space and assistance he required and tentatively explored the possibility of compensation for the mice: "I wish that I could offer him this material without any expense. Unfortunately, however, the Laboratory finances are handicapped and since it would cost something to raise the animals which he would need I am afraid that the expense of that work would have to be deferred by the Public Health Service." Not having sold mice before, Little had no fixed prices at hand, so he requested a list of the number, age, and sex of the mice Andervont needed, in order to make an estimate of the expenses involved.[48] Schereschewsky did not object in principle to a financial exchange; rather, he responded as a potential consumer and bargained to get exactly what he wanted out of the deal. He gave Little exact specifica-

[46] H. B. Andervont, "J.W. Schereschewsky: An Appreciation," *Journal of the National Cancer Institute* 19, 2 (August 1957): 331–33.

[47] Schereschewsky to CCL, 2 May 1932, Box 739, CCL-UMO.

[48] CCL to Schereschewsky, 23 May 1932, Box 739, CCL-UMO.

tions on the biological commodities he wished to purchase: Andervont would need approximately five hundred female mice from a high-cancer strain, Schereschewsky said, and "it would save a great deal of time if they were all mature" to the age where the tumors develop, "say between four and six months old."

Little's surprise at the magnitude and specificity of Schereschewsky's request suggests that he was thinking of this opportunity not as a long-term solution but rather as a temporary measure to make some quick money. Indeed, his initial efforts to frame the mouse exchange in precise financial and temporal terms seem almost too transparent in this regard. For example, he implored Schereschewsky to give the Jackson Lab few months to come up with the mice Andervont needed, explaining in great detail why the time lag resulted from Schereschewsky's own mistaken husbandry calculations. It also appears, however, that Little had in mind a particular strain of mice that was available that he wanted to sell, for he told Andervont that the average age of breeding females in the only strain both "available" for sale and suited for Andervont's experiments was ten months. This strain (not mentioned by name) must have also been one the Jackson Lab was already breeding in large quantities for research (e.g., *dba*), because Little stated he could have the total number of mice of that strain by the end of the summer. Until then, Little said, Andervont could come to the Jackson Lab to do the work and breed his own. Or, he suggested, Jackson Lab's mice could be sent "to you in lots of twenty to fifty as they could be raised" at cost, which was "approximately fifty cents a piece." Little concluded by giving Schereschewsky a marketing pitch that would soon become inextricably attached to these mice: only Jackson Lab mice were inbred "under the controlled conditions which would be necessary to make them valuable for your work."

In the end, Schereschewsky rejected parts of Little's plan, but he accepted the longer time frame necessary to scale up animal production. He did not transport the mouse material to Harvard on the basis of his own research failures with relocated mice, which (as he explained to Little) "don't seem to maintain the same tumor rate when transplanted in a new environment as in the original laboratory." Instead, Schereschewsky and Andervont paid a visit to the Jackson Lab in early June to determine "the details of what we have in mind" for getting the experiments done over the summer. The PHS would then take financial responsibility for the continuing work by simply purchasing the inbred strains used under contract. Little agreed to this arrangement and, within ten days of Schereschewsky's letter, submitted a formal bid for the contract: 500 mice at 50 cents each for a total of $250.00. Schereschewsky immediately accepted, and although the lab's financial ledger does not record income from the outside sale of its inbred mice—$258.16—until late January 1933, the first mouse sales deal was effectively sealed.[49]

Several Jackson Lab staffers later recalled that, before Andervont arrived, John Bittner had floated the idea of selling and shipping their extra inbred mice to the Cambridge laboratory, so that Schereschewsky and colleagues could do their work in-house.[50] But the surviving minutes of the Lab Workers' Meetings show no evidence that Bittner was lobbying Little and the other staff members to sell mice this early.[51] Furthermore, Andervont himself has written about his first days in Bar Harbor, particularly about being scandalized at how Jackson Lab workers remained oblivious to the earning potential of their experimental materials. Scientists and caretakers were being instructed to dispose of all of the mice they did not need for their research so that the lab did not incur any needless maintenance expenses—a practice that Andervont had to tell them was both scientifically and financially wasteful. "When I saw boxes of mice being killed," he said later in an interview, "I suggested that the sale of mice could help with the financial crisis that was beginning."[52]

FROM RESEARCH TO PRODUCTION: AN UNEASY TRANSITION

Wherever the idea came from, Little's actions after making this initial deal suggest a kind of "seller's remorse;" he remained deeply ambivalent about doing anything that might be perceived as too commercial and thereby sullying the reputation of the lab's pure research. Among geneticists, in particular, the free exchange of mice had always been standard practice,[53] and at the same time that he was doing business with Schereschewsky, Little also made new efforts to court the attention of this other group interested in inbred mice. These efforts were not manufactured, since the most successful Jackson Lab experiments that summer focused on describing the genetic behavior of various mouse mutants that had turned

[49] Ledger sheet labeled "Mice February 1933," Box 739, CCL-UMO; also Schereschewsky to Little, 25 May 1932 and 31 May 1932; CCL to Schereschewsky, 27 May 1932; Government Contract—Bid for Supplies, signed by CCL, 9 June 1932; Contract Letter of Acceptance, signed by Schereschewsky, 27 June 1932; all in Box 739, CCL-UMO.

[50] Elizabeth Fekete, in Holstein, *The First Fifty Years*, p. 79; Elizabeth Russell, in *Origins of Inbred Mice*, ed. Morse.

[51] "Roscoe B. Jackson Memorial Laboratory, Minutes of Meetings of Workers," 8 July 1930 to 1 December 1930, Box 739, folder of the same name, CCL-UMO.

[52] Interview with Jean Holstein, as quoted in Holstein, *The First Fifty Years*, p. 79; also Andervont to Holstein, 12 December 1979, JLA-BH.

[53] See Oral History Interviews: Joan Staats, Elizabeth Russell, George Snell, and Tom Roderick, JLOH-KR. On the importance of materials exchange in the work culture of geneticists, see Kohler, *Lords of the Fly*, chap. 4, and Rader, "The Mouse People."

up in the colonies. For example, Little and Maine graduate student Byron McPheters successfully manipulated (via artificial selection) the development of strains with myencephalonic blebs (fluid-filled sacs, the first of which appears in eye development): two lines of these mice (originally described by Little and Bagg in Cold Spring Harbor) had limb abnormalities that appeared to be linked to eye abnormalities. Also, Bittner investigated cancer tendencies in F2 hybrids of *dba* high-cancer and yellow low-cancer strains: yellow offspring (about half of the 260 hybrids studied) were five times more likely to develop cancer, an outcome Bittner suggested could be related to a genetic linkage between "a gene determining high incidence of mammary cancer and the gene for non-agouti" coat color.[54] But earliest publications resulting from Bittner's work appeared in genetic or general experimental biology—not medical—journals, and Little was encouraging his researchers to re-embrace the complicated hereditary aspects of the work, rather than its other applications.[55] Bittner himself wrote in his 1931 summary report: "We believe that we have added important data in support of the genetic theory of the transplantation of cancer . . . [though] the geneticists have criticized [this] theory on the basis that we have been unable to determine linkage. . . . We hope that future experiments will overcome this view."[56]

Likewise, when International Eugenics Congress exhibit chairman Harry Laughlin approached Little with a request to develop a display showing the human eugenic implications of the lab's work on mouse cancer, Little ignored the suggested topic and worked with Joe Murray to put together a chart showing all the known gene mutants of *mus musculus* and their linkage groupings, which were partially determined for only thirteen of the twenty mouse chromosomes; each character was to be illustrated by a dried mouse skin specimen. Despite the many strains of mice the Jackson Lab and Murray himself possessed, their colonies (like most mouse colonies at the time) were relatively specialized to the experimental work. Little had to write away to his former Bussey colleagues L. C. Dunn (now at Columbia) and Clyde Keeler to obtain "black and tan" and "rodless retina" skins. What was remembered as "the Jackson Lab mouse mutation display" at the Eugenics Congress made a powerful im-

[54] These were first reported as Staff of the Roscoe B. Jackson Memorial Laboratory, "The Existence of Non-Chromosomal Influence in the Incidence of Mammary Tumors in Mice," *Science* 78 (November 1933): 465–66.

[55] D. B. Young to CCL, 1 July 1932; also, Byron McPheters (once a University of Maine student) and Stanley Warner (of Cambridge, MA, and affiliated with the Bussey) spent time at Jackson Lab during the summer of 1931: cf. C. C. Little and B. W. McPheters, "The Incidence of Mammary Cancer in a Cross between Two Strains of Mice," *American Naturalist* 66 (1932): 568–71; and Little, "Further Studies on the Genetics of Abnormalities Appearing in Descendants of X-rayed Mice," *Genetics* 17 (1932): 674–88.

[56] John J. Bittner to CCL, 31 December 1931, Box 732, CCL-UMO.

pression on even those outside of genetics research: the next week Little received a request to borrow the display for the "Century of Progress" World's Fair taking place the next year in Chicago.[57]

At the same time, there are signs that Little was testing the waters quite literally among his genetics colleagues regarding a plan to encourage more widespread sale and use of Jackson Lab's own inbred mouse materials. In August 1932, at a Plenary Meeting of the Sixth International Congress of Genetics, Little proposed a formal resolution that suggested adopting inbred animals as the gold standard of genetics research. The resolution, if passed, would state that: "In all experimental work involving investigations of a pathological . . . or of a physiological or psychological nature with animals or plants, the importance of genetically pure strains be emphasized." Laissez-faire impulses were strong among the mouse geneticists in attendance, including those who were formerly Little's closest colleagues, such as E. C. MacDowell of Cold Spring Harbor and A. F. Shull of the University of Michigan. "No formal action was taken," Little would later write to a medical research colleague, "because of fear that such would be taken as unwelcome interference on the part of one group of experimental scientists with the activities of another." Ultimately the motion was tabled, and Little could only concluded that, from a market perspective, large-scale efforts to distribute his institution's animals to mouse geneticists were premature.[58] True to the academic culture of their origins at the Bussey Institute, this group collectively resisted centralization, even though individually each of its members was hobbled by the scarcity of mutant material and lack of institutional resources to sustain large-scale breeding projects.

Meanwhile, as mouse production for the PHS project grew, the economy within which Little defined and regulated the value of the Jackson Lab's mouse materials and other resources was gradually transformed from one of research rewards to one of financial rewards. For example, Schereschewsky's Office of Cancer Investigations at Harvard contracted with Little three more times, supplying a total of 1,230 inbred mice for

[57] H. H. Laughlin to CCL, 27 October 1931; CCL to Laughlin, 9 November 1931; CCL to L. C. Dunn, 26 July 1932; CCL to Clyde Keeler, 26 July 1932; Jay Pearson to CCL, 1 September 1932; CCL to Pearson, 6 September 1932; all in Box 739, CCL-UMO. Abstracts of the mouse and rat exhibits at the congress can be found in Donald F. Jones, ed., *Proceedings of the Sixth International Congress of Genetics*, vol. 1 (Brooklyn: Brooklyn Botanical Garden, 1932), pp. 249–55; cf. a request for deer mice, which Jackson Lab also could not provide: D. O'Brian to CCL, 19 July 1932; CCL to O'Brian, 23 July 1932, Box 739, CCL-UMO.

[58] The minutes of this second plenary session were published in Jones, ed. *Proceedings of the Sixth International Congress*, vol. 1, pp. 17–19. Cf. CCL to Ludwig Hektoen, 22 November 1937, Box 6, Records and Minutes of the National Advisory Cancer Council, 1937–55 (NIH A-1,E-26) National Archives (College Park).

Andervont's work, and the language of these contracts reveals how the values informing Jackson Lab's participation in this process were shifting. From Little's perspective, cash was in such short supply that the money was more important than the science being done. His interest was in codifying the rules for its exchange, and he pushed for each contract to state that all mice would be "prepaid." But from Andervont's perspective, the standards set for the financial exchange did not mean as much as those set for the scientific quality of the materials. For some of his experiments requiring small groups of animals randomly selected from various inbred strains, the contract description of the mice and their care was vague and open-ended. But because his largest research project was premised on the use of mice reared under highly controlled environmental conditions that would predictably develop tumors, Schereschewsky asked that these standards appear in every relevant contract: namely, that the mice sold be female, of the "specified tumor stock," and "kept in substantially the same environment as the parent stock."[59]

Along the same lines, Little encouraged his caretakers to streamline their caretaking procedures in order to capitalize on economies of scale that the PHS exchange had created. In June 1932 he ordered an inventory of all mouse stocks by researcher, including the number of wire cages and wooden boxes held by each. This revealed that the lab had 7,774 mice of at least twenty-five strains on hand at the start of the summer. Rather than seeking to reduce this number, Little tried to consolidate the mice into a lesser number of boxes and cages to make feeding more efficient as well as less costly.[60] Because feeding was still the biggest expense, however, Little moved to decrease costs even more through rationalizing the food itself. In July 1932 he decided to switch the inbred mice from expensive hemp and canary seeds to Purina Fox Chow (a commercially produced feed, very similar to Dog Chow), which the lab could order in great bulk (usually two and a half ton lots) from a distributor in Missouri. Furthermore, though each box of mice—regardless of the number of organisms in it—already received only one ounce of milk a day, Little suggested that the researchers try plain water instead. No one reported any significant differences in development or general health of the animal, so everyone made the switch. Finally, with the help of caretakers Clarke and Oakes, glass water bottles were created. These were fitted with purchased rubber stoppers and metal tubing and then attached to the mouse box lids. Now

[59] Series of PHS Contract Bids by the Jackson Lab, together with PHS Letters of Acceptance, 18 June 1932 to 21 July 1932; all in Box 739, CCL-UMO.

[60] Inventory of mouse stocks with Little's notes, 1 June 1932, Box 735, CCL-UMO.

the mice could drink whenever they wanted, and there would be no spills to dampen the bedding materials, which meant fewer box changes.[61]

By the end of 1933, the shift at Jackson Lab from research to production remained an uneasy transition that Little was not entirely willing to make. He made one last-ditch effort to gain foundation patronage for Jackson Lab on the basis of its genetics from the International Cancer Research Fund (ICRF). The ICRF, founded by industrialist William Donner in 1932 in memory of his son, defined its mission broadly but with a definite clinical twist: funds were to be used "to assist research work to discover the cause or cure of cancer in any part of the world." Other researchers who had interacted previously with this foundation's administration gave Little the impression that a successful application was determined "through the good impression one made on the executive secretary," Dr. Mildred Schram, who was described as "a southern lady with a honeydew accent, completely surrounded by . . . aspiring young and not-so-young men hoping for a munificent handout or perhaps a few thousand dollars."[62] So in September 1932, Little wrote a letter to Schram and told her that he had already discussed the possibility of an ICRF grant for Jackson Lab with James Ewing, a widely respected cancer researcher at New York Memorial Hospital. Little outlined three projects on mouse cancers that he believed worthy of funding: (1) transplantable tumor studies, "consistent with the hypothesis that susceptibility to cancer depends on . . . a number of mendelian factors," (2) studies in hybrid crosses in spontaneous tumor strains of mice, "of special interest at least by analogy in a country made up of such diverse racial mixtures as our own," and finally (3) studies of the effects of X-rays on the "general nature of the germ plasm." In both (1) and (2), he noted that Jackson Lab's "unique stocks of inbred mice" were vital for the work, but in outlining a budget, he simply requested stipends for his researchers.[63]

[61] CCL to Gladys Garland, 25 July 1932; also "Orders for the Jackson Lab, 1932–34" Folder; both in Box 739, CCL-UMO, especially Jackson Lab Secretary (Mary Russell) to Mr. H. Hopkins, 23 July 1932. Other letters in this folder reveal that the lab had some difficulty purchasing the right-size tubes and stoppers for the water bottles. For a general discussion of the feeding procedure changes, based on the recollections of Joe Murray, see Holstein, *The First Fifty Years*, pp. 77–78. Oakes and Clarke were often referred to as "mouse box changers," indicating that their primary responsibility was to wash out the mouse boxes weekly, change the bedding, and replace the mice: see interview with Earl Green, June 1993, JLOH-KR.

[62] On the ICRF's founding, see "William Henry Donner," *National Cyclopedia of American Biography*, vol. 44 (New York: White and Company, 1962), pp. 247–48; Michael Shimkin, *As Memory Serves: Essays on a Personal Involvement with the National Cancer Institute, 1938–1978*" (PHS/NIH: NIH Publication no. 83–2217, 1978), esp. p. 10.

[63] CCL to M. Schram, 24 September 1932; Little also enclosed c.v.s of all his researchers and a bibliography of the work published so far: all in Box 735, CCL-UMO.

Three months later, Schram replied by requesting more information about Little's institution. Specifically, she asked what portion of the total funds Little would apply to research and what portion to "experimental material." In his own desperation for funds, Little noted straightforwardly that the amounts "can be naturally shaped to meet the policy which your organization most approves of." But, consistent with the increasingly complicated results his researchers were obtaining, Little made no specific arguments that Jackson Lab research would contribute to discovering a cause or a cure for cancer.

Two weeks later, when the ICRF officially denied Little's application for funds,[64] Little fired back an angry letter to Schram that, in retrospect, offers a window on his mounting frustration as well as the lab's shifting institutional priorities. Little demanded to know if the application was considered by all members of the Scientific Advisory Committee, "for Drs. Ewing and [Burton] Simpson knew much more about the work than did Dr. [F. C.] Wood." Donner himself replied, insinuating that the real problem with the application was that it had been vague with regard to the value of the Jackson Lab cancer research program. Six months later, ICRF advisor Burton Simpson explicitly decoded this comment for Little: "a request for continued genetic study of the different stocks . . . would be the best way to present your request."[65] Little had founded his institution on this line of work, and yet in practice he did not invoke the group's cancer genetics research in a critical foundation negotiation, which suggests the extent to which mouse sales and production had already overtaken research at Jackson Lab.

Perhaps learning from the ICRF application, Little enclosed a letter in Murray's and Strong's new applications to the Fuller Foundation that touted the primary scientific value of the institution as its interdisciplinary cancer and genetics research program: "The group at the Laboratory is one which has worked together for years. It has been built up with a specific program and purpose in mind."[66] He also invited the prestigious geneticist J.B.S. Haldane (who was touring the United States from England) to visit Jackson Lab, and Haldane editorialized Little's same glowing arguments about this research in the American and British press.

Yet because the Jackson Lab had produced only interesting pure research leads rather than results with obvious clinical applications, this

[64] CCL to M. Schram, 10 December 1932; M. Schram to CCL, 21 December 1932; Box 735, CCL-UMO.

[65] W. H. Donner to CCL, 14 January 1933; CCL to Donner, 16 January 1933 and 25 January 1933; Donner to CCL, 5 May 1933; CCL to Burton Simpson, 8 June 1933; Simpson to CCL, 17 June 1933; all Box 735, CCL-UMO.

[66] Holstein, *The First Fifty Years*, p. 20.

argument remained unconvincing to the public and to potential patrons.[67] Murray's and Strong's applications to the Fuller Foundation were denied. But Little wrote ruefully to a lab trustee in October 1932 of his institution's predicament: "it seems a highly grim paradox that in direct proportion with the increasing scientific value of the work and its recognition by contemporary scientists of distinction, support of it has diminished."[68]

In sum, to sustain his institution during the Depression, Little was forced to modify the accepted practice of exchanging results and animals for free—now, by selling animals for cash. Little's transaction with the Public Health Service was thus implicitly a tradeoff of research for production. Although he conceived it as a temporary arrangement, scaled-up production created powerful imperatives for embracing businesslike values of efficiency and organization. The new caretaking procedures that resulted certainly benefited the research work going on at the Jackson Lab. But by commodifying the mouse exchange relationship, Little also invited the possibility that market forces beyond the scientific needs of his own researchers could compete to determine the value of Jackson Lab inbred mice. Along these lines, separating mice materials from the research for which they were initially bred proved both easier and more difficult than Little had initially anticipated, and it had profound consequences for the institutional mission of the lab.

RENEGOTIATING THE RELATIONS AMONG MONEY, MATERIALS, AND RESEARCH

At summer's end, it became clear to Schereschewsky and Little alike that Andervont had exchanged more with the Jackson Lab than just money for mice. Still, because the relationship was framed by a formal "bill of sale" for organisms instead of as a scientific meeting of the minds, proprietary links between research materials and research programs had been

[67] "Bibliography of Papers Published 1926–1929" by Little's Group, Box 735; CCL to Dr. Homer W. Smith, 11 January 1932, Box 739; both CCL-UMO.

[68] On Haldane's visit, see George Dike to CCL, 7 October 1932, and CCL to Dike, 10 October 1932; J.B.S. Haldane, "The Genetics of Cancer," *Nature* 132, 19 August 1933: 265–67. Regarding the Fuller Foundation, see CCL to E. M. East, 30 November 1932, plus attached proposal, Box 735, CCL-UMO. Regarding the ICR, see Grant application materials, as well as CCL to Mildred W.S. Schram, 24 September 1932 and 10 December 1932; Schram to CCL, 21 December 1932; W. H. Donner to CCL, 14 January 1933; CCL to Donner, 16 January 1933; Donner to CCL, 5 May 1933; all Box 735, CCL-UMO. On Simpson, see "Burton Simpson," *National Cyclopedia of American Biography*, vol. 35 (1949), p. 109; on Donner, see "William Donner," *National Cyclopedia of American Biography*, vol. 44 (1962): 247.

largely ignored in the interest of furthering short-term goals.[69] For example, while working side-by-side with the small group of Jackson scientists, Andervont had introduced Bittner and the others to a new transplantation technique he had developed: caudal or tail inoculation of mouse tumors. To reflect the importance of this technique in the research experience, Schereschewsky now wanted to formally redefine his workers' relationship with the Jackson Lab to include more than simply the sale of mice.

After consulting with Bittner, Little was the first to respond—he resisted the possibility that the unique genetic uses of Jackson Lab mice could be subsumed under PHS intellectual categories. Little noted that he and the lab's patrons were "enthusiastic" about the prospects of cooperative research with the PHS workers but implied that this could only occur if the work were planned based on the already existing mouse strains and research of Jackson Lab scientists. Little described three possible topics for collaboration: (1) the genetics of tumor transplants in first-generation hybrid mouse strains; (2) the uterus as a site for tumor transplant; and (3) Bittner's previous work on tumor immunity and the transplantation genetics of certain standardized tumors. But he concluded with what had become his new bottom line: "the speed at which this could be done would be largely determined by the amount of resources [i.e., money] available." Here Little-the-administrator appears to have recognized the benefits of a stable income such a long-term collaboration would provide to his lab, especially in still-tough fiscal times—but Little-the-scientist was clearly searching for ways to ensure that any research done with Jackson Lab mice would not infringe on his scientists' own interests in the mouse strains they had developed and maintained.

After a week of deliberation, Schereschewsky replied, equally territorially, that he felt their collaboration should be defined as an outgrowth of Andervont's—not Bittner's—work on tumor immunity, despite the fact that Bittner's experiments (involving 4,300 mice of A [Albino] and dilute brown stocks) had been going on for a year.[70] Notably, Schereschewsky did not address the state of the Jackson Lab project on extra-chromosomal influences. Although most of the collaborative work could take

[69] Little had been previously exposed to such issues, except on a much smaller scale: cf. his correspondence with R. Kortweg on whether it would be appropriate to circulate descendants of the mice Kortweg had obtained from Little (Kortweg to CCL, 1 January 1932, Box 739, CCL-UMO; and 29 March 1932, Box 740, CCL-UMO), where Little agreed to the arrangement as long as they also shared results of the experiments with him (CCL to Kortweg, 1 February, 1932, Box 739, CCL-UMO).

[70] John J. Bittner to CCL (describing the protocols and the results: "the same parental cells [mammary gland, in this instance] may give rise to cancers which have different physiological characteristics"), 10 February 1932, Box 732, CCL-UMO.

place at Bar Harbor under Bittner's supervision, Schereschewsky wrote, he wanted to maintain some independent projects at Harvard using "a supply of inbred mice of the *C3H* and Dilute Brown strains." For both Little and Schereschewsky, then, the scientific issue had come full circle: from access to materials to intellectual property, and the complicated relationship between these two aspects of the scientific enterprise. Each man was now trying to protect not only his predictable supply of standard organisms, but also the proprietary interests of his scientists.[71]

Little had no cash on hand and was desperate for funds, so he appears to have blinked first. Upon receiving Schereschewsky's letter outlining his terms of collaboration, Little accepted it implicitly by writing back and discussing only the practical matter of financial compensation. For example, Little demanded to know how much per month the Jackson Lab could count on as income from the supply of mice and personnel for the PHS project. Schereschewsky responded in kind. For the experimental work to be done at Bar Harbor (not described in any detail), Schereschewsky then offered $300 a month for a nine-month period starting 1 November. This would cover Bittner's salary if he were appointed a part-time employee of the PHS ($125 per month) and pay for "the subsistence and care" of all the mice used in Bittner and Andervont's collaborative experiments (at a rate of one cent per mouse per day). In addition, Schereschewsky drew up an open-ended contract bid for the actual purchase of mice used. He assumed that the usual ten cents per mouse figure would apply to those mice that remained at Jackson Lab, and he suggested a figure of twelve cents per mouse for those sent to Boston, to cover railway express shipping charges. This government contract treated the mice like all other goods and services purchased by the federal government: it required Little to certify that all of the mice he supplied were commodities "of growth, production and manufacture of the United States."[72]

In the end, then, the more complicated issue of the connection between materials and experimental result was exposed but brushed under the table in Little's struggle to keep his institution afloat. The decision to do so effectively instituted—and upheld—a policy of compartmentalization of Jackson Lab's two resources—mice and research. For example, in sharp contrast with his initially tentative approach to mouse sales, Little now sought to redefine the financial terms of mouse upkeep and distribution to provide more money for the Jackson Lab upfront. For the mice at Bar Harbor, instead of separating their purchase and their caretaking

[71] CCL to Schereschewsky, 12 September 1932; Schereschewsky to Little, 21 September 1932; both in Box 739, CCL-UMO.

[72] CCL to Schereschewsky, 23 September 1932; Schereschewsky to CCL, 24 September 1932 and 14 October 1932; Contract Bid, 14 October 1932; all in Box 739, CCL-UMO.

costs, Little suggested that they might be able to reach "a composite figure for each mouse used" of seventy cents per mouse for seven weeks of food and care. This figure also included what Little felt was a fair eleven cents per mouse "allowance for those [mice] . . . that have to be re-inoculated because they are negative [for tumor acceptance] in order to be certain of their reaction." Furthermore, on the basis of his practical experience in mouse husbandry, Little also quibbled with Schereschewsky's method of determining what it would cost to ship mice to Harvard. "To put the mice in a proper box to protect them from weather and get them to you safely," Little argued, "would cost about four cents a mouse for shipment." This extra expense was because the only material deemed suitable by Little for these boxes was wood (needed to insulate the mice against potentially fatal temperature changes during shipment), and these boxes would have to be custom-made.[73]

Ultimately, Little was unsuccessful in renegotiating the financial terms of the mouse commodity relationship on any counts. He allowed the PHS to purchase Jackson Lab mice for the collaborative project and pay for their care at Schereschewsky's suggested penny-per-day rate. Little also compromised and let Schereschewsky have the Jackson Lab mice shipped to Cambridge for twelve cents a piece, probably because Schereschewsky agreed to send the empty wooden boxes back to Bar Harbor to be cleaned and reused. Judging from the contents of Bittner's weekly reports to Schereschewsky concerning the project, Andervont also largely defined the intellectual terms of the project. The mouse experiments that Bittner reported on followed a structure and sequence very similar to that reported in Andervont's previously published work on tumor immunity.[74]

At the same time, Little did not give up on his researchers—indeed, the timetable and conditions for the transfer of mice to PHS research was compatible with sustaining the core mouse supply for the Jackson Lab's own work. A variety of widely used Jackson Lab stocks were employed for the tumor-immunity project, with the largest number of animals coming from the *A*, *X*, and *C57Brown* strains.[75] But Little promised Schereschewsky "to provide surplus animals when they become available," while the lab was in the process of "speeding up production in the stocks which you desire." In part, this was because the Jackson Lab mouse colo-

[73] CCL to Schereschewsky, 19 October 1932; Box 739, CCL-UMO, in which Little and Schereschewsky also argued over how to obtain the best rail shipping rate for the mouse boxes. Jackson Lab would later develop other low-cost means of keeping the animals healthy during shipping: cf. Maine Potato Growers Association to Beatrice Johnson Little, 27 March 1951, Box 727, CCL-UMO.

[74] See, for example, H. B. Andervont, "Studies on Immunity Induced by Mouse Sarcoma 180," *Public Health Reports* 47, 37 (9 September 1932): 1859–77.

[75] See Bittner's reports, Box 739, CCL-UMO.

nies had unexpectedly been hit by a devastating intestinal infection that badly depleted some of the stocks. But also, this strategy preserved the Jackson Lab's own mouse reserves by spreading out the supply of mice from different strains over the nine-month period. Little did not even begin sending mice to Boston until December 1932, and when he did, he sent large numbers only of very easy-to-breed, hardy strains such as Bagg albinos and dilute browns.[76]

Still, on the receiving end, Schereschewsky and Andervont encountered several problems by virtue of the long-distance nature of the mouse supply arrangement, the solutions of which suggest that Little was more focused on the concerns of researchers external to the Jackson Lab ranks. For example, to make his reporting procedure more efficient, Bittner developed a standard form for conveying all the relevant experimental information about the mice. Besides listing the number of mice used each week and the days of mouse care provided (for purposes of contract payment), Bittner employed the PHS's conventional notation for tumor types but connected them to Jackson Lab workers' own genetic shorthand notations for the mouse strains in which the tumors were contained. For example, the "tumor 180" inoculated progeny of an *A* strain crossed with an *X* strain were referred to as *AXF1/180*. These reports effectively standardized the names of the Jackson Lab inbred stocks to cancer researchers, who might not otherwise have been familiar with genetic nomenclature.[77] Designed to solve a communication problem between particular scientists, they would consequently remain popular (although not hegemonic) designations for mouse materials in medical and genetic research through the 1940s.

Likewise, when Schereschewsky and Andervont could not replicate the uses of Jackson Lab products in their own institutional setting, Little himself did the technical troubleshooting. Sometimes as many as 20 percent of the mice in a certain shipment would arrive in Boston dead, and although Jackson Lab assumed full cost for the delivery of dead specimens, dead mice in a shipment meant a substantial delay in the experimental work for which the mice were intended.[78] Also, soon after Andervont faced

[76] It was not until late March and April 1933, after production had already been stepped up for outside sales, that Little sent more than ten of the scarcer Leaden mice in his biweekly rail shipments to the Harvard lab. See series of letters: "We are sending today express prepaid the following mice," from the Jackson Lab to Schereschewsky, 16 February 1933 to 2 June 1933, Box 739, CCL-UMO.

[77] Schereschewsky to Little, 21 October 1932; Weekly Reports to Schereschewsky, "Respectfully submitted by J. J. Bittner, Special Investigator," 23 November 1932 to 14 June 1933; all in Box 739, CCL-UMO.

[78] Jackson Lab to Schereschewsky, 29 June 1933, detailing the contents of a 162-mouse shipment, with a handwritten note: "34 dead"; Dorothy Silverman (Schereschewsky's secre-

difficulties assimilating the Jackson Lab mice to a new caretaking regimen—his local animal feed distributor could only provide a close approximation to "Fox Chow" called simply "Purina Chow," which was more expensive. Andervont wrote Bittner to see about acquiring some of the food used at the Jackson Lab. Mary Russell, the lab secretary, replied that the only way they could get the Fox Chow cheaply would be to order it as the Jackson Lab did—in ten-ton lots, once every three months. A food purchase of this scale was beyond the financial and physical capabilities of Andervont's small lab.[79] Ultimately, in responding to Andervont's complaints, Little observed that the continued success of mouse sales would also depend on the widespread availability of caretaking methods for the scientific workers who used them. He began exploring options for standardized mouse food and instructed his researchers to get to work on preparing a manual on Jackson Lab mice that he could distribute to future consumers of the lab's animal products.

In the short term, however, the same economic factors that had initiated the focus on production in the lab's culture led to the demise of the PHS relationship. In late May 1933 Schereschewsky informed Little that government budget reductions in the coming fiscal year would mean the end of PHS funding for Bittner at Bar Harbor. Accordingly, Bittner submitted his final weekly report summarizing the tumor immunity work in July 1933. Over the nine months of the project contract, the Jackson Lab provided the PHS with 3,575 mice and received in return a total of $1,513.74. Together with the money provided for Bittner's salary, these revenues represented almost three times the amount provided for the Jackson Lab during the last six months of 1933 by Little's original Detroit patrons.[80]

The Jackson Lab's financial situation was not stabilized, and the outlook for research funding remained grim. By February 1933 Lab expenditures already exceeded income by almost $600. And as the banking situation in heavily industrialized Detroit worsened, his patrons' support tapered off to almost nothing: their original $50,000 a year budget was down to only $2,500 a month in 1932–33. During the last six months of 1932, the original benefactors would contribute a total of only $800.[81] Little redoubled

tary) to J. J. Bittner, 1 July (presumably 1933), requests a refund for 34 mice found dead on a shipment's arrival.

[79] Dorothy Silverman to J. J. Bittner, 16 May 1933; Mary Russell to Dorothy Silverman, 17 May 1933; Box 739, CCL-UMO. Russell said that for ten tons of Fox Chow, Purina charged the Jackson Lab $4.94 per hundred pounds. Thus one truckload shipment of feed, which would sustain the Jackson Lab mice for approximately three or four months, cost $988.

[80] J. J. Bittner, "Report for the Month of July 1933 to the USPHS, Boston, Mass."; see also Summary of "U.S. Public Health Service, J. J. Bittner Experiment," circa 1 September 1934; both in Box 739, CCL-UMO.

[81] "Income and Expenditures" for the months of January, February, April, and June 1933, Box 735, CCL-UMO; Webber to CCL, 6 January 1932; CCL to Webber, 9 January

AMERICAN—*A Paper for People Who Think*—SUNDAY, MARCH 5, 1933

ot Talk,
ck to Gold

Biological Research Offers Greatest Hope of Conquering Cancer—Dr. Little

New War on Ancient Scourge Being Fought on a Wide Scientific Front

By DR. CLARENCE COOK LITTLE,

Former President of the University of Michigan, Director of the Roscoe B. Jackson Memorial Laboratory at Bar Harbor, Maine, and a leader in the fight against cancer.

As Told to Earl Reeves.

IN the field of research a new type of battle is being fought today against cancer.

In this war there have been in the past many weapons used. "An unguent of green frogs," quince seed mixed with pig's blood, juice of the willow tree. Nor were all these merely ancient superstitions. Turpentine was used as a "cure" in the days of Christ, at intervals through the centuries, and with medical conviction behind it as late as 1904.

Within our own day the attack upon cancer has been from the "medical" side; that is, based on an assumption that the cause might be found to be some substance or virus, alive or dead, beyond range of the microscope, non-filtrable; but against which some form of chemical action would perhaps be found effective.

The research campaign on that basis might be described as a mass attack, an attempt to drive through to crashing victory with all the forces Science could concentrate on a given point. The effective research group, it was thought, must contain a variety of specialists,

laws governing tissue growth. It is becoming clearer that only the "progeny" of this original cell are abnormal; there is little or no "infection" of adjoining normal, healthy cells.

Hence, much of the promising research now is not that which is by "medical" approach. It has been necessary to change tactics. We must understand first the biological foundations: in a sense understand more fully the very basis of life itself, or the nature and functioning of the body cell.

We therefore come to have less faith in "Hindenburg Drives" in the field of cancer

Chamberlain, Chancellor of the Exchequer; Prime Minister Ramsay MacDonald, chief sponsor of the Conference, and Sir John Simon, Foreign Minister. International News Photo Service.

CRUSADERS—Two outstanding figures in the war on cancer. Above, Dr. C. C. Little, author of the accompanying article; below, Dr. Joseph C. Bloodgood, of Johns Hopkins University.

3.1. Article from *New York American*, March 5, 1933, authored by C. C. Little [Source Credit: C. C. Little papers, University of Maine].

his efforts to promote the lab's own research mission. He continued to approach various foundations to get money for the lab's survival. He even tried to parlay his service on the NRC Committee into a National Research Council Fellowship for one of his scientists. Shortly before the new year in 1933, Little had received the news that the International Cancer Research Fund again denied a last-ditch request for funds. He tried to appear stoic about his repeated failures in this arena. "Of course," he told Richard Webber in January, "the same general conditions that are shaking the foundations of all industry will make it difficult for us to do as well as possible in normal times." Meanwhile, he preached the value of JAX's mission to the media, explaining how (in the words of one headline) JAX's "Biological Research Offers Greatest Hope of Conquering Cancer" (see fig. 3.1).[82]

1933; Wm. Murray to David Rodick, plus enclosed financial statement, 17 January 1933; all Box 740, CCL-UMO; untitled series of budgets, page headed "1933/ June 1–Oct.1 / Oct.1–Dec. 1," Box 735, CCL-UMO.

[82] Cf. Clarence Cook Little, "Biological Research Offers Greatest Hope of Conquering Cancer," New York American, 5 March 1933, CCL Clipping File, CCL-UMO; Susie G. Barnes (secretary, NRC) to CCL, 24 March 1933, Box 735; CCL to Richard Webber, 9 January 1933, Box 740. Little did eventually obtain $400 from the American Academy of Sciences toward the work of Green, and he rationed it carefully: "I believe this will give him four months salary at $100." See CCL to Richard Webber, 27 May 1933; all CCL-UMO.

But by March 1933 the financial crisis at the Jackson Lab peaked, and Little was once again despondent. He wrote pleadingly to a Ford-affiliated lawyer, hoping to hear his original patrons soon expected relief from the Detroit banking crisis. But it was to no avail: "unless Mrs. Jackson had some money in banks outside Detroit," the counselor wrote, "she could get none here, and I think even the Ford family will feel rather hard hit by the wallop we all have experienced."[83] Shortly thereafter, Little wrote to one of his American Society for the Control of Cancer colleagues: "The financial situation here has rapidly gotten worse so that revenue from any possible source is a matter of critical need."[84] In a letter to the Macy Foundation that same month, he put the situation in broader perspective: "It would seem to me to be a tragedy if the unique material which we have carefully built up over twenty-five years should have to be discarded. Frankly, I don't intend that it shall be. . . . I am perfectly willing to stake the results so far obtained in three years of intensive work against the output of any other cancer research laboratory in the country over twice as long a period. We are getting interesting results and important leads that are too valuable to be dropped."[85]

In the context of this most recent crisis, as well as the expanding production venture, Little self-consciously decided that strong leadership (or at least, the *appearance* of it) was critical for Jackson Lab to retain credibility in experimental biology circles. Thus, he began circulating his intent to implement a strategic "three-point plan" for the long-term security of the Jackson Lab's original program in cancer and genetics. Little first described this plan in a letter to Harvard's E. M. East in April 1933. Although he would continue to apply for any small grants and fellowships that arose, Little emphasized: "My plans are: (1) Get enough to keep the crowd alive until autumn, (2) Request one of the big foundations to stake us a living minimum of say $12,000 per year for five years, . . . (3) Use those five years to try to build up an endowment."[86] Thus mouse sales, while critical for the immediate survival of the lab, were not explicitly on Little's agenda for the institution's future.

[83] James O. Murfin to CCL, 24 March 1933, Box 739, CCL-UMO.

[84] CCL to Frederick Russell, 30 March 1933, Box 740, CCL-UMO.

[85] CCL to Ludwig Kast, 20 March 1933, Box 739, CCl-UMO. Strong got very a small grant from the Macy Foundation for his own private research on the biochemistry of blood proteins in relation to metabolism in cancer susceptible (CH) and non-cancer-susceptible (N) mouse strains (see CCL to Kast, 27 September 1932, and attachment, also Box 729, CCL-UMO).

[86] East to CCL, 7 April 1933, Box 739, CCL-UMO; CCL to East, 11 April 1933, Box 740, CCL-UMO. Little reiterates this plan in CCL to Richard Ames, 14 June 1933, Box 740, CCL-UMO.

At the same time, however, the redistribution of resources that resulted from the scaled up PHS production mediated directly against achieving these goals. In practical terms mouse sales were Jackson Lab's only reliable source of income, and so when the PHS contracts ended, Little needed to make efforts to broaden the lab's potential mouse market. From where he sat, it is highly unlikely that Little would not have recognized the myriad ways in which mice were developing a presence in a variety of biology and medical laboratories. For example, several psychologists—most notably, Robert Yerkes at Harvard—used inbred mice in the 1920s to study the hereditary expression of behavioral phenomena, from simple neurological irregularities to more broadly defined conditions like "savageness." Bacteriologists used common wild mice to investigate immunological reactions—such as English bacteriologist Frederick Griffith's now infamous 1928 experiments on *pneumococcus*. Additional medical uses for mice ranged from research on the comparative physiology of blood pressure, epilepsy, and deafness to clinical pregnancy tests. By 1928 Carl TenBroeck and Leslie Webster of the Rockefeller Institute needed so many of these creatures that they were making plans to build a new "mousery" to support their own varied research needs.[87]

Little's initial act in the process of marketing Jackson Lab's mice to a wider audience was not to survey existing users but to develop a catalog of the features of the animals that he determined universally most useful. In mid-March 1933, Little drafted a primitive sales listing for all the inbred strains (there were, he said, ten "stocks" available for purchase). Little's specifications were loosely based on Schereschewsky and Andervont's feedback and other observations Little made (in retrospect, clearly dating back to the 1928 NRC Committee) on scientists' interests in mouse material. For example, the list noted each stock's known gene makeup, including the number of inbred generations as a sign of homogeneity; tumor incidence; and a name (e.g., strains *A*, *B*). Most names followed Bittner's previous conventions, although in some cases Little simplified

[87] Robert Yerkes, *The Dancing Mouse: A Study in Animal Behavior* (New York: Macmillan, 1907). On this project, see "Robert M. Yerkes," in *A History of Psychology in Autobiography*, vol. 2. ed. E. Murchison (Worcester, MA: Clark University Press, 1932), pp. 381–407, esp. p. 395. See also Charles A. Coburn, "Heredity of Wildness and Savageness in Mice," *Behavioral Monographs*, 4 (1921–22:1–32; Frederick Griffith, "The Influence of Immune Serum on the Biological Properties of Pneumococci," *Reports on Public Health and Medical Subjects*, no. 18 (London: His Majesty's Stationery Office, 1928). On Moore and other physiological research, see "Animal Studies," RIMR/RU Archives, RG 210.3, Box 22, Folder "N.Y. Society for Medical Research," RAC-NY. The Ascheim-Zondek Mouse Test for Pregnancy is described in chapter 4 of this book. On TenBroeck's and work on mice at the Rockefeller, see Corner, *A History of the Rockefeller Institute*, chap. 9. On the construction of a Rockefeller mousery, see Rockefeller University Archives, RG 210.3, Box 290, Folder "Business Manager/Subject Files/Mousery 1928–34," RAC-NY.

the language even further (e.g., *C57Black* appears abbreviated by one letter, "C"). Interestingly, Little solicited some small commercial mouse dealers to determine the "market price" for their non–genetically controlled mouse material, and he set the cost for his material below "industry" averages. All mice were priced equally: ten cents each plus the cost of delivery, with additional charges for specific sexes (as with the PHS agreement). And in the true spirit of consumer culture, researchers were offered a money-back guarantee in the rare event that mice might arrive dead. The target market: "heads of all medical schools and various departments in school and universities as are likely to use material of the sort."[88]

Geneticists and some cancer researchers who knew Little personally continued to write to him, seeking mice and word-of-mouth information about the status of various strains and mutants. Their letters suggest that the earlier culture of the "gentlemen's agreement" in exchanging laboratory materials remained in place in the early years of JAX mass production and sales. For example, in January 1932 Dutch cancer researcher Remmert Korteweg wrote about a Belgian scientist seeking *dba* and *C57Black* mice: "I informed him these were mice of the strain you sent to us the other year, and that I did not feel at liberty to send these mice to others without your consent." But Korteweg promptly vouched for his colleague's character, calling him "a good scientific worker," and demurred: "You would greatly oblige both him and me by permitting my sending him these strains."[89] Little responded effusively—"I shall be delighted to have Doctor Maisin or anyone else to whom you wish to send material to receive the two stocks of mice which I sent you." Still, the very fact that this communication took place suggests that many still believed the exchange of JAX's unique research materials should follow the same informal proprietary rules as the oral or prepublication exchange of JAX research results.[90]

The expanded efforts to sell Jackson Lab mice appear to have been very successful on the limited front to which they were directed. Cash orders

[88] CCL to Mary Russell and enclosed inbred mouse sales listing, 3 March 1933, Box 730; Irwin Wachtel to Mary Russell, 18 March 1933; C. D. Haedrich to CCL, n.d.; both Box 740; all CCL-UMO; see also Frederick Russell, 20 March 1933 and enclosed inbred mouse sales listing, "Animal Production Information" Box, JLA-BH. In May, Little would plead with the U.S. surgeon general to restore the PHS's funding for mouse materials: see CCL to Hugh Cummings, 25 May 1933, Box 739, CCL-UMO.

[89] See, e.g., C. W. Turner to Murray, 8 May 1933; Murray to Turner, 12 June 1933; both Box 739, CCL-UMO.

[90] Korteweg to CCL, 19 January 1932; and CCL to Korteweg, 1 February 1932, Box 739; Korteweg to CCL, 29 March 1993, Box 740: both CCL-UMO. By 1933 Korteweg's colony of *C57 Black* mice had grown to four hundred animals and probably formed the basis of his laboratory's experiments on the "milk influence" in mammary cancer (see chapter 5).

quickly began coming in from medical researchers at places like Children's Hospital in Philadelphia and Memorial Hospital in New York. One annual budget notes a figure of $2,519 in income from mouse sales during the first six months of 1933, and although it is difficult to extrapolate from this (due to the varying price of different strains), Little's notes on the extant archival letters requesting animals suggest that demands quickly exceeded the supply of inbred mice. In a 1937 report, Little noted that for the period from 1932 to 1933, the Jackson Lab supplied a total of six thousand mice to outside researchers. This number represents nearly 40 percent more than the total number of mice provided to the PHS.[91]

The lowest point of the Depression, then, marked a turning point in the scientific history of both the Jackson Lab and the mouse as a standard laboratory organism. Little founded Jackson Lab to pursue a revolutionary cooperative research program in cancer and genetics, and when initially faced with the financial collapse of his original donors, he defended this program with great conviction. But the meager successes of this research were untenable to foundations in the contemporary economic climate. Conversely, Little initially undertook mouse sales and distribution only to get money for the continuing operation of his institution, but selling mice now looked to be Jackson Lab's sole potential source of revenue. So far, even this move had been somewhat problematic, for it had meant staking the material future of Jackson Lab mouse stocks—the original cornerstone of Little's genetic research program—entirely on the availability of government funding for basic research. But ironically, a new institutional dynamic evolved out of the mouse-sales-versus-research dilemma that historical contingencies created—one that favored standardized mass production of animals over the innovative framing of experiments with them.

Jackson Lab's initial success with mouse sales, then, had a multiplicity of practical effects. In the short run, it literally saved the institution. But for the long run, continued success with mouse sales irreversibly transformed Little's attitude toward Jackson Lab's institutional priorities. The decision to sell—even as a stopgap measure—made the expansion of mouse production a necessity, both to meet the demand that Jackson Lab had created by offering mice to a wider set of consumers and (more immediately) to make the whole sales venture a maximally profitable venture for the financially troubled laboratory. In turn, despite Little's best efforts

[91] For replies from medical researchers, e.g., Halsey Bagg to CCL, 4 April 1933, Box 12, JLA-BH; Howard Andervont to Wm. Murray, 29 March 1933; C. W. Turner to CCL, 8 May 1933; CCL to C. W. Turner, 12 June 1933; all Box 739, CCL-UMO. See also Holstein, *The First Fifty Years*, p. 79.

at unifying research and production in this time of crisis, the broadened sales effort began to drive a practical wedge between Jackson Lab's research mission and its developing function as an animal material supplier. This structural dynamic, in particular, would be one the institution would continue to struggle with in decades to come.

By 1933 the departure of several original research workers underscored these developments. Strong was the first to decide that he could no longer endure the financial instability and declining research atmosphere. When offered a position at the Yale's Sterling School of Medicine, he jumped at the chance, and in April 1933 he packed up his mice in Jackson Lab's own wooden boxes and headed for New Haven. Little commented bitterly (in a letter to Strong's new boss) that Strong no longer belonged at the Jackson Lab since he was no longer interested primarily in genetic explanations: "The development of [his] research in the determination of possible environmental factors that may influence physiological constitutional types, although it recognizes fundamental genetic determination of biological variability, seems to be of such a nature that it ought to be of necessity conducted at another laboratory." Soon after, Joe Murray also left Jackson Lab to take a professorship at the University of Maine's biology department. And in a bizarre but accidental development, Charles Green drowned unexpectedly the following summer in a boating accident on a local lake.[92]

Looking beyond the immediate personnel situation at Jackson Lab, however, the commodification of inbred mice and the scaling up of production that resulted from the economic pressure of the Depression had bigger consequences. These early developments laid the groundwork for Jackson Lab to construct a larger, more elaborate system of production and circulation aimed at monopolizing the scientific market for mice. Jackson Lab set up the first formal experimental mammal distribution system that was sensitive to a wide variety of user needs in different disciplinary and institutional contexts. Over the next two decades, its development was negotiated jointly by Jackson Lab mouse producers and their scientist-"users," but from 1933 onward through 1960, the Jackson Lab's involvement in this process was primarily as a mouse factory, not a research factory.

[92] Little and Strong subsequently warred over who would bear the expense of the wooden boxes needed to move his mice, and over whether or not he would leave representatives of his strains for Jackson Lab production. See "Protocol in re: transfer of L.C. Strong," n.d.; CCL to Dr. Kast, 17 April 1933; both Box 739, CCL-UMO. On Murray, see Holstein, *The First Fifty Years*. On Green: See Little, "Charles Velmar Green," *Science* 83 (1936): 543.

CHAPTER FOUR

A NEW DEAL FOR MICE

Biomedicine as Big Science (1933–40)

In 1935 Little heralded the Jackson Lab's industrial production of rodents in a *Scientific American* article he called "A New Deal for Mice." First, he juxtaposed gains the mouse had made in science during the last decade with prevailing negative cultural images of the animal: "Do you like mice? Of course you don't. 'Useless vermin,' 'disgusting little beasts,' or something worse is what you are likely to think as you physically or mentally climb a chair." Then against this background, Little cast himself as "attorney for the defense" and argued that through their involvement with science, mice had been positively transformed. Inbred mice—as opposed to their "not very convenient" wild mice relatives—"provided a particular service" to both science and humanity. By contrast with those scientists who worked with other larger mammals (especially dogs), Little invited his lay readers to visit his place of scientific work: one of the Jackson Lab's "mouse laboratory 'cities' with its cleanliness, orderly arrangement, and activity." Such arrangements testified that "thoroughbred" mice (a concept Little acknowledged some people would find "amusing") had become "an integral part of man's helpers." Inbred mice, he wrote, were "the troops which literally by the tens of thousands occupy posts on the firing line of investigation" into the "nature and cure of cancer." Also, in contrast with the policy of some major medical journals in the 1930s, *Scientific American* included full-body photographs of these animals undergoing laboratory procedures (such as tumor injection). "Under these circumstances," Little concluded, "perhaps mankind will accept and develop his relationships with mice in a different light."[1]

Written for a popular audience, Little's essay eloquently captured the inbred mouse's scientific successes to date as well as foreshadowing developments that lay ahead. To rectify the dire economic situation at JAX (as the lab's name was now frequently abbreviated from its Western Union cable address—"JAXLAB"), Little offered them for sale to the larger biomedical research community in 1932. Initially inbred mouse materials did not easily separate from the genetic research for which they were created, but by 1935, mass production of mice threatened to overtake Jackson Lab's cancer and genetics research programs. Little wanted to sustain research at JAX, so in negotiations with foundations and private donors from 1933 to 1937, he successfully argued for the complementary relationship between production and genetics work at JAX. When JAX researchers achieved notoriety for a series of experiments on genetic and extra-chromosomal transmission of breast cancer in mice, Little decided to take the lab's work public in a broader sense. For an American middle class that was becoming increasingly interested in the socially transformative effects of scientific medicine, Little shaped the American Society for the Control of Cancer's educational publicity in popular magazines to feature inbred mouse research as the laboratory science most capable of curing cancer.

But Little's proposed "New Deal for Mice"—like Franklin Delano Roosevelt's eponymous legislative plan for America that unfolded alongside it—ultimately challenged the social relations that had enabled its existence.[2] In 1933 inbred mice were research tools tied to a narrow form of scientific practice: solving the genetic problems of tumor transplantation. By 1938 National Cancer Institute funding for JAX sought to ensure that these creatures would be the cornerstones of a new plan

[1] C. C. Little, "A New Deal for Mice: Why Mice Are Used in Research on Human Diseases," *Scientific American* 152 (1935): 16–18. Cf. Lederer, "Political Animals." Little showed some awareness that perceptions of pain inflicted on the animals photographed might affect the public's reception of the argument: thus, the photo of the mouse undergoing an operation had a caption indicating that the animal was anesthetized.

[2] Recent historiography of the New Deal highlights the ways in which the liberal policies it promoted were often contradictory and undermined prevailing market ideologies. See Steven Fraser and Gary Gerstle, eds., *Rise and Fall of the New Deal Order, 1930–1980* (Princeton: Princeton University Press, 1989), especially Alan Brinkley's essay, "The New Deal and the State" (pp. 85–121); Ronald Edsforth, *The New Deal: American's Response to the Great Depression*, Problems in American History Series (Malden, MA: Blackwell, 2000), esp. chaps. 8 and 9; Alan Brinkley, *Liberalism and Its Discontents* (Cambridge: Harvard University Press, 1998); Rodgers, *Atlantic Crossings*.

to build a federally funded system of biomedical research—not around shared experimental problems (as was traditional, and as Little had originally conceived JAX's own mission), but around the shared practical values of coordination and consistency that policymakers now believed should govern experiments across all types of life science research. Though still relatively small in terms of the number of animals and dollars generated, the increased production and circulation of laboratory mice enabled by the 1937 National Cancer Institute Act combined potently with the rhetorical circulation of Little's claims for the animals' broader moral utility. This synthesis defined inbred mice as suitable for a broad range of tasks—experimentally and organizationally, as well as culturally—and thus temporarily created a stable, but flexible, coalition among those interested in making scientific knowledge with mice, those interested in using this knowledge, and those interested in the animals themselves. Subsequent further investment in the material infrastructure of mouse production—by both Rockefeller Foundation and the federal government—buttressed the practical work of this coalition. When, in 1941, *Journal of Heredity* editor Robert Cook called JAX biology's own embodiment of "the multidependent nature of science," he recorded more than just a shift in the lab's own institutional priorities. Cook registered Little's success in using inbred mice to plant the seeds for biomedical Big Science—"a heterogeneous yet coherent constellation of conceptual, material, and social resources" that (in retrospect) was indebted to, and yet distinctive in its scope from, the more localized, less publicly accountable American biological and medical research practices that preceded it.[3]

THE RETURN OF GENETICS: PHILANTHROPIC NEGOTIATIONS AND PRACTICAL CONSTRAINTS, 1933–1935

In the immediate aftermath of his decision to sell JAX mice in 1932, Little wavered on the proper balance between the business demands of mouse production and the long-term survival of the original JAX scientific pro-

[3] Robert Cook, editorial: "Contributions in Mouse Genetics from the Roscoe B. Jackson Memorial Laboratory," special JAX Mouse issue, *Journal of Heredity* 36 (September 1945): 257, 271ff. On a new historiography of Big Science that recognizes its pre–World War II roots, see James Capshew and Karen Rader, "Big Science: Price to the Present," *Osiris* 7 (1992): 3–25, esp. pp. 22–25, and applications of John Law's concept of "heterogeneous engineering." Cf. Daniel J. Kevles and Gerald L. Geison, "The Experimental Life Sciences in the Twentieth Century," *Osiris* 10 (1995): 97–121, which offers an analytic framework that I would argue is amenable to understanding emerging intellectual and institutional values in the early-twentieth-century American life science as "Big Science."

gram. At the height of the Depression in 1933, he began intensive negotiations with the Rockefeller Foundation and other private philanthropists to obtain financing for expanding the lab's physical plant. But his deep reservations about straying from the lab's original commitment to genetic research combined with his growing awareness of programmatic funding priorities at the Rockefeller Foundation to result in a temporary retreat from the lab's experimental cancer studies. Ironically, this happened just as this work was beginning to yield some results of broad interest to other cancer researchers.

In April 1933—one month after he circulated the first JAX mouse catalog—Little contacted Alan Gregg, a former Harvard classmate, with news of the lab's mouse production expansion.[4] Little described the lab building as "three and one half stories of factory construction especially designed for the maximum utilization of living materials." Gregg headed the Rockefeller Foundation's Medical Sciences program, but Little hardly mentioned initial JAX successes with mouse sales to medical researchers. Instead he proposed that the foundation fund a large renewable grant of $12,000 per year (the "bare minimum survival" figure he noted in his strategic plan to JAX trustees) for pure genetic research. Specifically, Little noted the perfect unity of inbred mouse materials and genetics studies in mammals, which were evolutionarily closer to humans. Carefully recorded "pedigreed" mouse material at JAX, he wrote, was "very valuable" by virtue of its potential as a mammalian genetic instrument, comparable to Morgan's *Drosophila* and Emerson's maize stocks. As proof, Little noted that some JAX workers' published papers on mouse linkage studies and size inheritance had "been on genetics purely, without any reference to cancer research."[5]

Gregg's internal program diary suggests he took a slightly different message, but it was one of which Little would have approved. Gregg noted briefly the benefits RF funding of mammalian genetics would have for genetic research, and described JAX as "Little's Research Institute in Eugenics." Gregg had just returned from a special planning meeting, where Rockefeller trustees and officers gathered to inaugurate a new era of foun-

[4] Little also simultaneously approached the Carnegie Institution, which had funded his work at Cold Spring Harbor and Orono (see chapter 1), although he was not optimistic—ironically because he thought JAX would be perceived by the program officers there as primarily doing cancer research: see CCL to B.F.W. Russell, 16 May 1933, and CCL to John Merriam, 10 June 1933; both Box 740, CCL-UMO.

[5] Memo of Alan Gregg interview with CCL, 7 April 1933, RAC-NY; Little wrote to trustee Richard Webber in June that he did not expect to get the grant: "these organizations are extremely hard-boiled and can turn down a request with something of the same lack of mental upset that you or I would feel slapping a mosquito." CCL to Webber, 14 June 1933, Box 740, CCL-UMO.

dation funding. In what would come to be known as the Foundation's "Science of Man" agenda, the Rockefeller Foundation leaders articulated a vision that emphasized genetics as one of a handful of "fields of critical importance, fields that capitalize on the rich heritage of the foundation's long experience." By contrast, wrote Natural Science Program Officer Warren Weaver, lack of appreciation for laboratory research—especially of the kind that would illuminate important problems in "molecular biology"—plagued the medical sciences. Regardless of where each program officer stood on the question of eugenics, however, the newly appointed Gregg knew he needed to mobilize projects that promised more systematic, methodologically rigorous approaches, and Little's mouse factory certainly did that. That December the foundation agreed to appropriate $11,000 in Natural Sciences emergency funds to Little's group "for research in mammalian genetics."[6]

The initial foundation investment went primarily toward supplementing JAX researchers salaries and buying more supplies (food, bedding, and cages) to sustain the increasingly larger mouse colonies. As Frank Blair Hanson noted in a brief followup visit that October: "[Little's] immediate problem[s are] to maintain his stocks of animals and keep his group of investigators together." This funding stimulated a brief recovery for the JAX cancer genetics research program, which was at a critical stage. A few months after receiving the initial grant, "The Staff of the Jackson Laboratory" published a landmark paper detailing the results of a series of reciprocal crosses between three high mammary tumor strains (*A*, *D*, and *Z*) and three low tumor strains (*M. bactrianus*, *C57Black*, and *I*).[7] The paper noted a reciprocal crossing effect that JAX scientists had observed in genetically equivalent F1 hybrids: only those F1 mice derived

[6] Gregg diary entry, 7 April 1933; Natural Sciences Grant Resolution #33359, 13 December 1933, RAC-NY. On the "Science of Man," see Lily Kay, *The Molecular Vision of Life* (New York: Oxford University Press, 1993), p. 46; See also Kohler, *Partners in Science*, pp. 282–83. In lieu of Kohler's and Kay's differing interpretations (cf. Robert Kohler, review of *The Molecular Vision of Life* (1993), *Isis* 85 (March 1994): 183–84), historians of biology are still sorting out the degree to which the Rockefeller Foundation was influenced by eugenics by the mid-1930s. Kay's work extrapolates from the foundations commitment to Caltech molecular biology to eugenic ideals, whereas in the case of JAX, it is notable that Gregg himself makes that connection. Cf. Pnina G. Abir-Am, "Converging Failures: Science Policy, Historiography and Social Theory of Early Molecular Biology," in *Scientific Failure*, ed.Tamara Horowitz and Allen Janis (Lanham, MD: Rowman & Littlefield, 1994), pp. 141–66.

[7] FBH Diary excerpt, 5 October 1933, RF Archives, RG 1.1, Ser 200D. Box 143, Folder 1773; John J. Bittner, "Breast Cancer in Mice," *American Journal of Cancer* 36 (1939): 44; and "Relation of Nursing to Extrachromosomal Theory of Breast Cancer in Mice," *American Journal of Cancer* 35 (1939): 90. Cf. Walter Heston's review—"Milk Influence in the Genesis of Mammary Tumors"—in *A Symposium on the Mammary Tumors of Mice*, ed. F. R. Moulton (Washington, DC: AAAS, 1945), pp. 123–39, esp. table II, p. 127.

from high-cancer females were themselves prone to develop cancer. This, they concluded, "establishes the transmission of extra-chromosomal influence [though] it should not be taken as a denial of the existence of chromosomal influence."[8]

Rather than abandon inbred mice as cancer inheritance models, JAX researchers reframed the 1933 findings as a genetic problem for further investigation: namely, what are the relative contributions of chromosomal and nonchromosomal factors in cancer inheritance? Similar findings were independently reported in 1934 by Remmert Korteweg at the Netherlands Cancer Institute, which had obtained some inbred mice from JAX in 1932. Korteweg suggested the extra-chromosomal influence could be transmitted through the cytoplasm of the ovum.[9] The JAX staff met in early 1933 and agreed to divide up the necessary experiments to test what they believed were the possible sources of tumors: hormones, blood, and milk. Little would later tell a Rockefeller Foundation officer: "it chanced that Bittner, who had been assigned to study the maternal milk, was the lucky one."[10] Little and Murray performed a series of hybrid crosses between *dba* high-tumor and *C57Black* low-tumor strains, and a statistical analysis of the results (based on Mendelian laws) suggested to them that "the relative importance . . . of extra-chromosomal influence . . . is approximately six-times that of possible chromosomal influence." But in Bittner's initial experiments, female progeny of the high-cancer strains could be transformed into low-tumor strains by nursing them with low-tumor mothers. In an effort to define more precisely the nature of what he called the "milk influence," Bittner backcrossed these strains to create a series of inbred lines that were susceptible to conversion from foster nursing.[11]

[8] JAX Lab Statement of Receipts and Expenditures for 1934, RF Archives, RG 1.1, Ser 200D, Box 143, Folder 1773; FBH Diary excerpt, 5 October 1933; Staff of the Jackson Laboratory, "The Existence of Non-Chromosomal Influence in the Incidence of Mammary Tumors in Mice, *Science* 18 (November 1933): 465–66.

[9] R. Korteweg, "On the Manner in Which the Disposition to Carcinoma of the Mammary Gland is Inherited in Mice," *Genetics* 18 (1936): 350.

[10] Interview GWG with CCL, 19 January 1950, RF Archives, RG 1.1, Series 200D, Box 144, Folder 1779.

[11] C.C. Little and W. M. Murray, "The Genetics of Mammary Tumor Incidence in Mice," *Genetics* 20 (1935): 466–96; see also Little and Murray, "Further Data on the Existence of Extra-Chromosomal Influence in the Incidence of Mammary Tumors in Mice," *Science* 82 (1935): 228–30; and "Extrachromosomal Influence in Relation to the Incidence of Mammary and Non-Mammary Tumors in Mice," *American Journal of Cancer*, 1936, 27: 516–18. See also J. J. Bittner, "Mammary Tumors in Mice in relation to Nursing," *American Journal of Cancer* 30 (1937): 530–38; and "Foster Nursing and Genetic Susceptibility for Tumors of the Breast in Mice," *Cancer Research* 1 (1941): 793–94. As Lowy and Gaudillière note (in "Disciplining Cancer"), Bittner's attempts to "domesticate" the milk agent would ultimately be validated only by a "shared system of standards defining mice." Cf. Gaudil-

By 1935, then, JAX researchers looked as though they were about to follow successfully in the footsteps of Morgan's *Drosophila* group that had preceded them. Although they were not able to map mouse genes, by identifying transmission genetics as the key biological mechanism of cancer and then developing and using inbred mouse systems to "show" how Mendelian genetics explained this phenomenon, the JAX group appeared to be developing the kind of circularity in their problem definition and experimental practices that characterized all good model systems in classical genetics. Little himself still kept the eugenics faith about the social applicability of this knowledge. In 1933 he wrote a dismissive reply to a *Journal of Heredity* commentary entitled "Is Eugenics Dead?" Little answered "Not Dead But Sleeping!" and concluded that "It would seem to be a good idea to remember that regardless of what the body politic may accomplish, nature is very patient and very sure in her methods." But because no solid eugenic conclusions could be drawn from the genetic work on inbred mouse cancers, the declared goal of all the inbred mouse studies coming out of JAX was simply to use mice to solve an important biological puzzle: what role do genes play in determining cancer susceptibility and/or carcinogenesis?[12]

Bolstered by these apparent successes, Little sought more funds to expand cancer genetics work at JAX, but little came of these efforts. In October 1933 he approached Philadelphia industrialist Samuel Fels, whose philanthropic activities focused partly on medical research. Little told Fels: "The situation here is becoming desperate and I am hard put . . . to find bare necessities for its continued existence." He described JAX work on the genetics of cancer susceptibility in different strains as the lab's most "encouraging" scientific prospect and cited as proof the prominent mention of JAX work on transplantable tumors in British biologist J.B.S. Haldane's "unprejudiced and outstandingly frank" 1933 review of cancer genetics in *Nature*. Little proposed that Fels fund a small research project on the differential incidence of breast cancer among yellow- and nonyellow mice.[13] Since yellow mice had a higher cancer incidence and

---

lière, "Circulating Mice and Viruses," in *Practices of Human Genetics*, ed. Michael Fortun and Everett Mendelsohn (Dordrecht: Kluwer, 1999), pp. 89–124.

[12] C.C. Little, "Not Dead but Sleeping," *Journal of Heredity* 24 (1933): 149. The height of Little's eugenics leadership was probably in 1928, when he was president of the American Eugenics Society: see "Clarence Cook Little," *Who Was Who*, vol. 5, pp. 434–5. Lowy and Gaudillière have concluded that Little and the JAX staffers "produced cancer lines of mice as a part of a larger enterprise in the production of genetic purity." But although Little retained his eugenic vision, by the mid-1930s he had cut many of his professional ties to the active eugenics community and reinvested his administrative efforts in cancer research policy.

[13] There is no record of exactly what JAX stock Little was referring to, but yellow gene (*A y*) is one of the oldest recognized mutations of the *agouti* coat color locus, in which

also had "a marked tendency towards adiposity and increased sterility," Little reasoned, they would be the perfect model organisms for a large-scale study on the relationship among diet, metabolism, and carcinogenesis. Fels gave Little $2,000 to conduct a preliminary study, but six months into the project most of the yellow mice involved became so "emaciated and poorly nourished" from the low-fat diet that their failure to thrive competed with tumor growth. Little instructed the caretakers to "kill" the whole experimental colony during the autumn of 1934.[14]

Although Bittner and Cloudman were allowed to continue this cancer work throughout this period, when the lab's financial outlook worsened in late 1933, Little realigned official JAX Lab research priorities. At a time of plunging federal science budgets and public calls for a "science holiday," Little surely realized that only the Rockefeller Foundation could provide an institutional grant large enough to insure JAX's survival. But in turn, projects promising the union of pure and applied biology—such as those Little had lionized much earlier in his career—were more likely to attract the attention of multiple Rockefeller program officers.[15] Between October 1933 and August 1934, notes from the various program officer visits to Bar Harbor show that the foundation conveyed to Little that its continued interest in JAX lay squarely in the lab's contributions to mammalian genetics. When Hanson toured JAX in October 1933, for example, he used industrial language to describe the facility, but he linked this very tightly to the necessary work of generating more murine mutations for linkage studies: "[to] L[ittle]'s statement, FBH adds that the whole arrangement is admirably designed for its purpose and is probably the most complete plant

---

alleles associated with the synthesis of yellow pigment in hair follicles are dominant to those associated with the synthesis of black pigment (reviewed in W. K. Silvers, *The Coat Colors of the Mouse* [New York: Springer-Verlag, 1979]). Since in the homozygous state it is lethal, the *a x* allele is usually propagated by backcrossing inbred C57BL to albino (a/a) animals.

[14] CCL to Samuel Fels, 23 October 1933, 28 October 1933, 6 December 1933, 3 January 1934, 24 September 1934, 9 October 1934, Samuel Simeon Fels Papers (collection #1776), Box 5, Folder 5, Historical Society of Pennsylvania, Philadelphia, PA. I am deeply indebted to Jordan Marche for making me aware of these letters, which he found in the course of his own research. On Fels's philanthropy, see his obituary, *New York Times*, 24 June 1950, p. 13.

[15] Bud, "Strategies in American Cancer Research." Robert Kohler argues that the influence of the "Science of Man" agenda on Rockefeller Foundation programs, especially Warren Weaver's, had waned by the mid-1930s: see Kohler, " Warren Weaver and the Rockefeller Foundation's Program in Molecular Biology: A Case Study in the Management of Science," in *Managing Medical Research in Europe: The Role of the Rockefeller Foundation (1920s–1950s)*, ed. Giuliana Gemelli, Jean-François Picard, and William H. Schneider (Bologna: CLUEB, 1999), pp. 51–90. On the politics of science funding during the Depression, including "science blackouts," see Peter J. Kuznick, *Beyond the Laboratory: Scientists as Political Activists in 1930s America* (Chicago: University of Chicago Press, 1987), chaps. 1 and 2.

in America for mammalian genetics." Little asked Hanson about the possibility of expanding production, so that JAX could supply all the necessary mice to the Rockefeller Institute for Medical Research. Hanson immediately nixed the idea: "The maximum income from mice . . . is really a minor item in the total budget needed and would not solve the problem of saving the stocks to science in case L. has to close up, for other investigators are chiefly interested in getting a supply of animals suitable for their specific problem. . . . This is genetic material of great value and ought to plan an important role in the increasing interest that must be given to mammalian genetics over insect genetics." When Warren Weaver visited in October 1934, he noted that Little began their meeting with a"[g]eneral discussion . . . of the development of genetics. L says that the three great fields of *Drosophila*, maize, and mouse genetics stand in order of increasing difficulty and complication but in decreasing order as regards definiteness of results, amount of result to be obtained from a given amount of research, and general stage of development. A fruit fly is almost completely what he is genetically predetermined to be. A mouse, with his higher degree of organization and with the profoundly important influence of internal secretions, presents a more complicated problem." Weaver himself concluded there were humanitarian reasons for funding this work: "For this very reason, however, and because of its relative nearness to human genetics, mouse genetics presents a most important opportunity."[16]

Little's formal proposal for the renewal of the original Rockefeller Foundation grant in 1935 framed the laboratory's research program broadly: "long-term research on the genetics of physiological factors that influence continuing processes such as normal and abnormal growth." Furthermore, Little argued that large-scale breeding of murine organisms was in itself a key to success in this research. JAX's carefully supervised mouse breeding for sales purposes, Little noted (albeit vaguely), had increased the observation of rare mouse mutants necessary for this research.[17] It had also increased JAX's production of genetics papers: "L has sent FBH the last 51 papers published by the laboratory; of these 19 are on cancer and 32 deal with various general genetics problems. . . . FBH is of the opinion that for this work to stop . . . would be a loss to the science of genetics." Foundation trustees and officers voted to give the lab tapering grants totaling $35,000 over the next four years.

[16] WW diary entry, 20 October 1934, RF Archives, RG 1.1, Series 200D, Box 143, Folder 1773, RAC-NY.

[17] Jackson Laboratory Statement of Receipts and Expenditures for 1934, 31 December 1934, RF Archive, RG 1.1, Series 200D, Box 143, Folder 1773, RAC-NY. On "growth" as an emerging research classification for cancer in postwar biological research, see NAS history of the Committee on Growth: Rexmond Cochrane, *The National Academy of Sciences: The First Hundred Years, 1863–1963* (Washington, DC: The Academy, 1978), pp. 433–74.

Little immediately went to work looking for a mammalian geneticist to replace Charles Green, who had just died unexpectedly in a boating accident. Ultimately, he hired George Snell, a Bussey Institution of Harvard alumnus who had done his dissertation on mouse linkage studies and had just completed an NRC postdoc at the University of Texas.[18] Little wrote to mouse geneticist L. C. Dunn of the renewed hopes he had for mammalian genetics at JAX: "Those of us who believed in the old Bussey . . . may perhaps come to feel that the Jackson Lab is its child, and as such, may merit some of the affection we felt for the older institution."[19] But these gains would be short-lived, for although in theory mammalian genetics research offered important payoffs, in practice it remained a relatively intractable field. A small mammal like the mouse produced somewhere between ten and twenty litters of progeny in a year (depending on breed and conditions),[20] but the mouse had twenty chromosomes as potential linkage groups, as compared to *Drosophila*'s four. In contrast with corn seeds (which could be stored on a shelf) or flies (which could be housed economically in test tubes), laboratory mammals needed expensive provisions: special cages, substantial food, and caretakers to clean up after them. Perhaps even more importantly, even when done on a large scale at JAX with the fastest-breeding organisms like mice, naturally occurring mouse mutants arose infrequently. Mutant organisms were the stock-and-trade materials on which mouse geneticists, like all geneticists, depended to help them study normal genetic arrangements and processes. Unlike *Drosophila*, which had become abundant "breeder reactors" by 1910, mouse mutants (even in JAX's large colonies) were still rare and precious finds. In 1925, for example, only nineteen naturally occurring mutant mouse stocks were recognized by the genetics research community; by 1931 this number had only grown by six, to twenty-five. Compared with Morgan's *Drosophila* lab where at least thirty-four mutants appeared in the two-year period from 1911 to 1912 alone, mouse genetics remained a slow business.[21] Ultimately, the pace of work in mammalian genetics led Weaver to conclude, in the internal resolution for JAX's renewal grant: "although a major portion of the work of the laboratory falls within a field of special interest of the 'Science of Man' program in experimental biology, the officers nonetheless judge it to be un-

[18] On Snell's biography, see interview with George Snell, in both JLOH-APS and JLOH-KR; see also Snell's c.v., provided to author by Snell.

[19] Cf. Robert Cook, "The Roscoe B. Jackson Memorial Laboratory Celebrates Tenth Birthday," *Journal of Heredity* 30 (1939): 448–52.

[20] Gruneberg, *Genetics of the Mouse*, chap. 2.

[21] Kohler, *Lords of the Fly*, p. 48; Keeler, *The Laboratory Mouse*, chap. 4. Cf. analysis of mouse mutant reports (by year discovered) in folder "Mouse Linkage Group Map Info," c. 1950s, JLA-BH.

wise for the Foundation to become involved, over an indefinite period, in its general support."[22] JAX Lab may have embodied the cooperative approach that Weaver cherished, but as an institution it was what the foundation's trustees called "retail business," and they would rather fund wholesale ventures.[23]

THE NATIONAL CANCER INSTITUTE ACT: THE SCIENTIFIC AND MORAL VALUE OF THE INBRED MOUSE AS A LABORATORY ORGANISM

The mid-1930s marked a turning point for the inbred mouse as a laboratory organism, as well as for the American political fortune of cancer research. Anticancer sentiments reached a fever pitch both in the media and on the floor of the U.S. Congress, and earlier deadlocks between the executive and legislative branches gave way to a near-unanimous agreement among federal officials that cancer was a problem requiring immediate government attention. Until the Depression bottomed out in 1933, little private or public money was available to contribute to the cancer cause. But historian Stephen P. Strickland notes that, by 1935, the fact that cancer was a national problem requiring immediate governmental action was not even debated. The political and cultural forces battling cancer eventually converged during the summer of 1937. Three different cancer bills were introduced in the Congress on 8 July of that year.[24]

For legislators this tide of events uncovered a persistent policy anxiety in Washington: if the federal government wanted to commit funds to a program to fight chronic disease, how should such a program be structured and administered?[25] One basic answer in all of the proposed bills was the importance of funding research for cures, rather than simply treatments for sick people. The note was sounded in many public forums. In a March 1937 article detailing the cancer discussions at the American Association for the Advancement of Science medical section's annual meeting, for example, *Fortune* magazine reported that about $700,000 would be spent

[22] RF motion #35416, 15 November 1935, RF Archives, RG 1.1, Ser 200D, Box 143, Folder 1773, RAC-NY.

[23] This framing of the JAX case probably also reflects frustration with Weaver's propensity to fund small projects, deemed "retail business" by the trustees in comparison with the multimillion dollar program grants at places like Caltech. See Kohler, *Partners in Science*, pp. 289–90.

[24] Cf. Strickland, *Politics, Science, and Dread Disease*, p. 7.

[25] On the various political maneuverings leading up to the passage of the 1937 NCI Act, see ibid., pp. 1–15; Patterson, *Dread Disease*, chap. 5. On similar struggles over the nature of government intervention for the case of polio research, see Creager, *Life of a Virus*.

on cancer research by private organizations that year: $250,000 by universities and $450,000 by foundations. By contrast, *Fortune* noted, the federal government (through the Public Health Service) was expected to contribute only an additional $140,000 to the cancer cause. Criticizing national priorities, the article remarked that "the public willingly spends a third of that sum in an afternoon to watch a major football game."[26] In comparison with financial support for other medical research, these numbers were not outstanding: research into infantile paralysis, for example, received $479,000 that same year from the foundations.[27]

As director of the American Society for the Control of Cancer, Little had from the very beginning emphasized research. In his first "Report of the Managing Director," he harshly admonished his predecessors for directing all its efforts toward educating the public about early cancer detection: "It would seem unnecessarily stupid to fail to include encouragement and practice of research as an integral part of the Society's program." As an experimental scientist and the head of a cancer research lab, Little himself had a personal stake in this claim. But his "trickle-down" views on the social value of scientific research also had roots in his Progressive vision of the power of scientific knowledge to cure social problems.[28]

As the decade progressed, Little increasingly promoted cancer research—over clinical treatments—and he drew more and more on inbred mouse work to make this point.[29] To medically trained audiences, Little

[26] "Cancer: The Great Darkness," *Fortune* 15 (March 1937): 112–13.

[27] Columbia University physician Dr. Henry Simms's report on U.S. medical research expenditure gives comparative statistics dating back to the mid-1930s: see U.S. Congress, Senate Committee on Education and Labor, Wartime Health and Education, Statement by Henry Simms, *Hearings before Subcommittee pursuant to S.R. 74*, 78th Congress, 2d session, 14–16 December 1944: *Medical Research I*, 1945, part 7, pp. 2238–41. For Rockefeller funding, see p. 2248. Medical research in general was not widely supported in the interwar years: the PHS's total appropriation in 1938 was only $2.8 million—compared with the $26.3 million the Agriculture Department (with its powerful political lobbies) received. See Starr, *Social Transformation*, pp. 339–40. Kevles has noted that by 1951, when federal support for medical research was $76 million, contributions from foundations and other nonindustry sponsors still amounted to nearly $45 million, or about 25 percent of the total spent on medical research in the United States; see his "Foundations, Universities, and Trends in Support for the Physical and Biological Sciences, 1900–1992," *Daedalus* 121, 4 (Fall 1992): 195–235.

[28] The ASCC gave small denominations to researchers during Little's tenure, but it appears the Depression-induced lack of funds prevented a more organized grants program. See Strickland, *Politics, Science, and Dread Disease*; Shaughnessy, "Story of the American Cancer Society," pp. 155–56.

[29] In more public venues, Little also worked with the Women's Field Army of the Works Progress Administration to promote cancer education, through increased awareness of symptoms and earlier reporting to doctors. See Marsha Hurst and Jane Nusbaum, "Advocating for Women's Health: Breast Cancer and Models of Advocacy in the Mid-Twentieth Century," paper presented at the Berkshire Conference on the History of Women, Chapel Hill, NC, June 1996.

emphasized that *only* a cooperative research program between experimental genetics and clinical medicine could ultimately find a cure for the *causes* of cancer, and that inbred mice provided the material link for such a partnership. In a 1936 presentation before the Mayo Clinic staff, for example, Little remarked that although: "most medical men . . . are skeptical concerning its actual practical value . . . experimental genetics is one of the most promising of all the lines of cancer research in the laboratory. The reason for this is to be found in the remarkable extent to which such rodents as mice are now adapted for work of this sort."[30] Six months later, in the *Journal of the American Medical Association*, Little would expand on this point by noting that "the technic [*sic*] of keeping genetic constitution constant" in a research material—a control embodied by the JAX strains—was far superior experimentally to "the accepted method of making human matings, viz. by uncontrolled outcrossing combined with inadequate records and small numbers of progeny . . . in human families."[31] In short, he told a medical congress, the usefulness of inbred mice can be analogized to the usefulness of pure-grade chemicals: "The chemist would be helpless in his attempt to analyze unknown material if he did not have known chemical agents on the shelves of his laboratory to which he could turn for assistance in analyzing the unknown. The biologist, in the past fifteen years, has been given the tools by which he could approximate [his unknowns]. . . . In experimental medicine today, . . . the use of in-bred genetic material . . . is just as necessary as the use of aseptic and anti-septic precautions in surgery."[32] Thus defined, inbred mice were neither market-dependent commodities (as he had implied to his medical researcher consumers) nor mammalian genetics research tools (as he had told the Rockefeller Foundation). These animals were methodologically accurate tools for pursuing the socially responsible approach to all biomedical research.

In late 1936 these same arguments took a public turn as Little led the ASCC into its first national media blitz. At the same time that the foot soldiers of the new "Women's Field Army" (WFA) was spreading the word about cancer locally, in towns and cities throughout the country, Little was recasting the scientific and ethical usefulness of inbred mice in the more dramatic and popular rhetoric of warfare. Cancer, he told *Good*

[30] Little, "Genetics in Relation to Carcinoma," *Proceedings of the Staff Meeting of the Mayo Clinic*" (1936): 782–83.

[31] Little, "The Present Status of Our Knowledge of Heredity and Cancer," *Journal of the American Medical Association* 106 (27 June 1936): 2234–35, esp. p. 2235.

[32] C.C. Little, "Symposium on Cancer: The Biology of Cancer," *Proceedings of the Annual Congress on Medical Education and Licensure* (Chicago: The Association, 1941), 15–16 February 1937, 12–14. On the use of this same language to promote the rat as an experimental organism, see Clause, "The Wistar Rat."

The Conquest of Cancer

Begins With Wiping Out Ignorance, Fear, Neglect. Here's How You Can Help

By Dr. Clarence Cook Little

Dr. Little, Managing Director of the American Society for the Control of Cancer, whose article is based on extensive research and clinical work

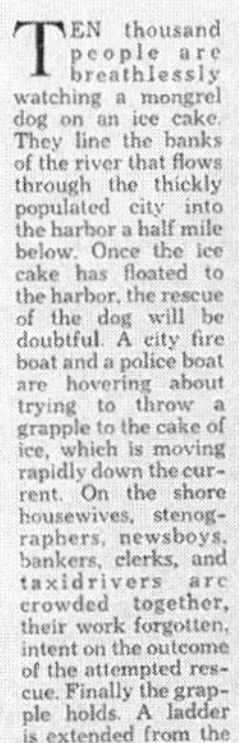

TEN thousand people are breathlessly watching a mongrel dog on an ice cake. They line the banks of the river that flows through the thickly populated city into the harbor a half mile below. Once the ice cake has floated to the harbor, the rescue of the dog will be doubtful. A city fire boat and a police boat are hovering about trying to throw a grapple to the cake of ice, which is moving rapidly down the current. On the shore housewives, stenographers, newsboys, bankers, clerks, and taxidrivers are crowded together, their work forgotten, intent on the outcome of the attempted rescue. Finally the grapple holds. A ladder is extended from the deck of the fire boat; a man crawls out —reaches the dog, reassures it, and gently carries it to the boat. A great cheer goes up. The crowd begins to melt away.

Among the crowd are perhaps twenty-five persons with early cancer. They are far more in need of prompt assistance than was the dog. Their peril is much more immediate. Through ignorance, fear, or neglect, however, they are allowing themselves to drift to the harbor and to the open sea of certain disaster beyond.

Let us see why this is the case.

Close by the rail on the river bank is Pat Crowley, retired fireman. It is a good ten years since he reached the sixty-year age limit, but his eyes are bright as he turns to leave. He becomes conscious of the fact that in his excitement he has allowed his clay pipe, an old and honored friend, to slip over to the left-hand corner of his mouth, from which favorite resting place of years' duration a small persistent sore has for the past month driven it. Reluctantly he moves it over to the other side, feeling, in so doing, as strange as if he had put on the wrong coat sleeve first.

Daydreaming of the time when he would have been on the fire boat, he almost bumps into Dr. Mann, an old acquaintance at whose hospital Pat's oldest boy is an orderly.

"Good job, Pat, don't you think?"

"That it was, Doc," says Pat, fingering his pipe.

The doctor's keen eye detects the small sore. "How long have you had that, Pat?" he asks, indicating it.

"Oh, about a month," replies Pat. "'Tis nothing. I've often had them, but they go away in a short time."

"But this one hasn't," counters the doctor. "Come up to the office and let me have a look at it."

"O. K. if you say so, Doc."

The next day at Doctor Mann's office a bit of the tissue in the sore area, removed under local anæsthetic and examined under the microscope, shows that cancerous changes have taken place.

"We'll fix that up at once, Pat," says the doctor. "You're lucky we caught it when we did. It would have meant serious trouble." He could have added "even death" and have told the truth.

"The saints be praised!" Pat is impressed. "If you hadn't chanced along, Doc, I'd never have known."

"That's the worst of it," thinks Doctor Mann.

A block or two away is Mrs. White, the mother of two children. She is forty-five years of age. In spite of capable obstetrical care, she received during childbirth injuries to the neck of the womb through which the child must pass during normal childbirth. No one has ever told her that such injuries should receive careful surgical attention and repair. No one has ever told her that unrepaired injuries of that sort are an unnecessary cancer risk.

Now, at her present age, certain irregularities and unusual symptoms of discharge are considered by her to be the normal circumstances attendant on

78

IN NO field is the proverbial "ounce of prevention" more important than in the early diagnosis of cancer. The cure of this dread disease in advanced stages still baffles the skill of the medical profession the world over; but if taken in time, many a potential victim can be saved.

I commend Dr. Little's article to GOOD HOUSEKEEPING readers on this basis. The story he tells should be known to every one. The advice he gives, if followed, may avert untold suffering. —DR. WALTER H. EDDY, Director, Good Housekeeping Bureau

4.1. Clarence Cook Little's 1936 article in *Good Housekeeping* [Source Credit: C. C. Little papers, University of Maine].

Housekeeping readers in 1936 (Fig. 4.1), could be conquered, but it was (as he later put it in a *Time* cover story), "a hard task requiring patience: trench warfare against a ruthless killer."[33] A few weeks later, 274 of JAX albino mice appeared on the cover of *Life* magazine, introducing readers to the story "U.S. Science Wars against Unknown Enemy: Cancer" (fig.

[33] C. C. Little, "The Conquest of Cancer," *Good Housekeeping* (June 1936): 78–79, 108, 110, 112. Quote from a published excerpt of a Little speech in "Cancer Army," *Time* 29 (1937): 40–41. In late 1936 he created the new office of publicity director in the ASCC, and he appointed media veteran Clifton Read to fill the position. Together, Little and Read

4.2. 278 JAX albino mice featured on the cover of *Life* magazine, March 1, 1937 [Source Credit: Time-Life].

4.2). Tapping into existing cultural fears of scientific cancer treatments, the "biggest gun" in the clinical treatment of cancer was presented first: Crocker Laboratory's 1.25-million-volt X-ray. Readers were cautioned that X-ray machines such as the one pictured were only in "experimental" use so far and that, in clinical terms: "Only the Surgeon's Knife or Radiation Can Cure Cancer."

---

engineered a virtual media blitz in many of the popular magazines of the day: see Patterson, *Dread Disease*.

By the comparison of mice to radium and its associated heavy machinery, the animals emerged as far more humane and effective weapons for the cancer war: they first appeared on a page headed "Mice Replace Men on the Cancer Battlefield." This animal, wrote the author, is "an ideal laboratory for the propagation and study of cancer," and that is why tens of thousands of them a year must "lose their lives on the battlefields of cancer research."[34] The only explicit connection of Little to this piece was a brief mention of the Jackson Lab's "valuable discoveries" about cancer through mouse research. But similar stories involving some combination of Little, inbred mice, and cancer research appeared (often with some of the same pictures) in several Henry Luce–owned magazines, such as *Newsweek* and *Fortune*, at around the same period.[35] Ultimately, the ASCC media campaign successfully articulated the meaning of inbred mice as morally acceptable animal stand-ins for humans in the cancer laboratory and, in the process, generated broader social support for cancer research and (to a lesser extent) the NCI Act.[36]

Little's public alignment of inbred mice and cancer research held some strategic risks, in light of the greater attention being drawn by contemporary animal advocates to medical experiments. Some medical research advocates singled mice out to argue for the overwhelming social benefits of medical research. For example, a 1928 piece entitled "What Science Owes to Animals" argued first that the value of animal experimentation "is the value of biologic therapy itself" and then noted that the "importance of the quiet little white mouse" simply could not be refuted in this regard: "his service to mankind . . . has been noteworthy."[37] Notably, however, antivivisectionist complaints about mice used in cancer research rarely focused on the mice themselves but rather on the amount of money that mouse cancer research drew from efforts to ameliorate the effects of the disease on patients.[38]

Just as the NCI Act was coming up for a vote, American antivivisectionists took on the cause of laboratory mice—not in Little's work, but in that of his long-time nemesis, Maud Slye. In 1937 *Starry Cross*, the journal of

[34] "U.S. Science Wars against Unknown Enemy: Cancer," *Life* 2 (1 March 1937): 11–17.

[35] "Cancer: The Great Darkness," *Fortune* 15 (March 1937): 112, 114, 162.

[36] Patterson, in particular, argues that because Little did not articulate a clear vision for a national institute, and he lent his support to the cause rather late, the 1937 ASCC media campaign must be read as an "unintended and roundabout" force in the passage of the NCI legislation. See *Dread Disease*, pp. 124–30, esp. p. 126.

[37] John A. Kolmer, "What Science Owes to Animals," *Hygeia* 6 (1928): 618–622.

[38] "Heard and Read," *The Starry Cross* 45 (1937): 55; "Editorial Correspondence" and "Heard and Read," both in *The Starry Cross*, 1938, *46*: 156 and 39, respectively. Susan Lederer found these materials in the course of her own research, and I am indebted to her for sharing them with me.

the American Anti-Vivisection Society, published a scathing critique of Slye's "rodent racket." Slye was not singled out for inflicting harm on mice per se;[39] instead, the editorial named her as a powerful counterexample to medical scientists' claims that their knowledge would have clinical application. Slye was described as operating "under the banner of medical research," but the writer argued that her vast body of work had "proved of no use whatever to humanity" because cancer was not yet cured. A year later, a commentary and a letter to the editor in the same journal echoed this criticism, with special reference to the mice. Slye's efforts were described as "futile": "When the smoke of battle has cleared away, and the sixty thousandth little mouse has uttered his last expiring squeak, the candid ones of the medicos will . . . admit that the medical profession knows little of the cause of cancer, and much less of the cure. . . . Therefore, even a vivisector could find no *honest* excuse for repeating these experiments." Other than this brief attack on Slye, however, mice received little antivivisectionist attention, either in internal or mainstream media—animal advocates were generally too busy attacking medical work with more popular animals like dogs and cats. Most contemporary promedical authors writing about animals in research simply debated the issue in general utilitarian terms, with the mouse presumably subsumed under the argument as a particular case, or they explicitly wrote in defense of cats and dogs. Only when these animal experiments "failed" to produce clinical results did antivivisectionists object to the utilitarian assumption, and this was the key ethical premise underlying Little's arguments for mouse use.[40]

[39] Because, as Sue Lederer has argued, "gender was an essential feature of . . . discussions on medical moral sensibility," it seems likely that antivivisectionists were capitalizing on Slye's lack of feminine qualities (in both her personality and her choice of scientific vocation) as an especially potent resource for public debates. See Susan Lederer, "Moral Sensibility and Medical Science: Gender, Animal Experimentation, and the Doctor-Patient Relationship," in *The Empathic Practitioner: Empathy, Gender, and Medicine*, ed. Ellen Singer More and Maureen A. Milligan (New Brunswick: Rutgers University Press, 1994), pp. 59–73, quote on p. 59.

[40] Cf. John Dewey, "The Ethics of Animal Experimentation," *Atlantic Monthly* 138 (September 1926): 343–46. Although Dewey advocated some form of animal protection in laboratories, he wrote: "To prefer the claims of the physical sensations of animals to the prevention of death and the cure of disease . . . does not rise even to the level of sentimentalism. . . . Scientific inquiry has been the chief instrumentality in bringing man from barbarism to civilization." Cf. W. W. Keen, "Anti-vivisectionists' Methods," *Hygeia* 5 (1927): 36–38; F. C. Cross, "What Animals Know About Medicine," *Popular Science* 127 (September 1935): 24+; Harold Ward, "The Sciences and Society: Prejudices against Animal Vivisection," *The Living Age* 348 (August 1935): 549–52. My generalization is made based on a search of published records via the *Readers Guide* from 1920 to 1937, a search through the Rockefeller University Archives holdings on "Anti-Vivisection" (RG 600–1, Box 21, Folders 2 and 3, RAC-NY), and in consultation with Lederer, who has done exhaustive research on the American Anti-vivisection Society's journals: *Journal of Zoophily*, *The Starry Cross*, and *The Anti-Vivisectionist* in preparation for a project on the history of animal experimentation in the United States.

Little took these arguments seriously enough to confront them head-on, by encouraging scientists to examine and act on stereotypical understandings of where mice lay on the "socio-zoological scale" of human and nonhuman animals.[41] Concurrent with the 1937 media blitz, he wrote an article in *American Naturalist* outlining "Some Contributions of the Laboratory Rodents to Our Understanding of Human Biology." He began by negatively stereotyping the antivivisectionists' cause: "We are all aware of the emotional bubbling and irrational babbling which pours in a stream over the paralyzed or bewildered brains of legislators [and] . . . has recently become a menace to scientific research." But, Little observed, the mouse and other rodents could easily confound the ethical categories and strategies taken up by these same activists because mice were *not* always viewed positively and with emotional attachment by your average 1930s American: "What we do not perhaps realize, . . . is that . . . it has been difficult to keep at fever heat a sufficient level of sympathy for the rodent similar to that which the dog or cat engenders. Were it not for the appearance of rabbits upon Easter cards and its resultant heart throb, the rodents would be a completely constructive element." Instead of ignoring the mouse's place in human culture, Little suggested, biomedical researchers should exploit "the age old enmity of woman and the Muridae" to "save the day" against the antivivisectionist threat. As he summarized: "The division of forces and the resulting restriction of objective . . . caused by the cold and impersonal eye of the rat, mouse, guinea pig, or rabbit, is a real contribution. It may make the rank and file of citizens understand more clearly the abnormal nature of the unbalanced and starved emotions that lie at the base of organized attempts to destroy scientific research with mammals. Since without that research human biology is doomed, the issue at stake is a truly great one."[42] In Little's view, then, mice could and should be actively used to confront ethical controversies over animal use. Working under the assumption that there would be almost no question in the public mind that knowledge obtained from these creatures was a long-term benefit for humanity, Little deployed inbred mice to define the scientific value of cancer research at the same time that he argued for more federal support of this research.[43]

[41] Drawing from Arthur Lovejoy's work on the "great scale of being," the "socio-zoological scale" is a concept introduced by Arnold Arluke and Clinton Sanders in *Regarding Animals* (Philadephia: Temple University Press, 1996), chap. 1.

[42] Little, "Some Contributions," 135–36.

[43] Cf., again later, C. C. Little, "The Value of Research with Animals," *Bulletin of the American Society for the Control of Cancer* 24 (1942): 7–10.

Public debates reached a rhetorical apogee when Little testified during the final federal hearings over the compromise bill.[44] To members of Congress he pedaled inbred mice as powerful solutions to the problems ailing both cancer research and cancer policy making. Against the predominant lay rhetoric of moral outrage over cancer's human death toll, Little juxtaposed an image of controlled animals in the laboratory: "Let me point out that research in the cause of cancer is not entirely a medical problem in any sense of the word. . . . The emphasis is entirely shifted from working with the slow, unsatisfactory human material to the material that is easy to handle, rapid breeding and conveniently controllable. That has been the biggest change in cancer research."[45] By simultaneously underscoring the practical usefulness of inbred mice in research and the moral status of inbred mice as human stand-ins for that research, Little's appeals to scientific and public audiences proved powerful for the passage of what is now known as the first National Cancer Institute Act in June 1937. The inbred mouse's moral suitability for cancer studies became instantiated through these public appeals. Experimental work with mice was the means through which biologists could produce publicly accountable science (i.e., cures for disease). But just as importantly, the bureaucratic power of these arguments is further suggested by the role that Little played in the initial federal cancer research policy-making under the new law.

COORDINATION AS A VALUE: JAX MICE AS A TEST CASE IN FEDERAL BIOMEDICAL RESEARCH POLICY-MAKING

The landmark quality of the NCI legislation lay less in its dollar-value scale than in its policy-making scope. The 1937 act authorized $700,000 annually in federal funds to support "grants-in-aid" for cancer research projects, and an additional $750,000 to build and run the National Can-

[44] The first two bills, which were identical, were introduced into the Senate and House in April 1937 by, respectively, Senator Home Bone and Representative Warren Magnuson; the third was introduced three weeks later into the Senate by Maury Maverick, and three months later into the House by Representative John Hunter (this time as a House joint resolution). Among health policymakers, the so-called Maverick Bill was preferred because Maverick consulted with the PHS in its formulation, so its terms were much more specific. The Maverick Bill also called for greater funding: an initial appropriation of 2.4 million and 1 million dollars annually thereafter (as compared to the "Bone-Magnuson Bill," which called for only 1 million dollars annually). See J. R. Heller, "The National Cancer Institute: A Twenty-year Retrospective," *Journal of the National Cancer Institute* 19, 2 (August 1949): 147–90.

[45] House of Representatives Committee on Interstate and Foreign Commerce, "Hearing on the National Cancer Act," *Congressional Record*, 75th Congress, 1st session, Report no. 1281, p. 4.

cer Institute in Bethesda, Maryland. This was not the first renewable federal appropriation for biomedical research: by 1938 the Public Health Service enjoyed a budget of $2.8 million dollars for research. Several years earlier, Roosevelt's top health advisor, Harry Hopkins, had pushed through a new amendment to the Social Security Act that earmarked several hundred thousand additional dollars from the national budget each year specifically for laboratory investigations of diseases. But in terms of other federal funding and other contemporary life science funding sources, these figures were relatively small. In 1938, for example, the U.S. Department of Agriculture received nearly twenty times the total 1937 NCI appropriation for research, and in 1930 Caltech alone received a single grant of $1.1 million from the Rockefeller Foundation to develop its biology division. What the NCI Act represented, however, was a symbolic breakthrough in the relationship between the life sciences and the federal government. As the first extramural grant program for biomedical research, it offered politicians the opportunity to claim a visible role in health issues at the same time that it offered scientific leaders the chance to create a system of peer-reviewed laboratory experimental work that they felt most rationally attacked cancer as a laboratory problem.

To this latter end, the law stipulated that Surgeon General Thomas Parran assemble a group of "leading scientific and medical men" to determine how this money should be spent. This organization, known as the National Advisory Cancer Council (NACC),[46] functioned both as the clearinghouse for government extramural grant applications and the agenda-setting body for the field of cancer research more broadly. Before the NACC, foundations that funded cancer research (such as the Childs Fund and the Fuller Fund) consulted with their scientific trustees as well as individual academic cancer researchers on particular grant-in-aid applications that fell within their field of expertise. When the NACC convened in November 1937, the occasion marked the first time a critical mass of cancer researchers were charged with formulating a set of general measures that could be used to guide funding in this field. Transcripts of the NACC's initial discussions, however, show their collective concerns lay less in promoting particular research areas than in promoting shared research values. In fact, a series of debates over whether and how to fund the production of inbred mice at Jackson Lab provided the NACC with a sort of a test case for working through what these values would be and how a policy-making body could encourage them to be practiced.[47]

Little's policy-making experience (as ASCC managing director) and long-time arguments for standardization and centralization of research

[46] Strickland, *Politics, Science and Dread Disease.*
[47] Ibid.

impressed lawmakers as well as Parran even before the NACC membership was named. One week after the passage of the bill, Little was called to testify at the Joint House and Senate Hearings on Cancer Research, where he spoke primarily to members of Congress and Public Health Service policymakers. All witnesses at these hearings were asked comment on how to best appropriate newly available federal funds for research. Little did not exploit this opportunity to preach the virtues of the genetic approach to cancer research that his JAX investigators were pursuing in their own experiments. Instead, he argued for JAX research materials on organizational grounds: inbred mice were the biological equivalents of standardized, interchangeable parts in a well-oiled machine of biomedical research production. Comparing inbred mice with industrially produced pure-grade chemicals, Little began: "[T]he first thing you should do . . . in spending your money would be to insure a constant and adequate supply of controlled, known animal material on which investigations could be carried out."[48] He also suggested that JAX was the only institution that could provide the necessary infrastructure to make this possible: "We can produce as nearly a chemically pure animal . . . as it is possible to produce . . . during this past year the little laboratory where I work has sent out over 65,000 such animals . . . for research in cancer and in other experimental medicine." (See fig. 4.3, which was likely a publicity photo because it features the iconically pure white albino mice, even though they were not JAX's only commodity.) Shortly thereafter he was asked to join the NACC, which held its initial round of meetings in November 1937.

In early NACC meetings, centralized organization and the coordination of federally sponsored research emerged as top priorities on which everyone agreed. Several members framed these issues by comparing the accountability of members of Congress to their constituency and scientists' accountability to their peers. Physicist Arthur Holly Compton, for example, noted that cancer researchers like Maud Slye were conducting "clinical-type" research on mice that promised quick applications to humans but agreed with Little that, in the absence of longitudinal human studies (such as an examination of cancer patients in Veterans' Administration hospitals), controlled laboratory research on animals provided the greatest potential for applied human cancer treatment or cure. At the very least, Little argued, since there is "no such thing as an expert on cancer," a coordinated but cooperative approach "centralizes local responsibility . . . and gives you a fixed agent whom you can hold responsible for the

[48] Minutes of the Joint Hearings on Cancer Research, U.S. Senate, Committee on Interstate and Foreign Commerce, 8 July 1937, *Congressional Record*. Cf. Kenneth Alder's work on interchangeable machine parts and the rise of the metric system in France—e.g., his essay in *The Values of Precision*, ed. N. Wise (Princeton: Princeton University Press, 1995).

4.3. JAX mice, prepared for shipping, c. 1935-40 [Source Credit: Jackson Laboratory Archives].

results obtained." As James Conant of Harvard concluded, these values should be what distinguished fundable projects from unfundable ones: could researchers give an affirmative answer to the question "is this work that could not be done except by government money, by virtue of its being mobile and centralized?" James Ewing agreed: only the big-project approach (where key experimenters and institutions are given lots of money over many years) stood a chance of producing results.[49]

The practical meaning of both centralization and coordination, however, was tested when Little himself submitted a proposal requesting federal support for the "stabilizing" work that JAX had already done with inbred mice. "The[se] animals," Little told his colleagues, "are generally used . . . in the American cancer field . . . [and] . . . are now recognized as standard material." But rather than cite the range of institutions that JAX now sold to as a result of its sales venture (see fig. 4.4, which shows boxes of mice going to the Public Health Service, Eli Lilly, Children's Hospital, the Merck Institute, and E. R. Squibb), Little instead noted only the two major consumers: the Rockefeller Institute of Medical Research

[49] NACC Minutes (meeting transcripts), Record Group 443, Box 6, Vol. 1, 9 November 1937, esp. p. 55 (Conant quote) and discussion on pp. 59–82; Little quote from February 1938 meeting (p. 596) of transcript, NACC-MD.

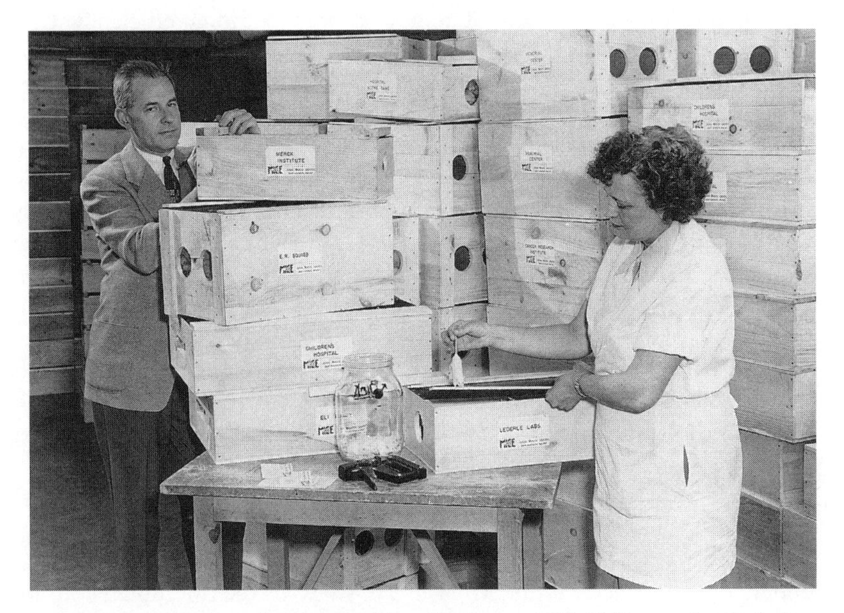

4.4. Boxes of JAX mice being prepared for shipping, c. 1935-40 [Source Credit: Jackson Laboratory Archives].

and the Public Health Service were each purchasing thirty thousand mice per year ("and both are constantly pressing us for more"). Little promised to double inbred mouse production at JAX (from around sixty thousand to one hundred twenty thousand mice per year) and to give government intramural and extramural grantees first priority on these materials. JAX, he concluded, would be providing an essential service to cancer research: "this has nothing to do with research work being done at the laboratory except that it stabilizes the supply of these animals and I think it is fair for me to say that it would cost approximately three times the amount [requested] to set up a separate institution, to raise those mice in an air-conditioned building in a climate like Washington or any place where the temperature in the summer goes much above 80."[50]

Drawing on industrial analogies, those who were in favor of Little's plan described it as way to make necessary materials available "with greater efficiency and . . . much *more economy*" [emphasis in original]

[50] 1937 NACC first meeting transcript, p. 271, NACC-MD; documentation exists to confirm this figure: the Rockefeller Institute's W. H. Sawyer and J. H. Bauer agreed in March 1937 to purchase 500–750 *C57Black* mice per week for their cancer research, and this adds up roughly to the figure Little is quoting here. See CCL to W. A. Sawyer, 19 March 1937, and J. H. Bauer to CCL, 24 March 1937; both RF Archives, RG 1, Series 100, Box 10, Folder 81, RAC-NY).

according to the general goal of achieving organizational efficiency. Carl Voegtlin labeled JAX's effort "essential" to the federal cancer research as a "standardizing job." Francis Carter Wood argued that it would "meet the situation we are in, like chemists doing work where everyone had to purify their own chemicals." William Middleton seconded this opinion, noting "we do not want to waste time" in either research material preparation or the unintentional duplication of results. But critics, especially Compton, worried that "this was not quite the right kind of bookkeeping for cancer work," given that JAX would be given a grant-in-aid of research but what they would be producing was not research but experimental materials. "The NCI would be subsidizing the [mouse] industry," Compton argued, by paying JAX a grant to produce mice and then again to buy the mice that JAX produced. Why, he wondered, could this centralized material resource not be made "self-supporting" on its own market terms by raising the prices of mice?[51]

Little acknowledged the desirability of self-sustaining material resources in principle, but he suggested that in the fast expanding world of early-twentieth-century experimental cancer research, production of standard animals was not such a straightforward business as the traditional conceptions of the scientific knowledge-making process would have it seem. Standardization, he argued, was equally a matter of demand as supply, and sometimes one could be used to create the other within particular user communities. When asked about the price of mice and how JAX determined it, for example, Little retorted angrily: "to be perfectly frank with you it has been a matter of missionary work." The subsequent discussion of what message Little had been preaching to create demand for JAX mice continued, but Little must have realized he was treading on hallowed ground for he insisted that they be off the record. Moments later, in response to Parran's and Conant's reservations about making a long-term commitment to the JAX project, Little drew on a market analogy to make the point that production and use of laboratory animals represent a "vicious circle"; "it is a question of business when you go into it, the needs and what will stand and what you can do . . . [but] you cannot expect to do very much in the way of making progress convincing more people, since the production of more mice that will convince them depends on getting the chance to produce them." Only after the motion to fund JAX was tabled did Little assert the inherent value of a genetic approach to the problem of cancer materials: "the main point is to . . . have it done by a geneticist, not a man of the medical school because they haven't adjusted to those

[51] NACC Minutes (meeting transcripts), 11 November 1937, pp. 28, 29, 131-B, 270/a–g (CCL's typed proposal to the NACC, addressed to Ludwig Hekoten, 22 November 1937), 278, 274–76, 272–74; and 27 November 1937, pp. 291 and 312; all NACC-MD.

things yet."[52] The perceived value of standardized organisms for cancer research turned, in large part, on how well policymakers could imagine a preexisting group of users—and yet, as Little noted, standardization was not a static but a relative value, and communities of experimental practices could be created, rather than just discovered, by materials suppliers.

Scientists from existing federally sponsored cancer research labs who already used inbred mice probably broke this deadlock on the question of supporting JAX's production expansion. The PHS cancer research unit at Harvard, considered the "bedrock of the national involvement in the problem of cancer," was headed by JAX's first mouse consumer: J. W. Schereschewsky. Many recall that Schereschewsky wielded great influence behind the scenes as a key advisor to the PHS's overall cancer research program. Beginning in early 1934, Schereschewsky's protegé, Howard Andervont, collaborated with the Jackson Lab geneticist J. J. Bittner on a long-term experiment involving tumor transplantation and induced tumors in inbred mice (see previous chapter and chapter 5). Also, by the mid-1930s, Halsey Bagg, now a Memorial Hospital cancer researcher and former collaborator of Little's from Cold Spring Harbor, ordered large numbers of mice from the Jackson Lab for his new line of research on the relationship between pregnancy and the development of mammary tumors. Bagg's research explicitly aimed at developing better criteria for geneticists to use in classifying inbred strains of mice as "high" or "low" cancer incidence.[53] On a small scale, then, the preexisting traffic in mice between JAX and cancer researchers validated strong ties between available materials and specific research problems.[54]

Three weeks later Little returned with a revised application that emphasized the current scope of inbred mouse production and use. JAX mice, Little argued, were already in great demand: six basic and ten secondary

[52] NACC Minutes (meeting transcripts), 27 November 1937, p. 282, NACC-MD.

[53] Quote regarding Schereschewsky from Shimkin, *As Memory Serves*, p. 10. See also Andervont, "J. W. Schereschewsky," cf. Halsey Bagg to Jackson Lab Secretary Mary Russell, 4 April 1933; Russell to Bagg, 18 April 1933; Bagg to Russell, 24 April 1933; Russell to Bagg, 2 May 1933; all Box 740, CCL-UMO; H. J. Bagg and J. Jacksen, "The Value of a "Functional Test in Selecting Material for Genetic Study of Mammary Tumors in Mice and Rats," *American Journal of Cancer* 30 (1937) 539–48. Bagg gained his research reputation primarily through his work on X-ray-induced mutations or other visible abnormalities in mice, and it was in this line of work that he and Little collaborated: see chapter 1.

[54] Jean Paul Gaudillière and Ilana Lowy, lectures given at the International History of Science Summer School, Paris, France, 10–15 July 1994; JAX Lab cites the 64 percent figure in its first *Handbook of Genetically Standardized JAX Mice*, ed. Earl Green (Bar Harbor: Bar Harbor Times Publishing, 1962), p. 53, JLA-BH. Ironically, Little felt himself chased out of his post at the ASCC by a new contingent of businessmen and doctors, with a "critical and aggressive attitude concerning the 'stupidity and inefficiency' of past efforts." See CCL typescript, n.d., "When in 1937 the National Cancer Act was passed . . .," Box 730, CCL-UMO.

strains of inbred mice were used in cancer research, and "these strains have been under careful genetic observation for a long period . . . 27 years in one case, 16 years in others, etc." Little presented revised sales figures: 40 percent of JAX animals produced are sold to the government, but an additional fifty to sixty thousand are already being distributed to "institutions all over the country at approximate cost [and] this demand now far exceeds supply." He asked the NCI for roughly $10,000 per year "to continue indefinitely" for supplies and equipment (food, bedding, racks, cages) to increase production of inbred mice. He also proposed that he would be responsible for obtaining the $40,000 in additional funds necessary to erect a new building in which these animals could be housed. A brief discussion ensued about what represented a "good animal" for cancer research: Ewing questioned whether there would be money to be saved by not using "pedigreed stock," and Little replied that JAX was bound to make a "surplus" of animals which made them, by definition, a good choice. A motion was made to fund the grant, and ultimately the decision turned on the practicalities of scale: there were no viable alternatives to JAX (either in-house or out) that could ensure the availability of large numbers of experimental animals, as quickly or as cheaply. When asked directly if supporting JAX would be worthwhile for cancer research, NCI Director Voegtlin said simply, "It *is* essential to provide mice." The motion passed immediately with no further discussion.

The NACC never issued a formal measure requiring that all NCI grantees use inbred mouse strains, but through this initial grant-in-aid arrangement with JAX, the organizational suitability of inbred mice for experimental cancer research became official federal policy. Before the committee could even agree on a research agenda (see next section and next chapter), the NACC deemed the coordination of laboratories around these standard mammalian materials a desirable configuration for future laboratory work. Because JAX was the only large-scale provider of these materials until well after World War II, the grant de facto promoted the use of JAX mice as an "industry standard." The final NACC resolution on the JAX grant bore the mark of the committee's structural vision for cancer research. It articulated the expectation that JAX would increase its facilities to produce the nearly thirty-five thousand inbred mice per year that NCI researchers would need and argued that this expense (one of the largest initial grants given by the NCI) was justified because only the inbred mouse's known genetic composition could ensure accuracy and efficiency at the benchtop, while also enabling better communication among all the various NCI-sponsored projects.[55]

[55] NACC Minutes (meeting transcripts), 11 November 1937, pp. 28, 29, 131-B, 270/a–g, 278, 274–76, 272–3; and 27 November 1937, pp. 291 and 312.

THE TRIUMPH OF PRODUCTION: THE ROCKEFELLER FOUNDATION MOUSE HOUSE AND "OPERATION BOOTSTRAP"

With the NCI grant in hand, Little immediately wrote to Warren Weaver, telling him of his intention to submit a proposal to fund a new building on the Bar Harbor campus. In 1933 this would have been a hard sell, but by the late 1930s Weaver had begun to show interest in technology—for example, ultracentrifuges and electrophoresis—as a means of creating a whole new sphere of life science knowledge he called "molecular biology."[56] Little himself did not explicitly make the instrument analogy, but in the context of initial NACC discussions, the plan to increase mouse production amounted to a promise that JAX would become financially independent while using their mouse distribution network to dispense inbred mice as standardized instruments of cancer research. On both these grounds—as a business proposition as well as a means through which accurately calibrated biological instrumentation in genetics could be obtained—Weaver and Gregg were now impressed with JAX.[57] But ironically, the NCI-sanctioned expansion of JAX that would attract additional Rockefeller Foundation support would effectively demote mammalian genetics to a discipline in the service of a broad range of biological and medical research projects.

When Gregg visited JAX in late 1937, he noted repeatedly (both in follow-up letters to Little and internal Rockefeller Foundation diaries) that inbred JAX mice were already in high demand. Over the period from 1933 to 1937, JAX Lab had gone from raising twenty thousand mice a year for their own use to distributing more than twice that number to outside workers, and as a result, their physical plant was at capacity.[58] Furthermore, the list of JAX mouse users had expanded from one to twenty and was now

[56] See Kohler, *Partners in Science*; Kay, *Molecular Vision of Life*. On the rise of instrumentation in the life sciences, see Nicolas Rasmussen, *Picture Control: The Electron Microscope and the Transformation of Biology in America, 1940–1960* (Stanford: Stanford University Press, 1997), and Pnina Abir-Am, "The Molecular Transformation of 20th-Century Biology," in *Science in the Twentieth Century*, ed. John Krige and Dominique Pestre (Amsterdam : Harwood Academic, 1997), pp. 495–524.

[57] See Kohler, *Lords of the Fly*, chap. 5. Cf. Jean-Paul Gaudillière, "Rockefeller Strategies for Scientific Medicine: Molecular Machines, Viruses, and Vaccines," *Studies in the History and Philosophy of Biology and the Biomedical Sciences* 31, 3 (2000): 491–509.

[58] JAX might have even been over its mouse housing capacity: Little told a newspaper reporter that the building was meant to house thirty thousand mice, but he told the Rockefeller Foundation's Hanson it was sixty thousand. See "Preparing Mount Desert House for 30,000 Mice to Be Used in Cancer Research Interests: Science to Play Role of Pied Piper for Humanity's Sake"; cf. FBH diary entry, 24 October 1933, and *Portland Sunday Telegram*, n.d. but probably January or February 1930, JLA-BH; CCL to WW, 13 November 1937, RF Archives, RG 1.1, 200D, Box 143, Folder 1774, RAC-NY.

TABLE 4-1
JAX Mouse Consumers, 1936-37

Carnegie Institute of Washington[G]
Michigan Department of Health
Rockefeller Institute for Medical Research[C]
J.H. Bauer
Ernest Stillman
James B. Murphy
U.S. Public Health Service[C]
Army Medical Center
Memorial Hospital, New York[C]
University of California[G]
University of Wisconsin
Johns Hopkins University
University of Minnesota[C]
University of Toronto
University of Cincinnati
Henry Phipps Institute
McGill University[C]
University of Chicago[G]
University of Georgia
National Institutes of Health[C]
Michael Reese Hospital[C]
University of Texas[G]
Harvard University[G,C]

LEGEND:
G = Institution with known genetics research program
C = Institution with known cancer research program
Unmarked = Unknown

Adapted from a list complied by C. C. Little, entitled "Roscoe B. Jackson Laboratory, Mice Supplied," in Minutes of the first Meeting of the National Advisory Cancer Council (letter of Ludvig Hekoten, 22 November 1937), NACC-MD.

highly diverse: it included genetics programs as well as a spectrum of medical research organizations from hospitals to university medical schools (see table 4.1). In internal discussions among foundation program officers, Weaver conceded that the Jackson Lab's function as a "wide and important" distribution center for the animals used in medical research now outstripped the value of its mammalian genetics research. "A highly standardized laboratory animal," he wrote in a 1937 summary of the JAX grants thus far, "can be obtained only by long-continued inbreeding so that recessive traits will come to the surface and be discarded."[59]

In his formal application to Weaver, Little worded JAX's needs in economic terms, rather than in relation to how his institution might build on its existing research program: "The demand [for mice] which is steadily increasing greatly exceeds the supply. . . . [JAX] needs to insure the maintenance of the stocks under circumstances and in sufficient quantity to stabilize and continue the present supply to various institutions at approximately the present cost." To support these claims, Little generated a numerical distribution projection for each of the six most popular strains: *A, dba, C57, C, C3H,* and *ABC*: *C3H* and *C57*, he claimed, could increase their physical plant. Little also included the latest version of the mouse stock list sent to potential scientific customers. Though he insisted that any plan for inbred mice supply stabilization must be "divorced from the individual fate" of himself or his institution, he clearly had some idea of the uses to which JAX mice were being put. The stock list evaluated each strain for its "Value for General Laboratory Work," as well as "Value for Genetic Research," "Incidence of Breast Tumors," and "Incidence of Internal Tumors." Little also touted the programmatic virtues of a JAX-type "single center" institutional model for solving the long-term mouse supply problem: "material produced under standardized conditions will eliminate variables and make more practicable comparison or repetition of work in or between different laboratories."

Weaver visited the JAX facilities in November 1937 and immediately grasped the business value of Little's proposal: "Should the proposed increase in breeding facilities be available, Little would . . . sell approximately 120,000 mice per year." He also seemed pleased to find evidence that JAX research benefited the mouse colonies, rather that the other way around. Cross-strain tumor transplantation experiments, for example, now interested Weaver not as research but as they were applied as routine controls to "assure the uniformity" of the stocks. Ultimately, the Rockefeller Foundation agreed to provide $40,000 in funds from the Division of Natural Sciences for the new building, and the most prominent members of the National Advisory Cancer Council attended the groundbreaking (fig. 4.5). Within a year, construction on the first JAX mouse house

[59] Staff Conference Discussion Notes, 7 November 1935; AG to Howard Adams, 22 April 1936; RF Grant Dispersal Summary #35416, 15 November 1935; all RF Archives, RG 1.1, Series 200D, Box 143, Folder 1773; WW's Summary of the JAX Lab grants, 20 December 1937, same except Box 1774, RAC-NY.

4.5. Snapshot at the 1938 groundbreaking for new JAX mouse production unit, attended by the National Advisory Cancer Council members. From left to right (faces visible): Francis Carter Wood, Little, Mr. Emory, James Conant, Thomas Parran, Ludwig Hekoten [Source Credit: Jackson Laboratory Archives].

was complete and the shipping room within was staffed to send out thousands more of the fifty to sixty thousand inbred mice it housed (fig. 4.6).[60]

By continuing to hire mammalian geneticists, many of whom were women, Little sustained a de facto commitment to this aspect of JAX's original research program. In 1936, George Wooley, a recent Ph.D. in genetics from Wisconsin, signed on at JAX. The following year, Little brought William Russell and his wife Elizabeth to Bar Harbor: both were finishing PhD work on mammalian genetics with Sewall Wright at the University of Chicago (see fig. 4.7, a staff photo from 1943, by which time Emelia Vicari and Katrina Hummel joined the staff.[61] JAX also gained a few cancer re-

[60] CCL to WW, 13 November 1937; handwritten letter from AG to WW, c. 11 November 1937; WW diary excerpt, 29 November 1937; CCL to WW, 4 December and 14 December 1937; WW Memo to Raymond Fosdick, 20 December 1937; all RF Archives, RG 1.1, 200D. Box 143, Folder 1774, RAC-NY. The Rockefeller Foundation didn't actually disperse the money for the building until JAX Lab's earlier tapering grants fully expired in early 1940. Mammalian genetics was mentioned in passing in this resolution, but the appropriation was not officially listed under the NS's genetics program: see Natural Sciences Grant Resolution #38011, RF Archives, 1.1, 200D, Box 143, 1773, RAC-NY.

[61] Emelia Vicari had diverse research interests, some genetic but more immunological: cf. C.C. Little and Emelia Vicari, " 'Lipid-Steroid' Fractions of Mouse Adrenal Lipids," *Proceedings of the Society of Experimental Biology and Medicine* 58 (1945): 59–60.

4.6. Judy Fielder and Watson Robbins in Mouse Shipping Room, c. 1940 (note wooden box marked "RUSH") [Source Credit: Jackson Laboratory Archives].

searchers through extramurally funded grants. In 1936 cancer geneticist Lloyd Law (a Castle student) came to JAX on a Finney-Howell Medical Research Fellowship, and in 1938 Walter Heston arrived with individual funding from the newly created National Cancer Institute (fig. 4.8).[62]

But as these scientists working at JAX saw it, production and research peacefully coexisted only when the former could serve the latter and not

[62] Jackson Laboratory Annual Report, 1939–40, JLA-BH. It seems to have been easier for women geneticists to get jobs working with mice than with flies (e.g., Kohler notes that Morgan's Drosophilists developed a "boss and boys" culture that was not friendly to women workers, while Salome Waelsch has said that Dunn's lab was much more congenial); but the existing historiography of gender politics of in early genetics (e.g., Amy Sue Bix's study of eugenics field workers and Marsha Richmond's work on Bateson's women students) has yet to be comparatively extended to the United States in the 1930s and 1940s. Although Elizabeth Russell had obtained a master's degree in 1937 under the direction of L. C. Dunn at Columbia, William Russell was the only member of this genetic research duo to be officially hired by Little in 1939. Elizabeth would continue to do independent research, sharing lab and office space with her husband, until 1946 when she was appointed to the JAX research staff. Shortly thereafter, William Russell left Bar Harbor for Oak Ridge National Laboratory, and in 1953, his now-ex wife was named senior staff scientist at JAX; see "A Biography of Elizabeth Buckley Shull Russell," Box 8, JLA-BH; Edna Yost, "Elizabeth Shull Russell," in *Women of Modern Science* (Westport, CN: Greenwood Press, 1984); interview with Elizabeth Russell, JLOH-APS and JLOH-KR; Provine, *Sewall Wright and Evolutionary Biology*, p. 189, for Wright's students and their dissertation topics. On Scott and Wooley, see *American Men of Science*, 8th ed. (1949): Wooley would later leave JAX

4.7. Jackson Laboratory scientists, summer 1943, informal photo. From left to right: George Snell, Emelia Vicari, Margaret Kelsall, Arthur Cloudman, Elizabeth Russell, Katrina Hummel, William Russell, George Wooley, Elizabeth Fekete.

the other way around. Elizabeth Russell, for example, would later argue that increased production—brought on by the need to sell mice during the Depression—actually advanced research in mouse genetics at JAX: "That period of financial stringency had another effect on mammalian genetic research, which I feel has been very beneficial . . . [this was] the system I sometime refer to as Operation Bootstrap." More mice were sold, from which JAX could use the money to support its research; secondly, mouse genetics research would become easier to do because a large-scale inbreeding program would generate more mouse mutants and therefore more genes with which to work. Or, as Little put it in the lab's Annual Report: "To increase by from 60–100% the number of breeding mice [would], thus increas[e] the opportunity to obtain and study new mutations and variations. This would add greatly to the strength of the Laboratory's research program."[63]

---

for Sloane-Kettering Cancer Center. See Jody Termial (handwritten signature) to June Smith, 6 March 1967, Box 2–6, JLA-BH.

[63] Elizabeth Russell, "The Origins and History of Mouse Inbred Strains: The Contributions of Clarence Cook Little," in *Origins*, p. 40; cf. Jackson Lab Annual Report, 1938–39, JLA-BH.

4.8. Snapshot of Walter Heston (far left) and Lloyd Law (far right), at a Jackson Laboratory "lobster bake," c. 1940s [Source Credit: Jackson Laboratory Archives].

In fact, however, Little bootstrapped JAX's genetic research agenda to the advancement of mouse production with only limited success, because great material investments were required. A stream of visiting scientists, such as cancer researcher Franscesco Duran-Reynals from Yale University, and the still-continuing summer student tradition suggest that JAX's mouse materials were highly valued by all types of biomedical experimentalists, and their reputation for research use remained mostly intact. By 1935, however, only thirteen loci had been mapped on five of the mouse's twenty chromosomes. From 1937 to 1942, seventeen new mutations were reported in mice, but only two of these occurred in JAX colonies. At the same time, according to the summer 1941 official photo of personnel (fig. 4.9), numbers of caretakers and support staff for these colonies (including now Henry Gordon, a business manager for the JAX production enterprise) rose at a rate faster than the number of staff geneticists.

During this early period of increased mouse production, Snell constructed breeding protocols for creating so-called congenic strains of mice (i.e., mice bred to be genetically equivalent to an inbred strain except for one selected differential chromosome segment). By virtue of controlling the genetic variables more precisely, these were strains that would stand to make murine genes much easier to manipulate and analyze experimen-

4.9. 1941 JAX Photo of staff, employees, and some students. From left to right, *first row*: Harry Gordon (business manager), George Snell, John J. Bittner, Arthur Cloudman, Elizabeth Fekete, C. C. Little, Elizabeth Russell, William Russell, George Woolley, Lloyd Law, Francesco Duran-Reynals. *Second row*: Ann Moore, secretary; Mary Simmons, assistant; "Mrs. Allen Savage," histologist; Elizabeth Keucher, secretary; Frank Clarke and Allen Salisbury (both caretaker/assistants, from the original JAX staff); Donald Harris and Everett Farley (both assistants); Katrina Hummel; Harold Woodworth, assistant. *Third row*: mostly students but sixth from the left is Lester Bunker, assistant. *Fourth row*: all students [Source Credit: Jackson Laboratory Archives].

tally, but Snell did not successfully produce these strains for another ten years.[64] Meanwhile, he took on projects that were designed to serve the needs of mouse production. In 1938, for example, Snell agreed to edit the first edition of *The Biology of the Laboratory Mouse*, an ambitious textbook-style synthesis of all available practical and technical information on laboratory mouse use. In Little's words, it was to be "a book of reference on the laboratory mouse for the use of the hundreds of research workers . . . who use this animal." The volume was authored by "The Staff of the Roscoe B. Jackson Laboratory," and virtually every chapter was written by a JAX Lab worker.

[64] "George Snell," *Current Biography*, May 1986, pp. 40–43; Cf. Margaret Dickie, "The Expanding Knowledge of the Genome of the Mouse," *Journal of the National Cancer Institute* 15 (1954): 679+; list of mouse mutants, arranged by year reported, in "Mouse Linkage Group/Mapping Info" Folder, JLA-BH.

In his introduction, Snell emphasized that the overriding aim of the book was not to argue for one type of research on the inbred mouse but to consolidate and "make available" all "established facts" on the practice of mouse work that cross disciplinary boundaries. Ideally, Snell wrote, such a "vertical cross-section of knowledge" about the mouse would encourage workers in different disciplines to pay more attention to one another: "It is a major purpose of this book, by gathering together the fundamental knowledge about the mouse from several fields of study, to make it easier for the research worker to traverse these interconnecting paths of science." Although there were literature review essays on the uses of mice in cancer and genetics research, the content of the volume was weighted heavily toward practical information for working with mice in laboratories. Such information ranged from the esoteric to the mundane. There was, for example, a pragmatic histological presentation of mouse tissues and mouse tumors, but also a general discussion of successful mouse husbandry arrangements. Observations included the statistical as well as the anecdotal. Thus Snell surveyed JAX colonies over a several-month period to determine what time of the day most mouse births occurred, while J. J. Bittner advised that he controlled mites or lice in his mouse boxes by sprinkling the organisms with a mixture of derris root and talcum powders. The only exception to these generalizations was geneticist William Russell's programmatic chapter, which elucidated the working assumption behind all of JAX Lab's practices: inbred strains of mice have widespread value for biological research.[65] Rockefeller Foundation's Frank Hanson heralded the publication of *Biology of the Laboratory Mouse* as an event "of value to all investigators who use mice." Indeed, the book served as what historian Steve Shapin has called a "literary technology":[66] conceptually complete and rhetorically rich in practical details, it codified a body of universal biological knowledge about laboratory mice at the same time that it increased access to formerly tacit knowledge shared only among JAX mouse workers and the people with whom they came in contact. Because nearly all of the information came from the hands-on experiences of JAX workers with JAX mice, *Biology of the Laboratory Mouse* also reinforced the equation of "JAX mice" with "standard mice." To use Bonnie Clause's computer analogy, the volume

[65] Little received outside funding ($5,800) for the preparation of the book from the John and Mary Markle Foundation of New York: on this and for Little quote, see Jackson Laboratory Annual Report, 1938–39, JLA-BH; George Snell, ed., *The Biology of the Laboratory Mouse* (Philadelphia: Blakiston Company, 1941), esp. p. viii. The only chapter not written by a JAX worker was the chapter on infectious diseases, by Harvard's John H. Dingle, JAX Lab's consulting bacteriologist for mouse production.

[66] Steven Shapin, "Pump and Circumstance: Robert Boyle's Literary Technology," *Social Studies of Science* 14 (1984): 484–520.

functioned as documentation but only for a specific type of software: for the information to be most useful, JAX mice had to be used.[67]

Another literary technology developed by Snell in the years following the NCI Act was *The Mouse Newsletter (MNL)*. The *MNL* had endured from Cold Spring Harbor since the 1920s, but it achieved more programmatic status in 1941, when Snell explicitly laid out its two primary aims and agreed to act as editor. The *MNL*, he wrote, would now strive to furnish a centralized listing of available mouse mutants for genetic study, in order to provide an official democratic forum for resolving questions about the standard nomenclature of these mutants.[68] It was to be "issued by the Roscoe B. Jackson Memorial Laboratory with the collaboration of investigators using mice in biological research," and not necessarily a permanent fixture: "future issues will appear when and if needed." The first volume of the news series, for example, was a mimeographed listing of all known inbred and mutant mouse strains, together with a list of laboratories and the specific stocks each maintained. For each inbred strain, Snell gave some information about its husbandry, as well as published references that detailed its specific uses. Ultimately, this hybrid format—something between an informal correspondence and a formal research presentation—proved useful for a wide range of inbred mouse users. JAX was deluged with twice the number of requests for copies that it had anticipated, and the first issue underwent a second printing a few years later.[69] Begun as a self-conscious attempt by mouse geneticists to standardize experimental nomenclature for their own research, the *MNL* also served to reinforce JAX's preeminence as a mouse supplier for cancer research (since, as *MNL* made clear, only JAX possessed stocks of every inbred and mutant strain).[70]

Medical researchers' perceptions of JAX as a "mouse house" were further bolstered by Little's apparent willingness to act as an animal provider

[67] FBH diary excerpt, 7 February 1939, RF Archives, RG 1.1, 200D, Box 143, Folder 1774, RAC-NY. cf. Clause, "The Wistar Rat," p. 345.

[68] Cf. form letter: "The Committee on Mouse Genetics Nomenclature" to "Biologists interested in Rodent Genetics," c. 1940, JLA-BH; "Rules for Assigning Symbols in Mutations," *Journal of Heredity* 31 (1940): 505–506. Relatedly, the first two volumes of the newsletter were officially titled "Mouse Genetics News."

[69] "Mouse Genetics News," no. 1 (ed. Snell), issued in November 1941 and again in November 1945, JLA-BH; Lloyd Law, "Mouse Genetics News," no. 2, *Journal of Heredity* 39 (1948): 300–307. The *Mouse Newsletter* officially became a separate trade publication again in 1949, although the listings of mouse strains and mouse mutants that it generated were periodically published in cancer research journals in the 1950s.

[70] Commercial mouse suppliers were not excluded from the early mouse newsletters, although the listings for such places as Carworth Farms in Rockland, New York, made it apparent that there was no scientist tending to the breeding of these organisms. See "Mouse Genetics News," no. 1, JLA-BH.

for even the most routine, nonresearch user communities. Starting in 1940, for example, JAX Lab contracted with area Maine physicians to conduct Ascheim-Zondek (A-Z) tests for pregnancy on JAX mice at 60 percent below what local commercial animal dealers charged. A-Z tests, which were devised by two German gynecologists, involved injecting large numbers of immature or virgin female mice with female patients' urine: if a mouse's ovarian follicles showed maturation within forty-eight hours, the test was positive. Little rarely (if ever) had JAX scientists doing A-Z tests, but in the political economy of the laboratory, he rationalized such work as a strategy to generate money that would benefit in-house research. In late 1941—just as Japan was invading Pearl Harbor—he told the lab's Board of Trustees that better marketing of JAX's A-Z testing program (along with JAX mice) would be a good way to ensure continued financial support for research during what he anticipated would be lean war years: "[W]ith the changing social and economic conditions throughout the country brought on by the present emergency there should be steps taken to make it certain we are not suddenly faced with a decrease in orders from our usual sources or lack of knowledge concerning emergency sources that might exist." In cancer research circles, however, such high-profile JAX activities arguably functioned to enhance the stature of the lab as a research-materials producer without bolstering its research reputation.[71] Perhaps the most symbolic culmination of this shift in institutional priorities came when in 1941 Little took the unprecedented step of registering "JAX Mice" with the U.S. Patent Office. JAX mice were now officially both a research material and a trademarked commodity: JAX Mice™.[72]

By 1946, Little wrote to Weaver that recent mobilization in medical research had affected JAX production positively, although inbred mice were not always needed for what Little vaguely described as "the war program": "As regards to the sale of surplus mice during the war, almost any stocks of mice were useful and we reproduced great numbers of cer-

[71] On the Ascheim-Zondek pregnancy test, see Selmar Ascheim and Bernhard Zondek, "Die Schwangerschaftsdiagnose aus dem Harn durch Nachweis des Hypophysenvorderlappenhormons," *Klinische Wochenschrift* (Berlin) 7 (1928): 8–9, 1404–11, 1453–57. Little's form letter describing JAX Lab's Ascheim-Zondek testing program (preserved only as CCL's scrap paper for the draft of a funding campaign letter!), c. 1940–44, Little Box #1 and Jackson Laboratory Annual Report, 1941, p. 4; both JLA-BH; Jackson Lab Financial Statement, 1943–44, RF Archive, RG 1.2, 200A, Box 133, Folder 1189, RAC-NY.

[72] "JAX Mice," Trade-Mark 387,519, registered 4 November 1940 and first used on 20 May 1941. Although the U.S. Plant Patent Act was passed in 1930 to protect the interests of commercial plant breeders, the JAX trademark predated the Wistar Institute's on its laboratory rodent—the WISTARAT—by a full year. See Glenn E. Bugos and Daniel J. Kevles, "Plants as Intellectual Property: American Practice, Law and Policy in World Context," *Osiris* 7 (1992): 75–104; Clause, "The Wistar Rat."

tain stocks which were not all of them closely inbred for forty or more generations." Little did not attach a description of where all these noninbred strains went, although he reported an income of $45,000 from "Sales of Supplies" and attached an endorsement letter from C. P. Rhodes, a cancer researcher at Memorial Hospital in New York. Frustratingly little of more precision can be said of the specific effects of World War II on JAX production and distribution. But Weaver later underscored Little's general conclusion about the war's positive effects in a report to Raymond Fosdick on JAX's production efforts (but again, not the specific research they enabled): "Their stocks were of critical importance during the war; and their activity [is] not only a first rate service to the whole of the country, but has grown to such dimensions that it returns the Laboratory a substantial net income per year." Little had cautioned Weaver not to read too much into JAX's wartime gains in the "surplus" animal category: "since the war," he wrote, "the pendulum is swinging strongly in the direction of the utilization of particular genetic strains of mice." But Weaver's take-home lesson was unclear and again reflected the institutional ambivalence with which JAX was viewed. Although Weaver told Little privately, "I am inclined to be more interested in the program than the plant," Weaver told his superiors that continued investment in the breeding facilities "constitutes a now essential part of their commercial supply service as well as their own basic research."[73]

CONCLUSION

By 1940 the extent to which JAX's mass production of inbred mice had begun to shape policy-making and experimental practices in life science research was already evident—notably, in biologists' discussions of the animal supply problem at the second NRC forum on the maintenance of biological research organisms in January 1940. Biologist and NRC Chairman Ross Harrison declared that the general purpose of the meeting, held twelve years after the first such gathering, was to once again "discuss problems involved in the production, maintenance, and supply of stocks of experimental animals and plants." But while mouse geneticists were well represented at the 1928 conference by Little and Dunn,

[73] CCL to WW, 3 May 1946; CCL to WW, 26 March 1946 ("marked rec'd May 20, 1946"), including 22 March 1946 letter from C. P. Rhodes to CCL, where Rhodes describes JAX's work as "complementary to Memorial Sloan-Kettering . . . [where] we must emphasize the application of highly technical procedures to the study of cancer in patients." Also, WW to CCL, 21 March 1946, and WW to Raymond Fosdick, 24 April 1946; all from RG 1.1, Series 200D, Box 144, Folder 1777, RAC-NY.

this time nongenetic mouse researchers from medical contexts were also invited. These scientists agreed with the geneticists that debates over what kind of standard mouse material should be preserved (inbred strains versus genetic mutants) were less urgent than questions about the proper infrastructure for distributing that material. As Little summarized the discussion midmeeting: "Both the geneticist and the man in experimental medicine . . . recognize what the other needs. . . . [T]he real handicap . . . [is] the machinery by which research can go ahead or plan in advance, knowing that the material it needs can be guaranteed to it with reasonable certainty."

Although there was some initial disagreement, Little persuasively argued that both *Drosophila* and mice workers had proven integration of material maintenance with the literary support technologies (like reference books and newsletters) was the most sensible approach. Zoologist L. J. Cole, for example, first noted that rigid centralization would be impossible: he argued "there is too much technique required" in keeping a standard strain "to expect it to be kept in a general laboratory for all purposes." L. C. Dunn then suggested more "regional agreements among investigators to see that certain stocks were maintained" that smaller groups found useful. But in contrast with his 1928 position, Little now told the committee that JAX's system worked precisely because it was primarily committed not to individual research projects but to being a centralized material repository: "a given strain [at JAX] . . . is not hitched up with the future of any one man. There is a good deal to be said in favor of more or less centralization under such adequate group control." Ultimately, Dunn and the others agreed that mouse material "continuity does not inhere in any individual," and that Little's approach appeared to be best in meeting researchers' needs.[74]

In 1941 the primary components of the "New Deal for Mice" were already in place. Over the next fifteen years, JAX's annual budget would remain relatively low in absolute dollars, but its mouse supply system would grow dramatically. By the mid-1950s, more than a quarter of a million mice were sold annually, or almost a 600 percent increase over the 1937 figure. The development of a user support system accounts, in large part, for the increased consumption of inbred mouse material during the latter era. Rockefeller "mouse houses" stabilized the material technol-

[74] "Conference on the Maintenance of Pure Genetic Strains," 27 January 1940, transcript of discussion (from which all quotes come) and supplementary materials in folder of the same name (incl. Division of Biology and Medicine), NAS/NRC Archives, Washington, DC. I am very grateful to Bonnie Clause for sharing this material with me. For a more detailed presentation of this debate from 1928 onward, see Rader, "The Origins of Mouse Genetics," pp. 2–4.

ogy of inbred mice production, which made inbred mice—by definition—more regularly obtainable and more reliable mouse organisms with which to work. At the same time, JAX was at the center of a rationalized network of practical support for inbred mice researchers, through creating (or, in the case of the *MNL*, sustaining) literary technologies to make their mouse products more easily obtainable, more accessible, and more useable for the general biological research community. The construction of a reliable mechanism through which researchers could obtain and use inbred mice helped transform the inbred mouse into a standard animal, both in terms of it being "widely available" and "widely used."[75]

Little himself recognized that inbred mice could be marketed as more broadly "useful" animals—for the American public, they were tools in the war against cancer, and for cancer policy, they were instruments of rational research coordination—and for this reason, their increased laboratory presence began to take on greater significance. America was about to enter World War II, in which science—in particular, physics—played a key role in both the military theater and the public imagination of the eventual Allied victory. After 1945, biology conducted according to the values of cooperation and coordination—the very values touted in the 1930s by science policy-making organizations from the Rockefeller Foundation to the National Advisory Cancer Council—would become an important resource for thinking about how to structure science and how to distribute the vastly greater resources available for laboratory work in the postwar world.

[75] Twenty-fourth Annual Report of the Roscoe B. Jackson Laboratory, 1952–53, p. 7; series from 1938 to the present is preserved in JLA-BH.

CHAPTER FIVE

RxMOUSE

JAX Mice in Cancer Research (1938–55)

In December 1938, after several months delay, the National Advisory Cancer Council published its agenda for a program of "Fundamental Cancer Research" in the *Public Health Reports*. The discussions leading to this document took place after the initial decision to fund the mass production of inbred mice, but final recommendations were made by a subcommittee, two members of which were JAX-affiliated scientists: C. C. Little and James B. Murphy. Not surprisingly, the report highlighted fields that would maximize consumption of JAX materials. But while the authors cautioned that "in any program for cancer research, patience and the adoption of a long-time point of view are essential," they also declared the unequivocal importance of genetics: "somewhere between the inherited or acquired cell tendency and the factor which releases this tendency lies the crux of the cancer problem." Three subfields, they argued, could best address this question: transplantable tumors, tumor induction, and spontaneous tumors. Of Bittner's landmark "milk influence" work, the subcommittee concluded, "it [is] considered established that mammalian cancer [is] not infectious" and grouped viruses with other micro-organisms as etiological agents that could be disregarded.[1]

[1] "Fundamental Cancer Research: Report of A Committee Appointed by the Surgeon General," *Public Health Reports* 53 (1938): 2121–30. My research indicates that there is no available transcript of this subcommittee's deliberations

5.1. Gathering of mouse genetics and cancer researchers at the Jackson Lab, c. 1938. *Front row,* second from left is Elizabeth Fekete, fifth from left is L. C. Strong; *Third row,* seventh from the left (with darkest suit and glasses) is Howard Andervont; *Fourth row,* fourth from left: Arthur Cloudman; *top row,* third from left: C. C. Little, then George Snell, William Russell, Walter Heston.

These recommendations appear to have had dramatically different institutional consequences at JAX and other federally sponsored laboratories. At the Public Health Service's Harvard Cancer Laboratory, mouse mammary tumor research flourished in the period between 1938 and 1948. In 1938 Howard Andervont began a series of experiments designed to replicate and elaborate on Bittner's results, and he was asked to set up an in-house transplantable mouse tumor lab at the new NCI compound under construction in Bethesda.[2] Andervont's program centered on foster-nursing cancerous strain and hybrid strain offspring, a procedure that assumed inbred mice embodied new disciplinary standards for tumor stability. By contrast, at JAX the sale and distribution

[2] "H. B. Andervont," *Annual Obituary* (New York: St. Martin's Press), 1981, pp. 171–72. Cf. Andervont, "The Influence of Foster Nursing on Incidence of Spontaneous Mammary Tumor Cancer in Resistant and Susceptible Mice," *Journal of the National Cancer Institute* 1 (1949): 147–53.

of inbred mice was at an all-time high (110,000 mice were distributed in 1939, a more than 45 percent increase over 1937), but morale had eroded among the remaining original group of researchers. Bittner, in particular, continued to receive small grants (from both the NCI and the Anna Fuller Fund) to support his continued research on the milk problem. In 1939 he put forth a multifactorial theory of three "influences" on the development of breast tumors in mice: (1) inherited susceptibility; (2) adequate hormonal stimulation; and (3) "an active influence in the milk." But by 1943 members of the NACC rated Bittner's projects of secondary importance to the federal cancer research effort and almost didn't fund him.[3] Thus while JAX mice seemed to be selling and Bittner's "milk problem" provided a fruitful area for other cancer researchers, JAX researchers were unable to make the most effective laboratory use of the inbred mouse materials and cancer research theories that they themselves had developed.

The "received view" of what Elizabeth Russell called JAX's "tiny and still struggling" research program in the period between 1930 and 1950 attributes this problem to Little's leadership—more specifically, to his stubborn refusal to engage in work that might overthrow the genetic theory of cancer. As geneticist James Crow summarized: "Little was interested in cancer and in genetics. He thought the way to study cancer was through heredity, the way to study heredity was in the mouse, and that the way to study mice was with inbred strains." Jean-Paul Gaudillière has suggested that Little prevented Bittner from fully developing a research program on "milk influence" because experimental work on the meaning of nursing-induced mouse mammary tumors stood to threaten JAX mouse production and sales. Contradictory results might challenge JAX marketing claims that its animal commodities were genetically stable, and the institution's reputation as a supplier could be ruined.[4]

But while genetics was "in" and viruses were "out" in cancer research throughout much of the 1940s,[5] the appeal of the genetic approach clearly extended beyond Little's circle of influence. Kenneth DeOme, a cancer

[3] "Report to Those Interested in the Work of the Roscoe B. Jackson Memorial Laboratory," enclosed in letter from CCL to WW, 7 July 1939, RF Archives, RG 1.1, Series 200, Box 143, Folder 1774, RAC-NY. On Bittner's recollection, see Holstein, *The First Fifty Years*, p. 93.

[4] James Crow, "A Century of Mammalian Genetics and Cancer: Where Are We at Mid-Passage?" in *Mammalian Genetics and Cancer: The Jackson Laboratory Fiftieth Anniversary Symposium*, ed. Elizabeth Russell (New York: A. R. Liss, 1981) pp. 309–24; Gaudillière, "Circulating Mice and Viruses."

[5] Daniel J. Kevles, "Pursuing the Unpopular: A History of Courage, Viruses, and Cancer," in *Hidden Histories of Science*, ed. Robert Silvers (New York: New York Review, 1995), pp. 69–114.

researcher trained during the 1930s, remembered: "Genetics was just the big thing to study. . . . People believed that genetics was the controlling factor in everything. No proof, but everybody believed it."[6] As director of the American Society for the Control of Cancer, Little had a long history of publicizing cancer's genetic links, and as founder of JAX, he surely would have wanted to spare his institution increased financial hardship from loss of mouse sales. Still, profound administrative and intellectual tensions shaped relations among JAX scientists, inbred mice, and the government-funded laboratory attack on the cancer in the years from 1938 to 1950. Seen in this light, Little's influence might be more accurately viewed as reflecting, rather than constituting, the new politics of work emerging in the field of cancer research.

Although inbred mouse usage flourished in cancer studies from 1938 to 1950, Little's envisioned "New Deal for Mice" failed to take hold during this period. Both at NCI and more broadly, inbred mice became standard laboratory tools: by 1940, the increased availability of inbred mice made possible by the NCI, as well as consumption of mice in wartime vaccine research, transformed tumor transplantation and foster-nursing protocols into dominant experimental methodologies. JAX researchers stressed particular laboratory and clinical interpretations for this inbred mouse work. Little focused on cancer education for the American Society for the Control of Cancer (ASCC), and public interest in cancer genetics grew, even while many inbred mouse researchers still pursued other theories of cancer causation (as part of a viral resurrection in experimental cancer studies).[7] NACC discussions of JAX's work, however, increasingly focused on the persistent tension in the relationship between animal production and cancer research at JAX: should the government invest directly in making mice, policymakers wondered, or should the NCI support JAX researchers as an indirect guarantee of the availability of mouse materials? Initially the NCI saw research support for JAX as a prudent quality control measure, but as federal cancer research policy gradually shifted toward valuing centralization, the government became increasingly reluctant to support any experimental work at JAX on the grounds that it duplicated their "in-house" efforts. To protect JAX research, Little attempted to develop long-term partnerships with other university research centers, but these efforts failed. Because the NCI still held significant fi-

[6] As quoted in Fujimura, *Crafting Science*, pp. 25–26.

[7] On the viral resurrection in cancer research—and the role of new lab instruments in hastening this development—see Angela N. H. Creager and Jean-Paul Gaudillière, "Experimental Platforms and Technologies of Visualization: Cancer as a Viral Epidemic, 1930–1960" in *Between Heredity and Infection*, ed. I. Lowy and J.P. Gaudilliére (London: Routledge, 2001), pp. 203–41.

nancial leverage over JAX, their decision to cut Bittner's research funding effectively wrenched inbred mouse materials from the institution that had given birth to them. At the same time, JAX lost many of its researchers to the NCI and other medical schools. Rockefeller Foundation came to the rescue. Little obtained funding to build new stock centers for other mammals (rabbits, guinea pigs, and rats), and once again to scale up inbred mouse production. But in the view of researchers and policymakers alike, this signaled at once the failure of Little's original biomedical research vision, even while it virtually guaranteed JAX's future as a materials supplier. Once again, JAX mice seemed to be succeeding, but not in the ways that Little had hoped.

After 1945, wartime models of medical treatments, public distaste for eugenics, and lay activism in the cancer lobby combined to change the social meaning of "successful" biological research—from esoteric basic investigations that illuminated disease mechanisms to experimental studies that quickly led to new clinical therapies. The sea-change in biomedical research policy that resulted represented the final blow to Little's Progressive vision of cancer research with inbred mice and represented an intensification of previous trends. In the wake of the 1947 Bar Harbor fire, the Jackson Lab was rebuilt through the efforts of a classic postwar trilogy of 'Big Science' actors—laboratory researchers, science policymakers, and members of the lay public. In 1952 Little commissioned an in-house publicity film entitled *R_xMouse* to mark its resurrection. As pictures of researchers working with inbred mice passed across the screen, Little's confident voice declared: "Just as the purity of the chemical assures the pharmacist of the proper filling of the doctor's prescription, so the purity of mouse stock can assure a research scientist of a true and sure experiment."[8] But despite some promising leads, no work on inbred mice had yet proven useful for doctors in the cancer clinic. Likewise, Little's chosen research materials had lost their metaphorical power as the "prescription" to treat American biomedicine's organizational and intellectual woes, and they would not recover it again until the mid-1950s when the government planned a massive chemotherapy screening program with these animals as its subjects. What Little had hoped would be an era of unfettered progress in the war against cancer, with heterogeneous coalitions centered around laboratory work with inbred mice, became instead a transitional period in American cancer research, in which the increased

[8] *Rx Mouse*, videotape of original ten-minute film. Special thanks to Grady Holloway for digging this out of the JAX archives and preserving it on videotape. There is almost no archival documentation on the making of the film, but because of its short length and popular message, it was likely designed to be shown to JAX Lab visitors before regular biweekly public tours.

5.2. Mouse Production Wing, c. 1940 [Source Credit: Jackson Laboratory Archives].

use of inbred mice did more to facilitate a simple convergence of experimental techniques than to consolidate the increasingly complex social and scientific goals of U.S. biomedicine.

INBRED MICE AS CANCER RESEARCH TOOLS: AN OVERVIEW OF USAGE, 1937–1947

In 1940 the triumph of mass production appeared to be a fait accompli at JAX. Little sent the Rockefeller Foundation pictures for their *Annual Review* that placed in the foreground the scaled-up, industrialized aspects of mouse breeding the building project embodied, including a view of the new production wing (fig. 5.2). But foundation officers and trustees saw more important evidence that their investment in mouse supply was paying off in the numbers. In the first six months the building was operational, mouse production soared 53 percent, while JAX made significant income relative its total operating costs. Annual profits from mouse sales doubled between 1936 and 1939, from $7,000 to $14,000 (nearly a quarter of the JAX annual budget). The March 1940 *Rockefeller Trustees Bulletin* gleefully reported that Little had "outwitted the old proverb that you can't eat your cake and have it too," but then explained that this meant simply that "the advantage" of the newly centralized mouse supply system was not "all on one side: the service rendered was highly appreciated by outside biologists."[9]

By 1947 JAX inbred mouse products were not universally used by all mouse researchers in the life sciences, but JAX's distribution network and product recognition had generated a reliable constituency, especially in cancer research. A survey of the "materials and methods" sections of papers published in the journal *Cancer Research*[10] indicates that mouse use (inbreds and noninbreds) was initially strong and rose slightly in experimental cancer studies in the decade immediately following passage of the NCI Act. In 1935, work featuring mice represented 27 percent of the total research articles published; by 1945 it represented more than 42 percent. But while this general upward trend held from 1920 to 1951, the percentage of mouse users (in comparison with those who did research on other experimental organisms like rats or plants) appears to have fluctuated greatly from year to year, probably reflecting random contextual factors in the journal's publication (e.g., editorial decisions, nature of submissions, and the logistical difficulties of doing research during wartime).

Perhaps a more significant measure of the JAX mouse's success, then, might be the number of mouse cancer researchers who used *inbred* mice in their studies rather than what NCI policymakers referred to as "mongrel strains."[11] Focusing on the "materials and methods" descriptions of their animals in the *Cancer Research* article texts, the historian can infer from such variables as whether researchers simply called them "white mice" or used JAX-standard inbred strain names. Here the numbers tell a more dramatic tale: the use of inbred mice rose from less than 1 percent of all studies published in this journal in 1932 to more than 30 percent in 1937,

[9] FBH to CCL, 3 December 1938; CCL to FBH, 13 December 1938; CCL to Warren Weaver, 7 July 1939; George Gray to CCL, 23 February 1940; CCL to GG, 25 February 1940, where Little notes: "Every one of the mice raised" is used by geneticists because each mouse "is used as material to observe whether or not mutations have occurred. We found *two* this past year" (emphasis in original); all RF Archives, 200D, Box 143, Folder 1774; "Eating Your Cake and Having It Too," excerpt from the *RF Trustees Bulletin*, March 1940, RF Archives, RG 1.1, 200D, Box 143, Folder 1775; all RAC-NY.

[10] Due to the unstable financing, *Cancer Research* (1941–present), the official journal of the American Association of Cancer Research, went through two previous incarnations: *Journal of Cancer Research* (1919–28) and *American Journal of Cancer* (1928–40). Since ASCR always maintained editorial control, my article count reflects my consideration of all three of these in series as the same journal. On the history of the AACR publications, see Victor Triolo's lively account in Victor Triolo and I. I. Riegel, "The American Association for Cancer Research, 1907–1940: Historical Review," *Cancer Research* 21 (1961): 137–67.

[11] Ora Marshino, "Administration of the National Cancer Institute Act, August 1937 to June 1943," *Journal of the National Cancer Institute* 4 (April 1944): 429–443.

and again from 1937 to 1947, to nearly 70 percent of all studies. Not all of these mice came from JAX—indeed, in the 1930s, many cancer researchers described where they got their animals, but just as many did not and chose to use simple generic descriptors, such as "laboratory stock." This rhetorical development is telling, however, for it suggests an incomplete transition in the scientific etiquette governing the exchange of biological research materials: from the "gentlemen's agreement" (where acknowledging the source of animals or tissue samples was as important as acknowledging the sources of data or theories) to a marketplace model (where the source of materials is relevant to, but less important than, their presumed standardized quality). The fact that JAX nomenclature was increasingly used to describe strains of mice used suggests that Little's decision to make progressively larger investments in mouse production had combined with NACC policy to recast both JAX's institutional identity as a materials producer and the meaning of inbred mouse mice as "the standard" cancer research animal.

Further analysis of what was being done with these mice once they got into the laboratory indicates that, although diverse approaches to studying mouse cancers persisted, inbred strain availability encouraged a convergence of laboratory techniques. At the same time that the use of JAX-type mice among mouse cancer researchers was on the rise, work on tumor transplantation and foster-nursing studies also escalated (as measured by the percentages of such experiments among mouse studies published by *Cancer Research*). In 1936, for example, these approaches combined to make up 66 percent of the work, while eleven years later, in 1947, they made up 98 percent of the mouse cancer work published. Taken together, the simultaneous rise of inbred mouse materials and tumor transplant or genetic methodologies demonstrates that federal investment effected a co-construction of animal tool and research problem in cancer laboratories from 1936 to 1947. Some diversity persisted in mouse cancer research: in any given year between 1937 and 1947, as much as 10 percent in of inbred mouse experiments published by *Cancer Research* were studies of chemical actions on tumors (for example, painting tumor-prone mice with tar, in order to induce cancer more quickly), and an average of 20 percent were observations of spontaneous tumor behaviors in mouse strains. Certainly, noncancer research on inbred mice was also on the rise: in a letter to Weaver in 1937, Little listed at least seven other kinds of "experimental work" for which inbred mice were used. These included research on pneumonia, yellow fever, and influenza by F. H. Bauer at the U.S. Army Medical College, and tissue culture studies by Warren Lewis at Johns Hopkins. Furthermore, working with albino mice of unknown origin at Harvard Medical School, Max Theiler and his colleagues proved in 1930 that the yellow fever virus could be readily transmitted to mice—a much more cost-effective organism with which to do the vaccine research in which the Rockefeller Foundation itself had shown great interest.[12] So while there was significant confluence of animal tool and experimental approach, this relationship was not deterministic: inbred strains simultaneously provided the means for developing and refining new techniques in cancer genetics, as well for better controlling experimental variables in the investigation of old problems, like the behavior of infectious diseases.[13]

During the period just before and during World War II JAX stock lists evolved to reflect just such a heterogeneous clientele of scientist-users. Above all Little strove to keep purchasing from JAX simple. The 1938 "Stock List" was divided into two major sections: "Inbred Stocks Available in Quantities," which included the five major cancer and noncancer strains: *C57Black*, Bagg albino, *A*, *C3H*, and *dba*, and "Stocks Available for Genetic Tests, etc.," which included some mutant variations obtained within cancer strains, like *L* (a nonagouti brown, from the *C57Black*s) and *W* (shaker-2, wavy-2 mice). The descriptive table format from the initial mimeographed JAX sales lists remained, with columns for possible experimental variables of interest to the scientist-consumer: "No. of brother x sister generations," "Color and Description," "Genetic Formula," "Fertility," and "Remarks" (which included cancer incidence). Ten years later, however, more frequently updated mimeographed listings replaced these tables, and the general descriptions of the possible experimental variables were replaced by a list of relevant published, peer-reviewed scientific papers on the experimental behavior of each of thirty-three strains available. Notably, now the JAX listing also included the "Swiss strain," a type of albino developed by Clara Lynch at the Rockefeller Institute, who obtained them from A. de Coulon of Lausanne, Switzerland around 1926. With virologist colleagues such as T. P. Hughes at the RIMR, she initially used these animals to develop a simple immunity test for the presence of yellow fever in a human population. Still, because the JAX stock lists were now indexed entirely according to the material (the particular JAX mouse strain) rather than by the material's specific disciplinary uses, such mini-bibliographies were an efficient means for inbred

[12] In 1951 Theiler was awarded the Nobel prize in physiology and medicine for this work: see his Nobel biography and lecture at www.nobel.sc/medicine/laurcates/1951/theiler-bio.html (accessed 15 June 2003).

[13] On the co-construction of tools and research problems, see the introduction to Clarke and Fujimura, *The Right Tools for the Job*, as well as Barbara Kimmelman's excellent essay on R. A. Emerson's experimental "ideology." CCL to WW, 4 December 1937, RG 1.1, 200D, Box 143, folder 1774, RAC-NY. Cf. Joan Fujimura, "Standardizing Practices: A Socio-History of Experimental Systems in Classical Genetic and Virological Cancer Research, ca. 1920–1978," *History and Philosophy of the Life Sciences* 18 (1996): 3–54, esp. 19.

mouse researchers to find out the whole range information about a given strain without having to search the specialized literature of scientific fields outside their area.[14]

MOUSE RESEARCH INTERPRETATIONS: GENETIC VERSUS VIRAL THEORIES OF CANCER IN THE LABORATORY AND BEYOND

Diversity in the approach of inbred mouse users at the biomedical research benchtop is less surprising if placed in the context of the mid-1930s' debate over cancer causation. This debate was more active than one might assume from paying attention only to JAX researchers, Little, and his circle of American cancer policymakers. Former NCI researcher Michael Shimkin argued that "[b]y 1940 cancer research was laboring under some premature and unfounded conclusions . . . one of which was that cancer was not an infection and that virus research in cancer was a waste of time." NACC members contributed to this general impression by urging researchers and the public to proceed with extreme caution toward any alternative hypotheses and results. For example, in 1911 Peyton Rous of the Rockefeller Institute of Medical Research reported cancer could be induced in healthy chickens by injecting them with a cell-free extract of the tumor of a sick chicken; in retrospect, this is often called the first laboratory demonstration of a cancer virus. But in a 1938 evening lecture to the National Academy of Sciences, James Ewing, director of the Memorial Hospital in New York, downplayed these results as suggestive but not definitive and stressed that Rous showed only that "multiple factors may be concerned in the origin of some cancers." In 1943, James Murphy, Rous's former assistant and colleague at the RIMR, summarized his own views on the matter in no uncertain terms: "There is insufficient indication at the present time that viruses play any important role in the general picture; therefore, no attempt will be made to discuss at length the possible relation of this group of agents to cancer."[15]

[14] 1937 stock list, enclosed in CCL to WW, 4 December 1937, RF Archives, RG 1.1, 200D, Box 143, Folder 1774, RAC-NY; 1947 stock list: "List of Inbred Strains Maintained at the Roscoe B. Jackson Memorial Laboratory," 1 January 1947, in Box of the same name at JLA-BH. On the history of the Swiss strain and Lynch's research on them, see "JAX Mouse Data Sheet: SWR/J; Stock Number 000689" at jaxmice.jax.org (acessed 14 February 2003); and Corner, *A History of the Rockefeller Institute,* pp. 222–23, 308. On Rockefeller Foundation's prewar interests in yellow fever research, see relevant documents from RG 5. *International Health Board, 1911-(1913–1927)-1951*, RAC-NY.

[15] Shimkin, *Contrary To Nature*, p. 35. On CCL at the International Congress, see *New York Times*, 30 August 1932, p. 10. Cf. James Ewing, "The Public and the Cancer Prob-

At the Jackson Laboratory, researchers initially focused on what Jean-Paul Gaudillière and Ilana Löwy have called "domesticating" the milk influence, with the goal of preserving a realm for cancer genetics research.[16] Nearly all of this work focused on transplantation, rather than on the spontaneous tumor studies with which the group had begun back in Michigan. With Little traveling, raising funds for JAX, and administering ASCC business, J. J. Bittner emerged as the team's leader. As a person, he was fondly remembered for "the exceptional brightness and intelligence of his eyes and his inimitable charm"; as a scientist, he was described as a "gifted researcher" with the patience for making "painstaking observation" and "keen intuition" for elegant experiments.[17] In 1938 he set out a comprehensive plan to analyze the multiple action and interaction of genetic susceptibility, hormonal stimulation, and milk agent in the development of mouse mammary tumors. The Jackson Lab, Bittner argued, should focus initially on the third, using three inbred strains it had developed—*dba*, *A* (both high tumor) *C57Black* (low tumor). One of Bittner's early projects demonstrated that the milk influence could be transmitted to noncancerous strains through normal tissue transplant (that is, inoculation with the normal spleen, thymus, or mammary tissue of cancerous strain animals), rather than through nursing.[18] Perhaps with an eye toward their ultimate marketability as a "standardized package" in JAX's production enterprise, another early project sought to create a series of stable inbred strains that were (as Bittner called them) "susceptible but agent-free": for example, *dba* progeny nursed on *C57* mothers. These presumed stable strains could then be used to test relative influences of genetics and the milk effect on tumor expression. Walter Heston, a recent Ph.D. who studied genetics with L. C. Cole at Michigan State, joined the JAX staff in 1938 as one of the thirteen original National Cancer Institute fellows, and he spent two years working on this and other projects with Bittner.[19]

Little also wanted JAX to pursue the two alternative influences, along with the hypothesis (detailed further in the work of Lacassagne in France)

lem," *Science* 87 (6 May 1938): 399–407; James B. Murphy, "An Analysis of Trends in Cancer Research," *Journal of the American Medical Association* 120 (1942): 107–11.

[16] Lowy and Gaudillière, "Disciplining Cancer."

[17] Lucio Sevari and Franz Helberg, "Obituary: John J. Bittner," *Nature* 197 (7 February 1963): 539–40.

[18] J. J. Bittner, "The Influence of Transplanted Normal Tissue on Breast Cancer Ratios in Mice," *Public Health Reports* 54 (1939): 1827–31.

[19] Bittner, "Breast Cancer in Mice," p. 44; and "Relation of Nursing to Extrachromosomal Theory," p. 90. Cf. Walter Heston's review—"Milk Influence in the Genesis of Mammary Tumors"—in *A Symposium on the Mammary Tumors of Mice*, ed. F. R. Moulton (Washington, DC: AAAS, 1945), pp. 123–39, esp. table II, p. 127. Cf. A. Lacassagne, "Rela-

that mammary cancer influence could be transmitted through hormones. Through private medical research foundations, he obtained funds for two postdoctoral fellowships and hired traditional mammalian geneticists who showed a keen interest in doing work with more direct applications to human health and illness: Lloyd Law (a recently minted Ph.D. from the Bussey Institute, where he worked on size inheritance in mice) and George Woolley (a 1935 graduate of the University of Wisconsin, who had written a thesis on genetic history of cattle). Woolley and Law worked together on the blood; their experiments showed that when foster-nursed JAX *C3H* mice—one of Bittner's "susceptible but agent free" strains—were injected with the blood of normal high-tumor *C3H* males, these animals had a significantly greater incidence of tumors.

Meanwhile, Elizabeth Fekete worked with Little to design a series of experiments to demonstrate the effects of the maternal physiological environment by the technically challenging procedure of creating mouse surrogate mothers. Fekete transferred fertilized low-cancer (*C57Black*) ova from the fallopian tubes of pregnant mice to the uterus of high-cancer (*dba*) mothers. This work showed a nearly 70 percent increase in tumor incidence among the low-cancer strains both born to and nursed by high-tumor mothers—an effect Fekete argued was significantly "greater than has been accomplished by foster nursing along in these two strains."[20]

Fekete also worked on the problem of spontaneous tumors, most notably by demonstrating the positive effect on tumor development of mechanically blocking the mammary ducts in the *dba* strain. Likewise, Arthur Cloudman—an original JAX staffer who returned to Bar Harbor in 1935—conducted a series of histological analyses of mammary tumors in *dba/C57Black* Hybrid, demonstrating the differential cellular origins of spontaneous tumors in F2 progeny (nonepithelial) from virgin female tumors (epithelial). "This finer analysis of the tumors at the site of the mammary glands," Cloudman concluded, "serves to strengthen the theory of extrachromosomal influence in the incidence of spontaneous tumors." Later Cloudman extended this approach to a study of the relationship between transplanted liver tumors in high-cancer "leaden" mice, which arose as a mutation in the JAX *C57* strain.[21]

---

tionship of Hormones and Mammary Adenocarcinoma in the Mouse," *American Journal of Cancer* 37 (1939): 414–24.

[20] George Woolley, Lloyd Law, and C. C. Little, "The Occurrence of Whole Blood of Material Influencing the Incidence of Mammary Carcinoma in Mice," *Cancer Research* (1941): 955–56; Elizabeth Fekete and C. C. Little, "Observations on the Mammary Tumor Incidence of Mice Born from Transferred Ova," *Cancer Research* 2 (August 1942): 525–26.

[21] Elizabeth Fekete and Charles Green, "The Influence of Complete Blockage of the Nipple on the Incidence and Location of Spontaneous Mammary Tumors in Mice," *AJC* 27 (1936): 513–14; A. M. Cloudman,"Gross and Microscopic Diagnoses in Mouse Tumors at

Cloudman was also put in charge of genetically typing all the transplantable tumors JAX held, and this led to a productive collaboration on a more theoretical genetics project with George Snell. In 1944 Snell began developing a series of what he called "congenic lines" of mice to study the genetics of tumor transplantation. Using repeated backcrossing of one inbred mouse strain (the donor) to another (the inbred partner), Snell wanted to developed lines of mice that differed at only one genetic locus. He planned to use Cloudman's typed tumors in what he called "immunogenetic" experiments that would allow him to evaluate a tumor incidence variation against a stable genetic background. As Snell remembered later: "I worked on this as a problem in basic research. I was asking what these genes are like, how many are there, and what their function is." Meanwhile, Snell also took on one of the most challenging, but routine, tasks related to the JAX colonies: "I became involved with problems of gene nomenclature and this, together with strain nomenclature, remained a concern for many years."[22]

Plans at the NCI initially expanded on only one theme: reciprocal foster-nursing studies, to explore the effect of the milk influence on a larger scale. Andervont appears to have taken some pains not to compete with Bittner. Notably, his foster-nursing protocols were designed around the *C3H* strain—rather than *dba* and A—and he obtained his mice from L. C. Strong at Yale, rather than buying stock *C3H* animals from JAX (from whom he was already buying other animals). He later explained this was because Strong's *C3H*s showed a more stable high-cancer incidence, but Andervont's behavior also suggests he was being exceedingly careful to distance use of JAX cancer research materials from replication of JAX cancer research results.[23] When Andervont and his group at NCI moved to Bethesda in 1940, insiders soon came to refer their operation as the NCI "mouse dairy": by 1941, Andervont's *C3H* and *C57Black* colonies numbered in the thousands.[24] The scale of the foster-nursing experiments grew accordingly: in 1940, for example, Andervont claimed to have replicated Bittner's conversion of *C57 Blacks* (this time, with the

the Site of the Mammary Glands, *AJC* 27 (1936): 510–12. Cf. Cloudman, "Organophilic Tendencies of Two Transplantable Tumors of the Mouse," *Cancer Research* (September 1947) 585+.

[22] G. Snell, "Studies in Histocompatibility," *Nobel Lectures* (8 December 1980): 645–60; Silver, *Mouse Genetics*, pp. 44–45. Snell quotes, respectively, from Robert Cooke, "A Persistant Pioneer in Research," *Boston Globe*, 11 October 1980, and George Snell, "Autobiography," *Les Prix Nobel* 1980, online at www.nobel.se/medicine/laureates/1980 (accessed 31 July 2002).

[23] Strong, "The Origins of Some Inbred Mice," *Cancer Research* 3 (1942): 531–39.

[24] Occasional references to the NCI 'mouse dairy" are made throughout NACC discussions; cf. Heller, "The National Cancer Institute," esp. p. 195.

C3H high-cancer strain instead of the A-strain) in a study based on approximately one hundred mice, where Bittner's original study was based on only fifty animals. Bittner, in turn, increased the scale of his experimental colonies at JAX: reports from 1942, for example, describe several hundred mouse matings per hybrid strain. Although presumably such projects competed with space allotted to mouse production for outside sales, Little initially supported Bittner's research, and for at least a few years, he encouraged Bittner's attempts to make it more efficient. In particular, several JAX staffers worked with Bittner to develop what was described at the NACC as a "mechanical nursing machine for mice." But mouse breast pumps never came to be, and consequently, neither JAX nor NCI researchers succeeded in fully industrializing large foster-nursing experiments.[25]

Other kinds of research with inbred strains went on at NCI research centers, with varying degrees of commitment to the necessity of a genetic inbreeding framework. Biochemist Murray Shear, for example, arranged a collaboration with Harvard chemist Louis Fieser to study the carcinogenic properties of polycyclic hydrocarbons isolated from tar, as evidenced by the incidence of various tumors they caused when applied to mice. Fieser and J. W. Schereschewsky (Andervont's titular supervisor at Harvard) reached a formal agreement to ensure the project would be done with Andervont's established strains, although Michael Shimkin recalled: "Fieser never did fully accept inbred mice for the work, considering them somewhat abnormal." Likewise, Floyd Turner, a commissioned officer of the PHS sent to head the Boston group upon Schereschewsky's retirement, had just witnessed his wife die of breast cancer and so was personally determined to direct the lab's work to therapeutic ends. For him, this meant rejecting inbreeding and transplanted tumor methodologies for work on individual animals that had developed spontaneous tumors. Descriptions of Turner's modus operandi in the laboratory evoke the same disdain for the impersonal, manipulative breeding trials and statistical analysis of JAX-type mouse cancer work that Maud Slye continued to express: "Transplanted tumors he used only grudgingly. [But] every source of mice with breast lumps was mobilized, and every mouse started on one chemical or some other preparation that same day, Sundays alone excepted. Turner handled every mouse and made every injection himself, with an assistant sometimes holding the beast."[26]

Despite the coexistence of competing modes of practice with inbred mice at the Boston research center, for nearly seven years JAX and NCI

[25] NACC transcript, 3 April 1939, pp. 117–18, RG 443, Box 7, NACC-MD.

[26] On Shear's and Turner's laboratory work in the Boston NCI center, see Shimkin, *As Memory Serves*, pp. 15–16 and 18. Cf. Murray J. Shear, "Studies on the Chemical Treatment of Tumors," *Journal of Cancer Research* (1935): 66+.

scientists appear to have proceeded in an almost idealistic cooperative fashion in analyzing the genetics of the milk influence. Indeed, even experimental anomalies proved easily accommodated within their shared theoretical and practical framework. In 1941, for example, Bittner reported the sudden appearance of tumor agents in some of his *A*-strain progeny nursed on *C57Blacks*. Bittner interpreted this variation in cancer incidence as a "mutation" in one of the genes modifying mammary tumor susceptibility.[27] Similarly, throughout the 1940s Andervont observed remarkably divergent tumor rates in foster-nursed, noncancerous mice, ranging from complete failure to induce tumors to conversion rates as high as 63 percent. But rather than challenge the genetic assumptions undergirding the model, he worked to establish the existence of "sublines" between the JAX (Little) and NCI (Strong) inbred mouse colonies. In 1941 Andervont demonstrated that his *C3H* strain showed that tumors arising in the NCI strain did not grow in *C3H* mice from JAX. "The results obtained with it," he concluded, "are not comparable with those procured in other lines of the strain *C3H*." Similarly, in 1945 when his group observed decreases in the incidence of breast cancers among the progeny of backcrosses with *C57/C3H* hybrids, Andervont suggested that different genes governed transmission and propagation of the milk agent, rather than challenge the genetic stability of the mouse strains themselves.[28]

Few other researchers had the mouse breeding capabilities to replicate JAX and NCI foster-nursing results, but they underscored the value of genetic inbreeding in other types of experiments with inbred strains. Halsey Bagg, for example, extended Fekete's work on the effects of mammary gland blockage on spontaneous tumors in four inbred strains of mice: low-tumor CBA and *C57* and high-tumor *dba*s (all of which he obtained from JAX), and low-tumor JK strains (available at JAX but obtained from L. C. Strong).[29] Also, Clara Lynch embarked on a program in the mid-1920s to investigate what she observed to be "the marked differences in susceptibility" to lung and skin tumors in Swiss mice and other inbred strains. Lynch's lung tumor strains bore her own idiosyncratic names—

[27] John J. Bittner, "Changes In The Incidence Of Mammary Carcinoma In Mice Of The A Stock," *Cancer Research*, 1 (1941): 113–14. Cf. A. Kisrchbaum and L. C. Strong, "Transplantation of Leukemia Arising in Hybrid Mice," *Cancer Research* 1 (1941): 785–86.

[28] H. B. Andervont, "Spontaneous Tumors in a Subline of Strain C3H Mice," *Journal of the National Cancer Institute* 1 (June 1941): 737–43; W. E. Heston, M.K. Deringer, and H.B. Andervont, "Gene-Milk Agent Relationship in Mammary Tumor Development," *Journal of the National Cancer Institute* 5 (1945): 289–307. For a more detailed analysis of the circulation of *C3H* mice, foster-nursing protocols, and their relationship to later NCI work on mouse leukemias, see Gaudillière, "Circulating Mice."

[29] Halsey Bagg, "Further Studies on the Relation of Functional Activity to Mammary Carconoma in Mice," *American Journal of Cancer* (1936): 542+.

e.g., "Strain no. 1194" or "no. 62"—but in her "materials and methods" sections she went to great lengths to describe the lineage of her other strains (e.g., Bagg albinos) with existing inbred lines, and her conclusion could not have been more amenable to JAX practices: "more than one pair of Mendelizing factors must be involved, though the exact number is underdetermined." Lynch also circulated her mice to women graduate students in biology at Bryn Mawr College, thus creating a subnetwork of inbred mouse circulation that reinforced (even if it did not interact directly with) JAX inbreeding protocols.[30]

In Europe, by contrast, cancer researchers' commitments to inbred strains were not as strong, and chemical as well as viral theories of cancer proliferated alongside American genetic studies.[31] In 1903, for example, Amédée Borrel of France's Pasteur Institute had observed tumors endemic to certain mice colonies (whose breeding origin was unspecified) and speculated that such tumors might be caused by a cancer virus propagated through a parasitic infection. Two decades later, Johannes Fibiger of Denmark was awarded the Nobel Prize in Medicine and Physiology for showing that stomach cancer in rats was caused by the roundworm *Spiroptera neoplastica*; he also fed healthy mice cockroaches containing the larvae of the *Spiroptera*, which produced cancerous growths. Drawing on this work, a significant number of less well known, but far from marginal, cancer researchers in the United States continued to explore the viability of the parasite infection model. In 1936, for example, American M. C. Marsh reported that albino mice from his own eponymous "high mammary-tumor Marsh strain" showed a threefold reduction of spontaneous cancer when placed on a strict regimen that reduced the infestation of nematodes

[30] Clara Lynch, "Strain Differences in Susceptibility to Tar-Induced Skin Tumors in Mice," *Proceedings of the Society of Experimental Biology*, 31 (1933–34): 215. Cf. John J. Morton et al., "The Effect of Visible Light on the Development of Tumors Induced by Benzopyrene in the Skin of Mice," *American Journal of Roentgenology* 43 (1940): 896–98; and Elizabeth Ufford Green, whose thesis research at Bryn Mawr was published as "On the Occurrence of Crystalline Material in the Lungs of Normal and Cancerous Swiss Mice," *Cancer Research* 2 (March 1942): 210+. Lynch came to RIMR after working in T. H. Morgan's "fly room," and some researchers believed her genetic conclusion was merely a rationalization of this program: for example, M. C. Reinhard and C. G. Candee of the State Institute for the Study of Malignant Disease in Buffalo, NY, noted that Lynch had more often than not failed to produced any variation in cancer incidence between inbred strains. See their "Influence of Sex and Heredity on the Development of Tar Tumors," *Journal of Cancer Research* (1930–31): 640–44. Cf. Kohler, *Lords of the Fly*, "Boss and Boys"; and Corner, *A History of the Rockefeller Institute*, pp. 218–22.

[31] Joost Lesterhuis and E. S. Houwaart (Vrije University, Amsterdam), "Bringing the Inbred-mouse to Europe," paper presented at symposium on 100 Years of Organized Cancer Research, Heidelberg, Institut für Geschichte der Medizin/DKFZ, 18–20 February 2000 (see www.uni-heidelberg.de/institute/fak5/igm/g47/eck_can1.htm; accessed July 2002).

in their cages.[32] JAX inbred mice themselves were sometimes invoked as the "cure" for such infectious cancers. Thus, William Murray reviewed Marsh's data and suggested that the Marsh strain had not been inbred enough to be genetically homogeneous and so variation in cancer incidence was probably due to genetic variability rather than infection.[33] But viral theories of cancer did not go away simply because inbred mice could offer alternative material explanations for experimental results. Many researchers believed instead that (as Charles Oberling wrote from Paris in 1944) "the science of genetics may be collapsing before our eyes."[34]

By contrast, even while "hereditary" and "infective" understandings were increasingly converging rather than competing in cancer research itself, genetic theories overwhelmingly swamped viral theories in American cultural understandings of this disease before World War II.[35] Ac-

[32] A. Borrel, "Parasitisme et tumeurs," *Annales de l'Institute Pasteur* 24 (1910): 778; J. A. Fibiger, "Sur le développement de tumeurs papillomateuses et carcinomateuses dans l'estomac du rat sous l'action d'un ver nematode," *Troisième Conférence Internationale pour l'Étude du Cancer, Bruxelles*, 1913; M. C. Marsh, "Evidence of Heredity among Mammary Tumor Mice," *Journal of Cancer Research* 8 (1924): 518, cf. Marsh, "Simple Experimental Cancer Research," *American Journal of Cancer* 26 (1936): 181. Because no one could replicate this work, not everyone at the time believed Fibiger's growths were produced by the nematodes: cf. Fibiger's entry in *Dictionary of Scientists* (Oxford: Oxford University Press, 1999); also Paul Weindling and Marcia Meldrum, "Johannes Andreas Grib Figinger," in *Nobel Laureates in Medicine or Physiology, a Biographical Dictionary* (New York: Garland, 1990), pp. 177–81. Still, Walter Heston noted in 1945 that this line of work was alive and well: see his review essay, "Genetics of Mammary Tumors in Mice," in *A Symposium on the Mammary Tumors of Mice*, ed. F. R. Moulton (Washington, DC: AAAS, 1945), pp. 55–84, esp. p. 57.

[33] W. S. Murray, "Genetic Segregation Mammary Cancer to No Mammary Cancer in the Mouse," *American Journal of Cancer* 34 (1938): 434.

[34] Oberling, *The Riddle of Cancer*, pp. 162–63. By the second (1952) edition, Oberling's view on genetics mellowed: he wrote: "The genetics of cancer had first to be unraveled and the etiological role of the hormones recognized. . . . Only then was the time ripe to search for an infectious agent and genetics led the way" (p. 174). See also Ludwig Gross, "Is Cancer a Communicable Disease?" *Cancer Research* 4 (May 1944): 293–303. Creager and Gaudillière have argued that "the resurrection of tumor viruses in the 1930s was . . . intimately related to the invention of new modes of visualization of *hidden* cancer viruses" and "debates over the etiological significance and materiality of cancer viruses hinged on the divergent practices and epistemological commitments associated with these." "Experimental Platforms," pp. 233, 234. To the extent that this interpretation neglects the surrounding social context (namely, the concern among policymakers of sparking public fear of a cancer epidemic), it gives disproportionate weight to the laboratory as a lever in this debate. Cf. Paul Keating and Alberto Cambrosio, "The New Genetics and Cancer: The Contributions of Clinical Medicine in the Era of Biomedicine," *Journal of the History of Medicine* 56 (October 2001): 321–52.

[35] Cf. Creager and Gaudillière, "Experimental Platforms," although this convergence did not occur early enough to have an impact on Bittner's research program at JAX. Andervont has noted that Little's own position on hereditary versus genetic explanations on this ques-

counts of cancer studies in magazines and newspapers show that, while Little himself did not easily cede the importance of genetics research, the American mass media independently embraced findings that promised to illuminate pre-existing popular beliefs in "cancer families."[36] In 1933, for example, the Jackson Lab's milk-influence work received almost no attention in the national press when it first appeared. But that same year, pathologist Maud Slye of the University of Chicago—marginalized among many cancer researchers since becoming Little's rival in cancer genetics two decades earlier—made the *Newsweek* cover and the front page of the *New York Times* for her claim—based on "116,000 mice examined, . . . 23 years of studying generations of mice"—that inheriting a susceptibility to cancer depended on "a single gene, only in the original germ plasm." In an editorial response a few days later, the *Times* concluded that while Slye's results may not be consistent with other studies, her genetic approach brought with it the much-needed rigor of basic science: "Much to be preferred are deductions from laboratory experiments under strict control. . . . The problem of cancer is the problem of life because it is a problem of the cell." Cancer research, then, was one of the earliest scientific venues in which Americans invested their faith in genetics as a cure for chronic disease.[37]

In the laboratory, several researchers in the emerging field of medical genetics pursued popular theories of cancer heredity by studying humans. Canadian Madge Macklin, for example, did twin studies of cancer, while American Alfred Warthin and Dutchman W. F. Wassink employed family pedigree studies. This work sometimes came to overtly eugenic conclusions. Warthin went so far as to recommend that "a man who has a history of multiple incidence of carcinoma in his family should not marry a woman who has the same kind of family history."[38]

---

tion was more nuanced than he is given credit for: "Little was never anti-virus: he was too much of a scientist to be anti-anything." Andervont to Jean Holstein (illegible month and day) 1979, JLA-BH.

[36] On the persistent belief in "cancer families" through the 1950s, see Patterson, *Dread Disease*, pp. 58–59, although Patterson also notes (p. 111) that in a public opinion poll in 1939, 41 percent of respondents thought that cancer might be contagious.

[37] Survey of *Readers Guide* from 1933 to 1941 for subjects "cancer" and then "cancer research." On Slye, see "Cancer Heredity Laid to Single Gene," *New York Times*, 15 July 1933, p. X; follow-up editorial "Is Cancer Hereditary?" 18 July 1933, p. 16. Cf. McCoy, *The Cancer Lady.*

[38] Madge Macklin, "An Analysis of Tumors in Monozygous and Dizygous Twins," *Journal of Heredity* 31 (1940): 277–90; A. S. Warthin, "Heredity of Carcinoma in Man," *Annals of Internal Medicine* 4 (January 1931): 681–96, quote on 693; W. F. Wassink,, "Cancer er Hérédité," *Genetica* 17 (1935): 103–44. On the emergence of medical genetics, see Diane Paul, "The Eugenic Origins of Medical Genetics," in *Politics of Heredity* (Albany: SUNY Press, 1998), pp. 133–56.

When Little directly championed JAX cancer genetics in popular forums, however, his remarks were neither eugenic nor aimed at the viral hypothesis. Rather, Little promoted what he saw as two complementary scientific and social goals: more inbred mouse use in the laboratory to unravel cancer's true biological behavior, alongside more education for the lay public to generate support for research efforts. In 1936, for example, Little went head to head with Maud Slye at the annual AAAS meeting in Atlantic City, where the *Times* covered the event. "A truce was reached," the reporter wrote, at a symposium "in which a large number of cancer authorities took part." Here Little argued that cancer's genetic behavior was multifactorial (that is, more than one gene was involved), and that in the case of breast cancer, "some extrahereditary factor was involved." Slye withdrew her original claim and announced that she, too, believed more than one gene was involved. Still, four years later, popular fascination with the genetic hypothesis lingered; when *Life* ran an article showcasing the Jackson Lab's work, the headline read: "Dr. Little's 1 Million Mice Have Proved Heredity Is a Factor in Cause of Cancer." And in early 1941, when *Collier's Weekly* ran an article specifically about the Jackson Lab as the "front line against cancer," it praised Little as "the major-domo of this Mousetown" and noted that "when the problem of cancer is finally solved we'll owe a gold statue to the mouse."[39]

Little's approach was undoubtedly shaped by the shifting wartime politics of the lay cancer advocacy and education, especially within the American Society for the Control of Cancer. As managing director, in 1934 Little had suggested that the ASCC redirect its efforts from doctors and medical men to "nationwide lay education." This led to the founding of the Women's Field Army, which broke new ground by distributing pamphlets door-to-door and encouraging more public discussion of cancer phobias. In this context Little first began to articulate for a broader audience how laboratory knowledge about mice informed scientific understandings of cancer. In 1939 Little published his first book, *Civilization against Cancer* (1939), which contained two full chapters on the use of mice as materials for cancer research, and mouse data and examples are inextricably woven throughout the other eight chapters.[40] Here, as in his later pamphlet, *Can-*

[39] "Mass Suicide Fate Conceived for Man: A Report on the AAAS Meeting," *New York Times*, 30 December 1936, p. 10L. Cf. "The Week in Science: Hereditary Cancer," *New York Times*, 12 July 1936; and "Cancer: National Research Center Explores Its Nature and Cause," *Life* 8 (17 June 1940): 35–38, esp. 36–37. Robert Cook, "Front Line against Cancer," *Collier's Weekly* 107 (8 February 1941): 21+, esp. 22 and 21.

[40] C. C. Little, *Civilization against Cancer* (New York: Farrar and Rinehart, 1939), esp. chaps. 4 ("The Hunt for Living Weapons") and 5 ("Making Cancer to Order").

cer: A Study for the Layman (1944), Little went to great effort to articulate why hereditary studies of cancer in animals were necessary, but not sufficient, to help combat the disease:

> Mice are thus remarkable reproductive machines providing a shorthand version of the life process . . . in humans. . . . To determine their individual potentialities to form or resist the formation of cancer would be very nearly impossible by any experimental study of the application of Mendelian principles of heredity. This means that for the present . . . the individual whose parents died of cancer or in whose family the disease is frequent, should be alert, cautious and intelligent. . . . It does *not* mean that he should take a pessimistic or fatalistic attitude . . . he is not doomed or condemned. . . . This does not mean that studies of heredity in animals or that the use of biological methods of analysis are without value. On the contrary, these methods both by themselves and in combination with chemical analysis are as promising a line of attack on cancer as now exists. For an indefinite period into the future this field should provide valuable information on the origin and nature of the cancer process. . . . [But] no biologist who overemphasizes heredity can or will be able to advance far in his understanding of growth as a process.

For Little, then, lay cancer education was a way to reinforce socially the scientific claims he made to cancer researchers about the usefulness of inbred mice, while at the same time underscoring the practical and moral necessity of laboratory genetic studies in the public imagination. Establishing support for research, Little realized, was especially important as both the growing lay cancer movement and the remaining NACC members began to press for funding for therapeutically relevant activities over experimental investigations in cancer etiology. An etching accompanying the chapter on "research" (fig. 5.3) summarized Little's view: mice are portrayed at the top and center of a group of experimental animals including dogs and cats, with the caption: "Friends of suffering humanity."[41]

Still, Little and the other JAX researchers remained cautious about overextending the inbred mouse genetic model to humans. When his audience was other experimental biologists, Bittner himself consistently took the position that "on the basis of the data available at the present time it

[41] C. C. Little, ed., *Cancer: A Study for the Layman* (American Cancer Society, 1944), quote from pp. 29, 43–44 (copy in SLC library). Cf. "Report on Clinical Congress of American College of Surgeons," *New York Times*, 22 October 1936, p. 36; Little, "The Conquest of Cancer," p. 77. On Little and the Women's Field Army, see Shaugnessy, "The Story of the American Cancer Society," pp. 162–63; and Hurst and Nusbaum, "Advocating for Women's Health." On the NACC funding of "theories of treatment," see table of grantees compiled by Kenneth E. Studer and Daryl Chubin in *The Cancer Mission; Social Contexts of Biomedical Research* (Beverly Hills: Sage, 1980), pp. 21–24.

5.3. From Little's *Cancer: A Study for the Layman* (1944) [Source Credit: Manal Abu-Shaheen].

is impossible to state if any one influence (genetic, hormonal, or milk) should be considered as the active mammary tumor inciter." But when he spoke to medical doctors, he conceded that breast cancer in humans "could be substantially reduced if the women of families with any tumors in their ancestry were to refrain from nursing their progeny." His JAX colleagues, by contrast, in a collectively authored 1936 piece argued that while the use of inbred strains was exceedingly important for laboratory determinations of the causes of breast cancer, other procedures—such as hormone therapy—might prove more valuable for eliminating the disease in the cancer clinics. Little believed the scientific issues made the matter complex: "If one asks the direct question concerning mother's milk in humans the answer is not so easy." He ultimately declined to recommend that women from so-called cancer families not nurse their children, for this conclusion, he believed, would be warranted only from "data by direct observation" in humans: "This does not mean that milk as such produces cancer. Certainly cow's milk can be given a clear bill of health for there is evidence that cows are singularly free of cancer of the breast."[42]

[42] The Staff of the Jackson Laboratory, "The Constitutional Factor in the Incidence of Mammary Tumors," *American Journal of Cancer* 27 (1936): 551–55; John J. Bittner, "Pos-

With regard to cancer education, then, JAX researchers showed little faith that the analogy between mouse and human cancers in the laboratory would be properly translated in the broader public sphere.

Worries over public accountability of cancer researchers also haunted the National Advisory Cancer Council, as evidenced in their early discussions about the grants-in-aid system. In an effort to fulfill their congressional mandate to coordinate cancer research and eliminate what they called "unnecessary replication of effort," the NACC met in 1939 with officers of private philanthropies and attempted to develop a plan to rationalize all extramural support, public and private. This group called themselves "The Committee on Cooperation in Cancer Research," and they produced a survey of research work broken down by subject matter for the NCI's first several years. Genetics research had consumed only about 6 percent of grants-in-aid funds, but almost all of it had gone to Bittner. As Little himself would crassly put it to the committee, "90% of grants-in-aid [are], if you please, apparently useless," but "I should hate to see them entirely eliminated from the picture because somewhere, sometime you might get a line on a problem or person that you wanted." What Little could not anticipate, however, was that within that same year he would be called on to save Bittner's genetics research and, by extension, federal support for all research work at JAX.[43]

WHAT IS "COORDINATION"? EVOLUTION OF A JAX POLICY ON PRODUCTION VERSUS RESEARCH

In what amounted to an obvious conflict of interest, at an early meeting the NACC asked Little to defend an additional $10,000 grant-in-aid proposal made by Bittner for an expansion of his milk-influence work at Bar Harbor. Chair Ludwig Hekoten recognized the situation was awkward, and discussion was put off until the very end of the February 1938 meeting. When Little did finally speak to the issue, his ambivalence was apparent. On the one hand, he believed in the importance of research at JAX, but on the other hand, he did not want to jeopardize the JAX-NCI mouse production arrangement by appearing greedy on behalf of his own institu-

sible Types of Mammary Gland Tumors in Mice," *Cancer Research* (November 1942): 755–58, quote on 757; "The Influence of Foster Nursing on Experimental Breast Cancer," *Transactions and Studies of the College of Physicians, Philadelphia*, 9 (1941): 129–43, quoted summary from Gross, "Is Cancer a Communicable Disease?" p. 301; Little, *Cancer: A Study for the Layman*, p. 38.

[43] "Report of the Committee on Cooperation in Cancer Research" appears in the NACC transcripts for the March 1940 meeting, pp. 17–34; Little quoted in NACC transcripts, December 1939, p. 84; RG 443, Box 6, vol. 2, NACC-MD.

tion. Little argued that Bittner's work was worthy of consideration—above and beyond the NCI grant for production—because it fell squarely within new NCI Director Carl Voegtlin's "outline of work of interest to the Government"; he also noted that Bittner was being considered for a longer-term appointment as research fellow to the Surgeon General. Those members of the committee who viewed the JAX-NCI group collaboration more broadly hailed Bittner's proposal as "the most important work in cancer research in 1937." Voegtlin himself noted that supporting Bittner would "complement" the work of Andervont's group in Boston on milk influence in different strains: *C3H* and I. In retrospect, however, Little's insouciance about Bittner's research stood in marked contrast to his earlier passionate arguments for funding JAX mouse production, and it is impossible to imagine that fellow NACC members would not have perceived the contrast. For example, in one exchange, Karl Conant proposed to reduce Bittner's funding to $3,200, an amount that would proportionately cover only the rest of the fiscal year, with future funding contingent on demonstrating the work's relevance to the cancer program. Little replied:

> *Dr. Little*: Okay for this year, but I will be back. . . .
> *Dr. Ewing*: How long will this problem require about $10,000 a year?
> *Dr Little*: As long as the Government is interested and no longer, I hope. The work is going to be done whether the government does it or not. The Childs Fund is interested in it. I can get money for it if the government does not want it. . . . I am not urging it in any way at all.[44]

Years later, Andervont would note that "Little's firm belief that the acquisition of knowledge is essential for the practical application of science to human welfare led to this Laboratory becoming one of the few organizations devoted exclusively to this objective." As his cancer education work suggested, in the early years of NCI funding, Little had few expectations that research on genetics of milk influence could provide clear, practical recommendations for women to avoid breast cancer. Another NCI grant would also prove awkward for JAX accounting—later described by Little as following a "beggar policy," because it showed that the lab's research support was cobbled together from many small funding sources. All these things combined to make his support for Bittner's proposed research seem lukewarm at best.[45]

The following year, in the context of debates over the efficacy of the grants-in-aid program, Bittner's application was revisited and more

[44] National Advisory Cancer Council Transcripts, 14 February 1938, esp. pp. 596–607; RG 443, Box 6, vol. 2, NACC-MD.

[45] Quote from "Resolution about C. C. Little," proposed by Howard Andervont to the JAX Board of Trustees, 17 February 1972, JLA-BH.

strongly challenged by the NCI leadership, who wanted to avoid charges that the NACC supported particular institutions as unofficial "satellites." Nepotism between the NCI and JAX was already practiced: for example, Andervont's group in Boston had started transplantable tumor work with Bittner and JAX mice in the early 1930s and was the first to move into the new NCI headquarters in Bethesda. More generally, however, concerns were now mounting among medical educators and researchers over what they perceived to be the NACC's too-narrow implementation of the extramural grants-in-aid program. The NCI already had two officially designated "cancer research centers" in Boston and Washington, many reasoned, so wasn't supporting other institutions doing similar work an inefficient use of federal funds, in light of existing in-house commitments?[46] This controversy reached its height just as the new Bethesda labs were opening, and Little pressed the committee in the June 1940 meeting similarly to formalize his institution's research status by appointing an NCI representative on the JAX Board of Directors. Some members considered this a good measure of quality control—the NCI gave JAX so much money to produce mice, governance support would merely be insurance on that investment. Surgeon General Parran and Conant, however, nixed the idea on the grounds that it would be more "desirable" to "divorce" the two institutions "while fostering proper cooperation." As Parran put it: "I am anxious for the impression not to become fixed that we have a continuing stock interest in that laboratory. . . . I am not willing to assume permanently the major commitment for the support of that institution."[47] Thus NCI's decision to effect what its research director later described as an "ironclad separation of the intramural staff and the extramural affairs of the institute" in the case of Bittner and JAX ultimately became policy for the grants-in-aid program that persisted into the 1950s. The NCI's first director, Carl Voegtlin, considered it "sacrosanct" and used to it maintain impression of the Bethesda group as a politically and scientifically neutral body: "Voegtlin wanted no help from outside with his direction, and at the same time wanted no criticisms that the intramural personnel would profit from intimate knowledge of grant requests from scientists from other institutions."[48]

[46] The NCI extramural grant program was under fire for other reasons: for example, many argued that its funding should be administratively centralized within the NIH. See Daniel Fox, "The Politics of the NIH Extramural Program,1937–1950," *Journal of the History of Medicine and Allied Sciences* 42 (1987): 447–66, esp. 459–61. But this criticism of the NACC's decision making was made within larger debate, taken up in earnest after the war, over the relationship between federal funding for biomedical research and government involvement in both health care and medical education: cf. Strickland, and the work of Harry Marks, "Leviathan and the Clinic," paper presented at the History of Science Society Annual Meeting, 27–30 December 2002.

[47] NACC transcripts, June 1940, pp. 43–44, RG 443, Box 7, vol. 2; NACC-MD.

[48] Shimkin, *As Memory Serves*, p. 29.

This decision on JAX was based in large part on a false assumption that the programmatic aspects of inbred mouse work held absolute sway at the benchtop. One could easily get this impression from looking at Andervont's work. He ultimately published over two hundred papers on the genetics of mammary tumor behavior in inbred mice, and he was, by his colleagues' accounting, "a master of the economical experiment to answer specific questions." But many other important discoveries that emerged from the use of inbred mice at NCI were more accidental. For example, in the mid-1940s Washington radiobiology group members Paul Henshaw and Anderson Nettleship discovered that the urethane they were employing to anesthetize their inbred strains to receive radiation caused lung cancer only after they autopsied dead animals and found them to have multiple tumors.[49]

False or not, Little recognized that this new standard of efficiency had hazardous potential for Bittner's work. At this point in the history of JAX, funding for inbred mouse production was administratively separate, but physically and practically, the routine work of supervising the large breeding colonies still fell to JAX scientists, each of whom took care for a few strains with the help of one or two caretakers and technicians. Little, therefore, saw the need to develop other institutional affiliations that would demonstrate to the NCI that while JAX mouse caretaking was routine, JAX's program of research was at the leading edge of academic work in the field. In communications with Warren Weaver, he boasted (falsely, based on NACC transcripts) that "the Jackson Laboratory has frankly been mentioned as the first" institution to be funded as a new NCI research center. At the same time, he also confided his efforts to get Bittner's work on a stronger footing by establishing a permanent collaboration between JAX and Yale University Medical School: "those in charge of cancer research and genetics at Yale . . . have indicated a lively interest in the possibility of . . . cooperative research and perhaps a certain amount of teaching by the staff members . . . equivalent to perhaps 25–33% of the laboratory's annual budget." But these efforts stalled repeatedly, as Yale struggled with reorganizing its entire division of basic sciences. As a result, Little told Frank Blair Hanson in 1941 that he was "now trying to consolidate in preparation for a possible storm," and he promised his staff he would travel less, so he could be more in touch with what was going on at the JAX benchtop.[50]

Bittner's research remained vunerable to the criticism that it was unnecessary on two counts: it was a duplication of research effort (milk influ-

[49] Ibid., p. 15, 42.

[50] CCL to WW, 7 July 1939, RG 1.1, ser. 200D, Box 143, Folder 1774; CCL to WW, 1 January 1940; and interview with FBH, 28 July 1941, both RG 1.1, ser. 200D, Box 143, Folder 1775; all RAC-NY.

ence work at Bethesda) being supported by a duplication of federal funds (JAX grant for mouse production). By 1941, with Little no longer present at NACC discussions, Voegtlin frankly acknowledged that the Bethesda group's lack of animal facilities made them "dependent on [JAX] indirect grants as a source of materials," and he registered his approval for funding Bittner as well as "subsidiz[ing] the business of selling mice." But Parran questioned the wisdom of the Bittner grant-in-aid, which (based on past funding decisions) appeared to be "permanent subsidy in half support of this institution" and its research. The $15,000 grant was eventually approved, but the political damage was already done.[51] In 1942, with the total NCI budget capped due to increased American military involvement in the war, the NACC decided to taper Bittner's research funding. C. P. Rhodes began the discussion of Bittner's renewal application by noting that he had begun to collaborate formally with, academic cancer researchers at the University of Minnesota. Rhodes described this work as "replicating and confirming" the NCI groups and complained that even Bittner's new institutional alliance would not bring his work fully up to speed: like JAX, Minnesota did not have an ultracentrifuge and therefore was "simply not equipped" to isolate and characterize the milk factor.[52] That year Bittner received 20 percent less funding, and the following year, led by long-time JAX supporter James Murphy, the committee savaged Little's final reapplication: why, Voegtlin wondered out loud, did the NCI give grants-in-aid to raise inbred mice and then pay again to purchase nearly 20 percent of JAX's entire animal output? Little, he commented, seemed to want "to use Bar Harbor as a Genetic Branch of NCI," whereas ideally, other members concurred, support for JAX mice would be based on "no specific project" but for making animals that lent themselves to multiple kinds of laboratory research. JAX's utility, like that of the animals it produced, was defined by the NACC as its ability to make a universal product that sustained many diverse user communities, rather than a specialized product that sustained only one. Compton eventually suggested that the amount be reduced to $9,000 but no one seconded it: the question, Murphy pointedly observed, was whether they cut the research funding without "killing" the lab. The grant was finally approved for the full amount, but Bittner appeared to recognize that support for his research at JAX was effectively dead. Unhappy with the prospect of a career spent stewarding the JAX

[51] Full cite for RF letter from CCL to WW; NACC transcripts, March 1941, pp. 62, 66–67, RG 443, Box 8, vol. 1, NACC-NA.

[52] NACC transcripts, April 1942, pp. 71–72, RG 443, Box 8, vol. 1; NACC transcripts, January 1943, pp. 62, 70–78, RG 443, Box 8, vol. 3; both NACC-MD. Cf. Sevari and Helberg, "Obituary: John J. Bittner." Earlier, Little sent a telegram to Wendell Stanley asking if he could purchase a centrifuge"; see Creager, *Life of a Virus*, p. 110, n. 122.

production colonies, he left Bar Harbor in 1943 for an appointment at the University of Minnesota Medical School.[53]

Bittner was not the only JAX staffer whose research funding nearly evaporated in this shifting policy climate. Elizabeth Fekete's ova transplant studies were threatened when the Rockfeller Foundation invoked its own clause about supporting people and their projects, not institutions. In 1940 Weaver agreed to fund ova transplantation studies at JAX under Little's supervision, but he later told Little: "As I think the matter over in a cold-blooded way uninfluenced by your convincing presence, it looks a good deal as though we were being asked to make a contribution . . . to the regular budget of the institution." Little explained that the foundation would be paying to replace Fekete's usual production-oriented labors: "The relative amount of time she can spend on research and the degree to which she can be freed from routine work in connection with the production of surplus animals will depend on whether her expenses can be met from a source independent of the production and utilization of these animals." Rockefeller Foundation granted JAX three years funding, in order to exploit what Little called "the first opportunity to measure . . . just what the nature of the foster mother influence may be."[54] Similarly, George Snell met with Rockefeller Foundation officers to describe a routine project to analyze the mutation-inducing properties of different amino acids in mice and *Drosophila* (which was an extension of his postdoctoral research at the University of Texas). This smaller project—"To determine the chemical nature of the gene," as Snell wrote—was not encouraged; on Little's urging, Snell withdrew the application.[55] Research funding, while not the sole barometer of scientific activity at JAX, represented an important resource in the practical ecology of the institution in the 1940s. Without targeted funding, some original research still went on, but this work was limited by the extent to which JAX staffers could be freed from their more routine scientific tasks related to the production colonies.

L. C. Strong has argued that Bittner's approach to milk influence work changed markedly after leaving JAX. Bittner himself has also implied that he felt pressured by the genetic "received view"; "If I had called it a virus, my grant applications would automatically have been put into the category of 'unrespectable proposals.' As long as I used the term 'factor,' it

[53] Sevari and Helberg, "Obituary: John J. Bittner"; see also Holstein, *The First Fifty Years*, p. 96.

[54] WW to CCL, 29 January 1940; CCL to WW, 1 February 1940, RG 1.1, 200D, Box 143, Folder 1774; and "Resolved, RF 38004," RG 1.1, 200D, Box 143, Folder 1773; all RAC-NY.

[55] CCL to WW, 16 November 1938; George Snell to FBH, 24 March 1939; both RG 1.1, 200D, Box 143, Folder 1774; cf. George Snell to RF, 31 January 1939, RG 1.1, Series 200D, Box 144, Folder 1781; all RAC-NY.

was respectable genetics."[56] Bittner's later publications suggest a renewed his enthusiasm for inbred mice as experimental tools after leaving JAX. In a 1948 review essay entitled "Some Enigmas Associated with the Genesis of Mammary Cancer in Mice," Bittner quoted generously from a speech called "The Geneticist's Contribution," which Little made to a conference of medical educators in 1937, just as JAX's institutional affiliation with the NCI was beginning. Inbred strains, Bittner argued, were necessary to the process of replicating and confirming experimental findings, and the only major obstacle to that process was a lack of awareness among researchers about the development of sublines.[57] At JAX throughout the late 1930s and early 1940s, Bittner dutifully conducted and published thorough encyclopedic studies on spontaneous and transplanted tumor behavior in individual cancerous mouse strains JAX produced, and he more sporadically published articles that reflected his "three influences" theory of mouse breast cancer. At Minnesota, his work ranged more freely over tumor transplantation, chemical studies, and foster nursing. In 1947 Bittner first described the mammary tumor milk agent in print as a virus (and he cited his own earlier papers as references for this point); two years later he returned to present a paper to the JAX twentieth-anniversary symposium and he described the same idea using the term "virus" in scare-quotes.[58] But by the mid-1950s, as Gaudillière has observed, an alternative animal model for cancer transmission—the *Ak* leukemic mouse—emerged at a liminal time in the field of cancer genetics: when work was stalled and "reviews on the causes of mammary tumors in mice had become a series of experiments organized in chronological order."[59] Bittner's shifting approach, then, had much to do with fulfilling the more expansive vision of his original research program.

[56] Bittner quote from George Klein, *The Atheist and The Holy City*, trans. Theodore and Ingrid Friedman (Cambridge: MIT Press, 1990), as quoted in Kevles, "Pursuing the Unpopular, p. 81. See also Strong, "The Origins of Some Inbred Mice" and "Inbred Mice in Science," pp. 45–68.

[57] J. J. Bittner, "Some Enigmas Associated with the Genesis of Mammary Cancer in Mice," *Cancer Research* 8 (December 1948): 625–39; cf. Little, "The Geneticist's Contribution," *Proceedings of the Annual Congress of Medical Education and Licensure*, Chicago, 15–16 Feb 1937 (photocopy found in Box 730, CCL-UMO).

[58] John J. Bittner, "Transplantability of Mammary Cancer in Mice Associated with the Source of the Mammary Tumor Milk Agent, *Cancer Research* 7 (1947): 741–45. Cf. "Mammary Cancer in Mice," in *Lectures on Genetics, Cancer, Growth, and Social Behavior* (Bar Harbor: Bar Harbor Times, 1949), pp. 51–57, esp. p. 57. For excellent discussion and bibliographies of Bittner's work, see Walter Heston and Howard Andervont in F. R. Moulton, ed., *A Symposium on Mammary Tumors in Mice* (Washington, DC: AAAS, 1945), pp. 55–84 and 123–39.

[59] Gaudillière, "Circulating Mice."

Little, in turn, relied on tried-and-true methods for closing the budget gap caused by the loss of Bittner's NCI funds. Having just found out that JAX had been willed the deed to the Hamilton farm, several miles from the original JAX site, Little began to explore with the Rockefeller Foundation the possibility of developing the new site into a series of inbred mammalian stock centers at JAX. Little sought to serve medical researchers; he wrote to several members of the Rockefeller Institute, for example, and asked them what kinds of animals JAX would have to produce if he wanted to get the institute's business. Plans quickly focused on the smaller experimental animals—rabbits, guinea pigs, and rats. Dogs were discussed but ultimately left out, since Little was developing a separate project on behavior genetics of dogs with Alan Gregg in Medical Sciences. In April 1941 JAX obtained guinea pigs from a well-known commercial animal breeder—Carworth Farms—and rabbits from the Rockefeller Institute's Harry Greene. For a time, Little explored the possibility of taking over the Wistar Institute's well-known inbred rat strains, during that institution's reorganization, but Weaver nixed this idea as overlapping too closely with the mouse stock center. After nearly two years of discussion and planning, the Rockefeller Foundation gave Little $35,000 annually for three years, during which time he was expected to make the enterprise (as Hanson put it) "self-supporting," as he had with mouse production.[60]

One measure of how federal cancer policy had affected JAX was the number of original cancer researchers still at Bar Harbor: with Bittner at Minnesota, Joe Murray at University of Maine, William Murray at the State Institute for Malignant Diseases in Buffalo, and Heston at Bethesda, only Fekete and Cloudman remained. NCI policy had effectively distanced JAX inbred mouse materials from the researchers whose work had given birth to them, and the Rockefeller Foundation seemed eager to finish the process. Neither geneticist T. H. Morgan nor the 1941 National Research Council Committee of Pure Genetic Strains—both of whom Weaver consulted—had any doubt that JAX could do this new standardizing job for genetic and biomedical researchers. "Little has been so successful with his mouse stock center," Frank Blair Hanson gushed, "that he now proposes

[60] Series of letters between Little and RIMR Researchers in RIMR Archive, RG 210.3, Box 2, Folder "Animals 1922–1962;" all February 1941; FBH to CCL 4 February 1941; CCL to FBH, 13 February 1941; Memo FBH to WW, 28 February 1941; CCL to FBH 23 April 1941; all RF Archive, RG 1.1, Series 200D, Box 143, Folder 1775. W.E. Castle to FBH, 26 February 1943, RF Archives, RG 1.1, series 244D, Box 1, Folder 14; CCL to FBH, 18 December 1942 (with handwritten note about urgency of Wistar situation), RF Archives, RG 1.1, Series 200D, Box 144, Folder 1776; cf. WW's diary description of his interview with CCL at Bar Harbor, 29 November 1937, RF, RG 1.1, Ser 200D, Box 143, Folder 1774; all RAC-NY.

to do the same for guinea pigs [and] rabbits." The following year the Rockefeller Foundation gave Little a small supplemental grant to hold a conference at JAX on inbred mice cancers and heredity, but Hanson's sentiments—and the far larger flow of foundation dollars to animal production—better reflected JAX's status as a maker of materials, not research.[61]

POSTWAR BIOMEDICAL RESEARCH AND MICE: PROBLEMS AND POSSIBILITIES

World War II mobilized scientific researchers—especially physicists—into a new relationship with the federal government, and Americans more broadly into a new relationship with science. The atomic bomb was said to have won the war, and the Manhattan Project, run by the U.S. Army and physicists out of secret laboratories in Los Alamos and Oak Ridge, was held up as an example of what science could do for the country if given enough resources. As news of Nazi race hygiene research on Jewish prisoners spread across the ocean, there was a public backlash against eugenics in the years immediately following 1945. But other wartime medical research, sponsored by the U.S. Committee on Medical Research of the Office for Scientific Research and Development, had had well-publicized successes, such as the development of penicillin, new antimalarial drugs, and the insecticide DDT. In this new context, debates about how to manage cancer research became part of a wider debate concerning the proper structure and function of federal funding for postwar science as a whole. Prewar cancer research policy values like "coordination" and "organization" took on new meanings. For the emerging lay leadership, they meant mobilization on the model of industrial research and development: doing "successful" biological research meant developing experimental studies that quickly led to new clinical therapies. As a *New York Times* editorial summarized: "We are convinced that industrial methods (organization, planning, competent direction) are more likely to teach us how cancer can be controlled than the kind of empiricism and floundering that passes for cancer research in the universities." Anticipating these developments, in 1944 Surgeon General Thomas Parran drew up a ten-year plan for the expansion of the NIH that used language that echoed Little's

[61] FBH to CCL, 21 April 1941; FBH interview with T. H. Morgan, July/August 1941; FBH Diary excerpt from visit to JAX, 28 July 1941; all RF Archives, Rg 1.1, Series 200D, Box 143, Folder 1775; FBH interview with CCL 14 January 1943 (where grant terms are discussed); CCL to FBH 1 February 1943; both RF, RG 1.1, Series 200D, Box 144, Folder 1776. Grant-in-aid for JAX signed by Alan Gregg, 28 June 1944, RF Archives, RG 1.2, Series 200A, Box 133, Folder 1189; all RAC-NY.

earlier appeals for NCI funds: "It is believed that the use of public funds is fully justified in developing the physical plant for health, in training professional personnel, in supporting both public and private medical and scientific research of a broad public interest." But as Nathan Reingold has pointed out, Parran assigned research a "decidedly secondary role" in the national biomedical expansion.[62]

By contrast, Little's conception of public health relied on an old-line Progressivist faith in scientific expertise, and this approach—along with personal animosities he developed with new lay leadership in cancer policy—alienated him from emerging currents of lay activism in the cancer control movement. Although Little's vision of industrializing inbred mouse production had brought JAX some measure of success with both biomedical scientists and business-oriented philanthropies, in 1944 he confronted what he perceived to be a disingenuous effort by businessmen to "reform" the Society. The influential Mary Lasker, wife of advertising millionaire Albert Lasker, sought to rid ASCC of the "impractical, ineffective, and unrealistic" efforts of medical men and scientists. In the spring of 1944 they met with the doctors who formed the Executive Committee—of the now renamed American Cancer Society (ACS)—and suggested that the existing organization's commitment to research needed revamping. Specifically, they sought to use broader public support for the anticancer movement—generated by publicity for the NCI Act—to raise an ACS endowment for awarding research fellowships that targeted the development of cancer therapies. The idea that nonscientists could properly evaluate any course of laboratory investigation galled Little, but he equally resented the implication that he had not been cognizant enough of core merchantilist values. He resigned his post that same year, but not before "speaking frankly" against what he perceived as a "violent and emotional . . . attack" on his ability to advocate and manage research. "The result," Little reflected later in his life, "was that the businessmen developed a very live personal dislike of me, which I think still blooms

[62] On the historiography of World War II and U.S. science, see A. Hunter Dupree, "The Great Instauration of 1940: The Organization of Scientific Research for the War," in *Twentieth Century Sciences: Studies in the Biography of Ideas*, ed. Gerald Holton (New York: Norton, 1977), pp. 443–67; Daniel J. Kevles, "The National Science Foundation and the Debate over Postwar Research Policy, 1942–45," *Isis* 68 (1977): 5–26; and Kevles, "Foundations, Universities, and Trends." Also Peter Neushul, "Science, Government, and the Mass Production of Penicillin," *Journal of the History of Medical Alllied Science*, 48 (1993): 371–95; Paul, *Controlling Human Heredity*; Robert Proctor, *Race Hygiene*; W. Kaempffert, *New York Times*, 26 August 1945, part 4, p. 9. Parrant memo quoted from Nathan Reingold, "Choosing the Future: The U.S. Research Community, 1944–46," *Historical Studies in the Physical and Biological Sciences* 25, 2 (1995): 301–28. On the prewar roots of this attitude toward medical research, see Robert Bud, "Strategy in American Cancer Research."

5.4. Little talking to lab visitors, c. late 1940s, with bulletin board in background: "Jackson Laboratory Mice Go Everywhere" [Source Credit: Jackson Laboratory Archives].

with full fragrance today." More immediately, his falling out with the Lasker coalition meant that he could never apply to what would become the Lasker Foundation, a major player in terms of research funding for cancer (especially milk factor work) in the 1940s.[63]

Apparently undaunted, Little returned to what he knew best: making mice. Little would later claim that "the role of basic research at Jackson Laboratory was just as important to medicine as nuclear fission was to the atomic bomb." But during and immediately after the war, the impact of federal medical and public health policy shifts on the fortunes of the JAX's production venture appear to have been quantitative, rather than qualitative. JAX mouse circulation had increased tenfold in the period from 1933 to 1941; a contemporary in-house bulletin board featured a map of the United States with the slogan "Jackson Laboratory Mice Go Everywhere" (fig. 5.4). But because Little almost never broke down the

[63] Quotes from C. C. Little typescript draft, "When in 1937 the National Cancer Act was passed," n.d., Box 730, CCL-UMO. See also Shaughnessy, "The Story of the American Cancer Society," chap. 11; and Richard A. Rettig, *Cancer Crusade: The Story Of The National Cancer Act Of 1971* (Princeton: Princeton University Press, 1977). For an insider account of the Laskers' takeover of the ACS and the development of the Lasker Foundation,

numbers of mice sold by consumer in his reports on the NCI grants, it is difficult if not impossible to tell precisely how much of this increase was due to wartime projects alone.[64]

Little himself did not let his old dreams for mammalian genetics go lightly. In early 1947, he approached the Rockefeller Foundation with a plan to build three more mouse production buildings at Bar Harbor. Little noted that the genetic research benefits of the increased mouse production would enable more mutants to be found, and he suggested that the buildings themselves could be a symbol of that relationship: "[they] should be connected either directly or by a passage to the existing laboratory structure." But the real value of such an investment, he emphasized, would be to bolster the "usefulness of the inbred stocks being used elsewhere." Weaver balked and said he was "inclined now to be more interested in the [JAX Lab's] program rather than the plant, though the two are inseparable of course." Yet when handed the opportunity to potentially recharge JAX's Lab's research program, Little reasoned that making and distributing mice could be seen as already having circumscribed his institution's total mission:

> [While] the Jackson Laboratory is on the threshold of becoming . . . a really outstanding center for mammalian research . . . more and more people have also begun to come to us with mouse problems and programs which send new roots of the Laboratory's contacts and influence into broader fields of biology and medicine, related to its original, narrow program. . . . The intellectual and spiritual gap which exists between the simple and quiet surroundings where resources are concentrated on a single objective and the complex and confused environments of great universities where competitive interests of schools, departments and individuals continually arise remains striking and compelling.

In short, Little argued that as long as JAX Lab was a centralized source of inbred mouse supply, it would represent a network of resources—material, social, intellectual, and practical—too valuable for mouse workers to ignore. Weaver quickly relented and approved the grant. When JAX got the go-ahead, Little sent Snell out on site visits to other relatively large-scale mouse breeding facilities to look for construction ideas. He also encouraged Snell to look at an actual industrial lab—Bell Telephone

---

see Judith Robinson, *Noble Conspirator: Florence S. Mahoney and the Rise of the National Institutes of Health* (Washington, DC: Francis Press, 2001).

[64] Quote from transcript of *R$_x$Mouse*. On production and circulation figures, see W. L. Russell, "Inbred and Hybrid Animals and Their Value in Research," in *Biology of the Laboratory Mouse*, ed. George Snell (Philadelphia: Blakiston Company, 1941), pp. 325–48. which cites an increase from 12,000 to 120, 000.

Company. Within a year, construction on JAX's second and largest mouse production facility was virtually complete.[65]

All work at Bar Harbor, however, was severely interrupted by the 1947 Bar Harbor Fire. JAX workers would later claim that some news reports about JAX's losses were mildly exaggerated, while others did not capture the scientific tragedy. For example, few accounts noted that a small group of original *C3H* descendants had survived the ordeal, having been boxed up for shipment in a room with a fire-proof door. But even fewer still mentioned George Snell's devastating loss of all his congenic strains, the crosses for which he had begun in earnest only the previous summer.[66]

New reports aside, all scientists and policymakers who had any connection with JAX understood that this was a turning point: should Little's group attempt to rebuild, or should the lab's mouse production and meager research work be dissolved and dispersed among already existing institutional outlets? Before anyone knew the full extent of the damage, the boards of the ACS and NACC both held special meetings and, together with the Rockefeller Foundation officers, decided to offer Little a new building for mouse production.[67] Before this formal decision was reached, however, inbred mouse workers themselves began a grass-roots effort to rebuild JAX: they sent back breeding pairs of strains they had received from JAX before the fire. Their desire to rebuild JAX Lab as a center of inbred mouse production and supply spoke volumes: it reflected a communal belief in the value of its unified system of mouse standardization, production, and distribution. Margaret Dickie, a relatively new mouse geneticist staffer, designed a new lab crest commemorating the fire (fig. 5.5): the lab is symbolized by a phoenix rising from the ashes, and it was intended to feature photographs of its Progressive founders, "Jackson" and "Little or [Paul] Revere." But this crest was never adopted, for in the end, regardless of the specific benefits each individual researcher accrued from JAX Lab's centralized enterprise, as a group they agreed with Little's assess-

[65] On making *Drosophila* a widely distributed research organism, see Kohler, *Lords of the Fly*, chap. 5; CCL to WW, 18 January 1946, 22 March 1946, and 26 March 1946; WW to CCL, 21 March 1946; WW to Raymond Fosdick, 24 April 1946; CCL to WW, 23 May 1946, 24 March 1947, and 21 August 1947, which notes that RF supplemented the original grant $10,000 for increased construction costs; all RF Archives, RG 1.1, 200D, Box 144, Folder 177. Snell to Sidney Strickland, 22 October 1946, Box 7–6, JLA-BH. cf. Kay, *The Molecular Vision of Life*, chap. 1, for a general argument about the inseparability of Weaver's own programmatic and practical goals.

[66] Joan Staats to W. G. Hoag, "Production Catalog, Comments," 6 September 1961, Box 2–6, JLA-BH: Staats cautioned about reports that all mice were killed in the fire: "I'm tired of this canard. . . . Let's not make the mistake of believing our own publicity." Snell, "Studies in Histocompatibility," p. 648.

[67] Telegrams contained in Box 735, CCL-UMO.

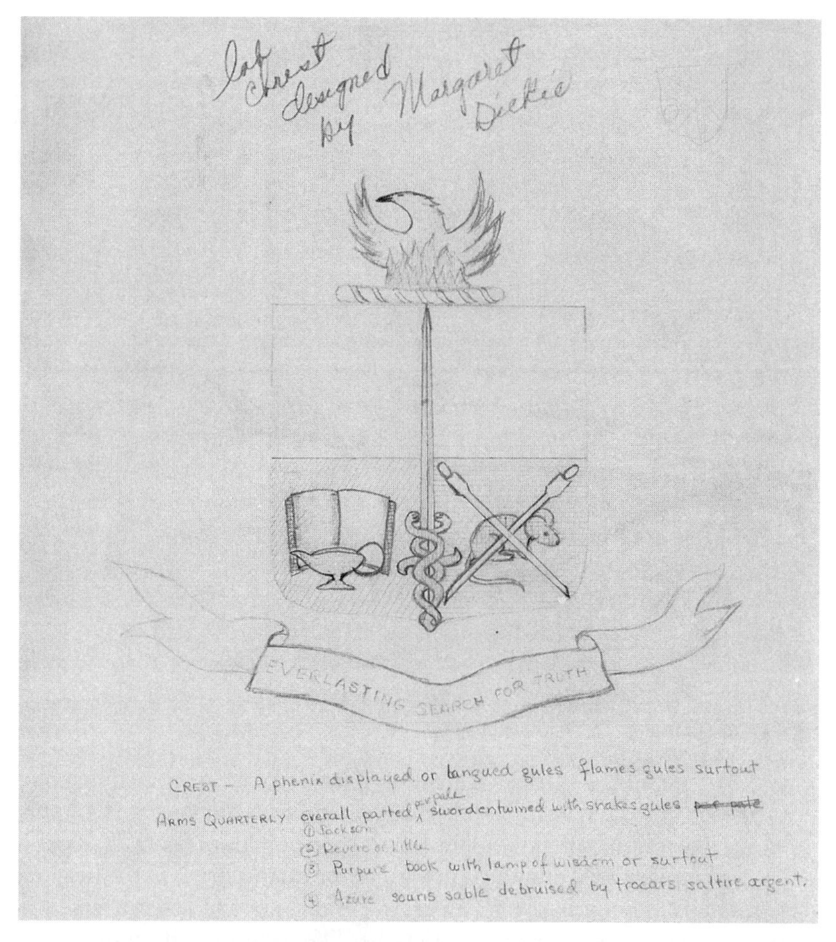

5.5. Jackson Laboratory crest, designed after the 1947 fire by Margaret Dickie [Source Credit: C. C. Little Papers, University of Maine].

ment: the JAX Lab and its inbred mice should be maintained "not only as a physical entity but as a spiritual force" to unify mouse researchers.[68]

Little took more deliberate precautions to ensure that his supply of animals could be replenished in the event of disaster. He created a so-called inbred nucleus unit, and he appointed Elizabeth Russell its head

[68] On responses to the JAX Lab fire, see WW to K. Compton, 3 October 1947; WW to RBF, 7 November 1947; both RF Archives, RG 1.2, 200A, Box 134, Folder 1191, RAC-NY; "RBJ Lab," *RF Trustees Bulletin*, November 1947; CCL to WW, 21 November 1947 and 28 November 1947; all RF Archives, RG 1.1, 200D, Box 144, Folder 1777, RAC-NY. For a specific example of a "mouse return," see CCL to Muller, 3 November 1947; Muller

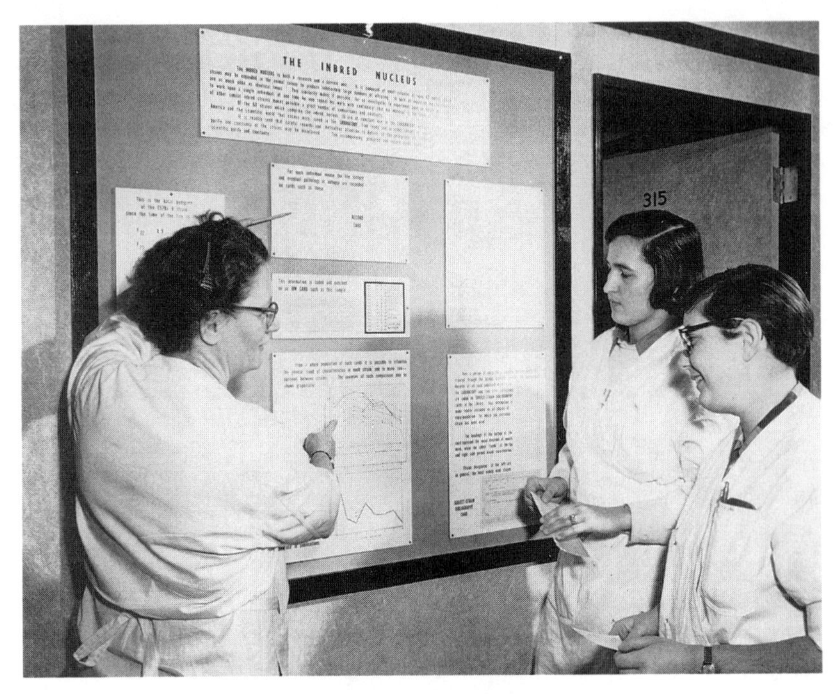

5.6. Elizabeth Russell explaining Inbred Nucleus, c. 1950 [Source Credit: Jackson Laboratory Archives].

(fig. 5.6). The inbred nucleus consisted of small, pure-line breeding colonies of each of the ten most widely used JAX inbred strains. It was initially housed apart from the new main laboratory building (fig. 5.7), itself a marvel of modernistic design compared to the quaint brick of the original lab. Russell's main job was to supervise a small staff of researchers who "collected data," as Little put it: "[R]ather like the information life-insurance companies study for different groups of humans, and collected for very much the same type of reason: to find out how good a risk each type of mouse is for some particular type of research work." Little himself soon took this metaphor more literally: in 1952 he pursued the possibility of purchasing life insurance for the JAX mice.[69]

to CCL, 7 November 1947; CCL to Muller, 17 November 1947; all in the H. J. Muller Mss., Manuscripts Department, Lilly Library, Indiana University-Bloomington; more generally, see 1948 JAX Annual Report (public version) and the film *Rx Mouse*, both in JLA-BH. For Little on JAX Lab as a "spiritual force," see Elizabeth Russell, "The Roscoe B. Jackson Memorial Laboratory: Leaders in Cancer Research and the World's Largest Supplier of Inbred Mice," *Purina Laboratory Animal Bulletin* (Spring 1955), in JLA-BH.

[69] Elizabeth Russell, "Inside the Inbred Mouse," *Bulletin for Medical Research* 9 (January–February 1955): 2–6; CCL to Lewis W. Lukens, 12 March 1952, Box 740, CCL-UMO.

5.7. Jackson Laboratory during rebuilding after 1947 fire [Source Credit: Jackson Laboratory Archives].

Meanwhile, Little did not fail to notice that the fire had once again pushed JAX and inbred mice onto the national stage. As former JAX librarian Joan Staats would note: "In many ways, it was the fire that put it (Jackson Lab) on the map . . . because then everybody heard of it—not just scientists. Previous to that time, it was famous within the scientific world—but that's all. But after the fire, there were a lot of fund-raising efforts and headlines in various newspapers about the fire and about the burning of the Laboratory." A scan of newspapers for the few years following the fire supports Staats's remembrance: editors from Boston to St. Louis lamented JAX's fate and lavished praise on Little's decision to rebuild and redouble his institution's cancer-fighting efforts.[70] The long-term stability of the JAX institution also had origins in the 1947 fire, for it was through the publicity surrounding the fire that Jackson Laboratory's mice became famous. In fact, in a development exemplifying the changing postwar organization of American science, public support was as much responsible for rebuilding JAX as any Rockefeller or federal government funds.

Little had exploited every opportunity to channel renewed public interest in scientific mice; in fact, the mouse's popular image underwent a significant transformation as a result JAX publicity after the fire. A great

[70] Interviews with Joan Staats and George Snell, JLOH-KR, and interview with George Snell, JLOH-APS; "Cancer Files, Mice Lost at Institute," *New York Times*, 25 October 1947, in RF, RG 1.1, Series 200D, Box 144, Folder 1777, RAC-NY. Other newspaper clippings can be found in Box 735, CCL-UMO; many date to April and May 1948.

5.8. Ladies Auxiliary of VFW laying a wreath on a commemorative tree at Jackson Laboratory, Memorial Day, 1955 [Source Credit: Jackson Laboratory Archives].

majority of articles about mice in the American popular press for the period from 1947 through the 1960s focused on the mouse's success as a "servant of science." For example, in 1949 *Science Newsletter* quoted cancer researcher Walter Heston's remark that "one of the greatest hopes for attacking the medical problems of the future lies in the development of mice that inherit many kinds of diseases." And a 1952 article on JAX in *Colliers* claimed to focus "On Important Mice and Men" in medical research: "in the tiered cages of this unique research institute may lie the answer to the mystery of diseases which kill two out of three Americans." Four years later, JAX mice were labeled "disease fighters."[71]

Several charity groups actively supported Little's "service to humanity" argument because it resonated with their own value systems. Ladies Auxiliaries of the Veterans of Foreign Wars, for example, donated money to immortalize their fallen family members or husbands who did wartime service, and this funding helped rebuild JAX buildings and support JAX educational ventures (fig. 5.8). Little visualized the metaphor of "mice-

[71] "Mice Exposing Man's Ills," *Science Newsletter* 56 (3 September 1949): 146; J. D. Ratcliff, "Of Important Mice and Men," *Colliers* 130 (8 November 1952): 13–15; "37,000 Disease Fighters," *Science Newsletter* 70 (18 August 1956): 103. See also "Unique Animals Aid Science," *Science Newsletter* 56 (2 July 1949): 10–11; "Of Fat Mice and Men," *Newsweek* 40 (4 August 1952): 52–54; "Of Mice and Men," *Newsweek* 68 (12 September 1966): 93. Cf. also Jean Paul Gaudillière, "Taking Mice for Men: The Production and Uses of Animal Models in Postwar Biomedical Research," paper presented at the Davis Center Workshop of Mice and Men: Animals and Medical Models, 14 March 1997.

5.9. Cover of "Project Mouse" (1952) [Source Credit: Jackson Laboratory Archives].

for-men" in R_x*Mouse*, which was co-produced by the Ladies Auxiliaries and the United Way of Michigan. This short piece, likely shown at the weekly public laboratory tours, sought to persuade viewers that mice at the Jackson Laboratory were not mistreated but appropriately sacrificed for humanity's benefit. Thus as one shot depicts the injection of a live mouse with tumorous tissue from another mouse, Little reassures the viewer: "This particular injection makes it possible for us to study the cause and effect of disease in our miniature human being."[72] With the mouse-animal explicitly acting as a stand-in for the human subject, the moral lesson of this scene is difficult to miss.[73] Likewise, in 1952 he initiated a campaign to educate high school students about the JAX mission by creating a book-

[72] *Rx Mouse*, at JLA-BH; on antivivisection activities at JAX, see interviews with Joan Staats, George Snell, Earl and Margaret Green, and Tom Roderick, June 1993, JLOH-KR.

[73] Cf. Anita Guerrini, "Animal Tragedies: The Moral Theater of Anatomy in Early Modern Europe," paper presented at the HSS Annual Meeting, Sante Fe, NM, 1993. For a modern fictional rendering of a mouse that draws the opposite moral, see Art Spiegelman, *Maus:*

let called "Project Mouse" (fig. 5.9). The final product included a dizzying array of often conflicting uses for mice: for example, instructions for dissecting a mouse (instead of the traditional frog) for an anatomy lesson, as well as for organizing your own wild mouse hunt.[74]

In 1949 Little had approved the consolidation of the local Bar Harbor fundraising effort with those of several major donors from the United Way of Michigan to create the Jackson Lab Association (JLA), a network of local fundraising organizations dedicated to spreading the JAX gospel. By 1950 JLA had active chapters throughout New England as well as in Cleveland, Omaha, Philadelphia, and Southern California.[75] Since many of Little's administrative strategies remained deeply steeped in the personalized managerial patronage approach of the 1920s, Little himself never became entirely comfortable with relinquishing control of the lab's funding to lay leaders, and this created some tension. The satellite JLA organizations did not receive much direct leadership from Little, yet when they showed initiative, the local leaders felt that they could do nothing properly in Little's eyes.[76]

As JAX received more national media attention, control over this and other aspects of JAX public relations escaped Little's tight grasp. By 1952, for example, because of the JLA's work the public perception of JAX was broader than simply a "mouse house." On the assumption that JAX was a clearinghouse for all animals used in research, many people sent Little letters in the early 1950s, asking if he was interested in selling or buying species ranging from guinea pigs to canaries.[77] Not all of the letters Little

---

A Survivor's Tale (New York: Pantheon, 1986). On Little's attempts to arrange to pitch this film to Walt Disney, see the introduction to this book.

[74] C.C. Little, ed., *Project Mouse: Rx Mouse and X Mouse*, with drawings by Francis J. Rigney (New York: The Paulist Press, 1952), by the New Hampshire Chapter of the Jackson Laboratory Association, JLA-BH. The copy of *Rx Mouse* in the Jackson Lab Archives belonged to George Snell's son, Tom, who went on to a scientific career.

[75] "Jackson Laboratory Association, Chapters Chartered" and "Jackson Laboratory Association, Memorandum of Procedures for Chairman," 8 September 1950, Box 735, CCL-UMO.

[76] This was best expressed by an amusing poem called "The Association Secretary," composed by a local JLA worker:

> If he asks for advice, he's not competent.
> If he doesn't ask for advice, he's a know-it-all.
> If he talks on a subject, he's trying to run things.
> If he remains quiet, he has lost interest completely.

"The Association Secretary," undated typescript, no author indicated but probably Robert Gantt, national chairman of the JLA, c. 1950–55 (see E. F. Chinlund to R. S. Morrison, 19 November 1952, RF Archives, RG 1.2, 200A, Box 134, Folder 1193, RAC-NY), JLA-BH.

[77] Gerry Canning (CCL's secretary) to Mrs. Kenneth Beach; William Murray to Mr. Jon Rood of Putney School in Vermont, 24 October 1952; CCL to Mrs. Glenn Decker, 27 October 1952; CCL to Mr. H. R. Gwilliam, 27 October 1952; all Box 740, CCL-UMO.

received agreed with the ethical assumption that the mouse was otherwise worthless to society. One anonymous author, for example, noted that she and her friends objected to the well-publicized JAX breeding experiments on inbred rabbits, which aimed at trying to create "good-" and "bad-tempered" strains. But Little refused to respond to this argument of individual animal integrity. Instead, he attacked the antivivisectionist woman ad hominem, writing back: "Dear Madame: The members of your club seem much more bad-tempered than the rabbit," and reiterating that animal experimentation represented the only true noble ethic: scientific progress in service to humanity.[78]

Little's paternalism persisted in other venues, including the lab's administration. In 1953 he hired a new associate in his own Boston brahmin mold: Robert Buell, a graduate of Harvard and Phillips Exeter Academy. Buell was highly aware of the personal touch Little still had with the local donors and, like Little, sought to maintain this link to JAX's scientific and social past. For example, in 1953 Buell objected to a staffer's idea of sending mimeographed invitations for a JAX party to the Bar Harbor summer colonists because "it might create an unfavorable impression, in view of their social prominence." Instead, Buell made sure they received engraved invitations.[79] Internal bureaucratic conflicts aside, however, JAX presented a unified identity to its potential patrons. In a 1949 foundation endorsement letter, the Board of Trustees enthusiastically noted that the institution had rebuilt and reclaimed its status as the "Bureau of Biological and Medical Standards."[80]

At about the same—for what reason, it is not clear—Little moved to reclaim the high ground for JAX research. Little and JAX staffer Katrina Hummel had continued working to characterize what he now called 'the mouse mammary tumor agent' even after Bittner's departure.[81] But instead, he submitted to the Rockefeller Foundation an experimental program designed to investigate the nature of the mammalian egg, arguing that JAX's history of research on maternal inheritance of cancer susceptibility was extremely productive and important to its genetic program. But

[78] Bartlett, "The Big Mouse Man."

[79] Buell to CCL, 31 July 1953, in Tom Roderick's personal collections of Little letters; on Buell's biography, see Little memo, c. 1 July 1953, Box 740, CCL-UMO.

[80] Richard W. Jackson (Roscoe's son) to Warren Weaver, 7 October 1949, plus enclosed endorsement, Rockefeller Foundation Archive, RG 1.1, Series 200D, Box 144, Folder 1778, RAC-NY.

[81] K. P. Hummel and C. C. Little, "Studies on the Mouse Mammary Tumor Agent": "I. The Agent in Blood and Other Tissues in Relation to the Physiologic or Endocrine State of the Donor," "II. The Neutralization of the Agent by Placenta," III. "Survival And Propagation of the Agent in Transplanted Tumors and in Hosts That Grew These Tumors in Their Tissues," *Cancer Research* 9 (1949): 129–34, 135–36, 137–38.

when Natural Sciences program officer Walter Loomis reviewed Little's application, the numbers told a story more in line with the trustees' interpretation. Loomis noted that the Natural Sciences division had already given JAX $256,000: "slightly over 88% of which has been used for the establishment and maintenance of his inbred mouse colony and only about 12% having been for research." Furthermore, he disparaged JAX's comparatively low output of research papers—in relation both to other labs and to its own output of mice:

> It is probably significant that Little sends out vast quantities of typewritten reports rather than sending reprints. . . . It can probably be honestly said that the valuable export of the Jackson Memorial Laboratory is in terms of boxes of mice rather than scientific publications. . . . The genetics of "maternal influence" at Bar Harbor just can not compare to the high quality of the [Sonneborn] Indiana group. . . . Leaving aside many of Little's slogans, such as "the development of inbred strains ranks with the discovery of microbes" as well as all the other various gildings of various lilies, the hard core [of JAX] seems to consist in a supply service of standardized mouse strains.

The Jackson Lab, Loomis ultimately concluded, was earning its scientific keep by producing mice, not research—and the grant was denied. Postwar developments at JAX had permanently recast its identity as the "bureau of mouse standards."[82]

A NEW ERA FOR BIOMEDICINE AND FOR JAX

In 1955 the practical relationship between federal cancer policy and JAX inbred mice got a dramatic boost when Lasker's newly christened American Cancer Society pressured the federal government directly to achieve better coordination of NCI cancer research labs and cancer treatment clinics. This led to the establishment of the Cancer Chemotherapy National Service Center (CCNSC), basically a "drug development program" for which Congress allotted nearly seventy million dollars over the period from 1956 to 1958. The CCNSC's sole purpose was the screening of anti-

[82] William Loomis Diary excerpt, 28–29 March 1951, RF Archives, RG 1.1, Series 200D, Box 144, Folder 1780, RAC-NY. On Sonneborn's research and research on maternal inheritance more generally, see Jan Sapp, *Beyond the Gene: Cytoplasmic Inheritance and the Struggle for Authority in Genetics* (New York: Oxford University Press, 1987), chap. 4. Cf. Schloegel, "Life Imitating Art."

cancer therapeutics, an enterprise that relied heavily on standardized tumors and standardized mice.[83]

While driven by a different set of actors with different priorities, the development of the CCNSC relied heavily on the pre-existing strategies for organizational and cultural justification that had led to the inbred mouse's success in the prewar period. Animal breeding for the CCNSC began under the supervision of the National Research Council at the Jackson Lab, and inbred mice were the accepted "gold standard." Because of the enormous numbers of animals needed—one scientist remembered: "it required more mice than were available for all of cancer research at that time"—breeding was soon also contracted out to commercial animal suppliers who agreed to follow established JAX inbreeding protocols. As a result of the CCNSC, then, inbred mouse production at JAX and beyond literally became part of what one NCI leader called "an industry-government cooperation . . . as effective in the pharmaceutical industry as it is in some of the defense areas."[84]

For benchtop researchers, the significance of this development ranks alongside other contemporary developments in the infrastructure of scientific instrumentation. At the same time that JAX was focusing on the standardized production of its living research tools, the separation of material production and research had begun to permeate the realm of biological instrumentation more broadly. By the late 1940s, the ultracentrifuge, electrophoresis apparatus, and the electron microscope all became mass-produced equipment that scientists could order from commercial companies that often had little or no connection to promoting the work of the scientists who first pioneered the tools' use. JAX inbred mice were now more valued by cancer researchers for their reliable standardization—each one, as Little put it, "as identical as newly minted coins"—than for the laboratory work of their developers.[85]

[83] In a short and often-neglected article, historian Donald Swain makes a similar point about the general relationship between pre- and post-war biomedical policy at NIH: see Donald C. Swain, "The Rise of a Research Empire: NIH, 1930 to 1950," *Science* 138, 3546 (14 December 1962): 1233–37.

[84] Shimkin, *As Memory Serves*, p. 55; Kenneth Endicott, "The Chemotherapy Program," *Journal of the National Cancer Institute* 19 (1957): 275–93. Cf. Bud, "Strategies"; also Ilana Lowy, *Between Bench and Bedside: Science, Healing, and Interleukin-2 in a Cancer Ward* (Cambridge: Harvard University Press, 1996), chap. 1.

[85] Cf. Creager, *Life of a Virus*, pp. 130–38; Rasmussen, *Picture Control*, chap. 1; Lily Kay, "Laboratory Technology and Biological Knowledge: The Tiselius Electrophoresis Apparatus, 1930–1945," *History and Philosophy of the Life* Sciences 10, 1 (1988): 51–72. For an overview of the development of scientific instrumentation, with an excellent treatment of the "Mouse" by Jean-Paul Gaudillière, see Robert Bud and Deborah Jean Warner, *Instruments of Science: An Historical Encyclopedia* (New York : Science Museum, London, and

The cultural significance of the CCNSC program, however, lies in large part with the historical transition it marks in the criteria by which biomedical research was judged by the public and policymakers as successful. The era of "rational therapeutics," begun in the late nineteenth century in U.S. medical schools, had come full circle; now members of the lay public were using the laboratory as their lever in reformist efforts to put medical research more in touch with the clinic. Policymakers, in turn, responded to this call with a new willingness to develop and support mission-oriented, applied projects in experimental biology and medicine. By both these criteria, JAX and its mice still showed promise, but funding decisions reflected the prevailing opinion that this promise could be best fulfilled if the animals were removed from the very experimental genetics context through which their laboratory use had been developed. JAX mice were, as the author of a 1952 *Colliers* article on Jackson Lab put it, "The World's Most Important Mice," because "in their tiered cages of a unique research institute may lie the answer to the mystery of diseases which kill two out of every three Americans" (fig. 5.10).[86]

For many mammalian genetics researchers, however, JAX's twenty-five years of existence nevertheless deserved a celebration, and many descended on Bar Harbor in the summer of 1954 to join in the festivities. Little planned a program that downplayed for his contemporary audience JAX's reputation as a mouse producer.[87] Instead, the event was designed to draw attention to the lab's founding research mission. The official title of the twenty-fifth anniversary conference was "Twenty-five Years of Progress in Mammalian Genetics and Cancer." Little invited Castle to give a talk on the links between JAX and mammalian genetics, and the session was organized around a Castle student reunion. His students even recreated a "live family tree" of the Bussey Institute's descendants in the field. But while Castle discussed some early JAX research on size inheritance in mice, he ultimately emphasized that Little's inbreeding program was the "cornerstone" of the institution and its scientific contributions. Furthermore, when Little extended Michigan State's H. R. Hunt—a Bus-

National Museum of American History, Smithsonian Institution, in association with Garland, 1998), with "Mouse" entry on pp. 403–4.

[86] Cf. Harry M. Marks, *The Progress of Experiment: Science and Therapeutic Reform in the United States, 1900–1990* (New York: Cambridge University Press, 1997), chap. 1. Also Timothy Lenoir and Marguerite Hays, "The Manhattan Project for Biomedicine," in *Controlling Our Destinies*, ed. P. Sloan (Notre Dame: University of Notre Dame Press, 1995), pp. 29–62.

[87] JAX secretary, Gerry Canning, joked to a correspondent in August 1954 about the tightly controlled discussion about the anniversary conference among Little and the rest of Castle's students: "the whole reunion has a startling resemblance to the establishment of a second front, a violent, bloody business, shrouded in secrecy." Canning to Diane Kelton, 10 August 1954, Box 12 (Little letters); JLA-BH.

Collier's, November 8, 1952

The World's Most Important Mice

By J. D. RATCLIFF

In the tiered cages of a unique research institute may lie the answer to the mystery of diseases which kill two of every three Americans

"WITHIN the next 24 hours, 4,000 people will die in the United States. Two out of three of them will die of constitutional diseases—cancer, heart and circulatory trouble, kidney disease, diabetes and others. These diseases aren't passed man to man. They are, in effect, 'built-in' weaknesses. For the most part they don't come from outside stresses or from known infectious agents, but represent a breakdown of the individual's body machinery. They are the greatest challenge facing the research man today. We must find *why* these failures occur, and what can be done to prevent them."

The speaker is Dr. Clarence Cook Little, head of one of the world's most remarkable research institutions—the Roscoe B. Jackson Memorial Laboratory, of Bar Harbor, Maine. Most of the Jackson lab's studies are conducted on mice. It has the largest collection of bizarre mice in the world. There are 60-odd strains: fat mice and dwarf mice; mice with corkscrew pigtails and mice with bobbed tails; some with curly hair and some with no hair.

Dr. Clarence Cook Little

One strain has cleft palates, and another clubfeet. There are shaking mice and waltzing mice, who waltz because of an inherited inner-ear defect (the inner ear controls body balance). One strain occasionally bears babies with no teeth—freaks that perish when they are weaned because they can't eat solid food. Others have extra-long teeth which constantly irritate lips and may give rise to cancer. Many of these mice are predisposed by inheritance to disease: cancers of all types, anemia, leukemia, glandular disturbances and other disorders.

Dr. Little says: "It is almost impossible to study the 'built-in' diseases in human beings. Man lives too long—a patient may outlive the research man. We have, therefore, turned to short-lived animals—mainly mice.

"In humans, for example, we rarely see cancer until it is well developed and possibly on the way to killing the victim. In mice, we can follow the de-

13

PHOTOGRAPHS FOR COLLIER'S BY RALPH ROYLE

5.10. Collier's article: "The World's Most Important Mice," 1952 [Source Credit: C. C. Little Papers, University of Maine].

sey alum—an open-ended offer to join the JAX staff, Hunt turned it down because he doubted that the lab's small genetics research program could garner permanent grant support for his work.[88]

Little also devised a publicity-oriented presentation of the anniversary event, which he called "Genes, Mice, and Men." This booklet illustrated

[88] The proceedings of the conference were edited by Elizabeth Russell and published as *Journal of the National Cancer Institute* 15, 3 (December 1954); Hunt to CCL, 23 July 1954, Box 12 (Little letters), JLA-BH.

JAX's successes with statistics: in twenty-five years, JAX had gone from 8 assistants to 118, from a $56,000 budget to an $800,000 one, and from 40,000 mice raised and 500 shipped to one million mice raised and 300,000 mice shipped. But while insisting that mouse production alone was a laudable achievement, the text was organized and largely devoted to JAX research, especially cancer research and its applications: "The Jackson Laboratory both pioneers and develops the living 'pay dirt' of biology."[89] This message was not lost on the readers. Education leader Arthur Adams wrote to Little that the "Genes, Mice, and Men" report "puts in popular and dramatic form the story of the intensely scientific yet intensely human objectives of the Laboratory."[90] Ironically, however, by the mid-1950s many cancer policymakers were more willing to put their faith in the inbred mouse than the group of human researchers that had given birth to and nurtured it. Little said of his institution's achievements: "If the Jackson Laboratory had done nothing more than create and continue to produce these inbred 'pure' strains of laboratory mice, all of the labor and enthusiasm expended by it would be justified." In light of the expansive visions for reform biology of his earlier career, it is difficult not to wonder whether this opinion was something Little himself believed or merely mimicked.[91]

[89] *Genes, Mice, and Men: A Quarter-Century of Progress at the Roscoe B. Jackson Memorial Laboratory* (Bar Harbor: The Jackson Laboratory, 1954), pp. 6–7, JLA-BH.

[90] Adams to CCL, 4 March 1954; see also Beulah France, R.N., to CCL, 5 March 1954; and Carl Fenninger to CCL, 8 March 1954; all in Box 12 (Little letters), JLA-BH.

[91] *Genes, Mice and Men*, p. 8.

CHAPTER SIX

MOUSE GENETICS AS PUBLIC POLICY

Radiation Risk in Cold War America (1946–56)

For two consecutive summers before she started graduate school at the University of Chicago, biology student Liane Brauch visited the Jackson Laboratory. There she met William "Bill" Russell, a former Sewall Wright student at Chicago and a JAX scientist since 1938. Russell's work was far from marginalized but certainly dealt in the less politically dramatic aspects of the lab's now signature organism. "I was more interested," he later said, "in the theoretical genetics and other applications, rather than the cancer side." While Brauch stayed on at JAX, working as a technician for most of 1945, Russell's projects spanned from ovarian transplant studies in mice to tissue transfer experiments in guinea pigs, but his goal remained the same: to study variability within inbred strains, "trying to break down variability into genetic and environmental components." JAX was an ideal place for this work. With many inbred and mutant mammal strains available, Russell had a wealth of material, and in turn, his projects proved beneficial for characterizing the genetic behavior of the JAX production stocks (fig. 6.1). But 1947 marked a transition for Russell, both personally and professionally. Shortly after his divorce from fellow staffer Elizabeth Russell became final and the Jackson Lab fire destroyed all of his mice and his research, he decided it was time to leave Bar Harbor.[1]

[1] Oral History interview with William and Liane Russell by the author, 8 December 1995, Oak Ridge, TN, JLOH-KR. Complete bibliographies for William Russell and Liane Brauch

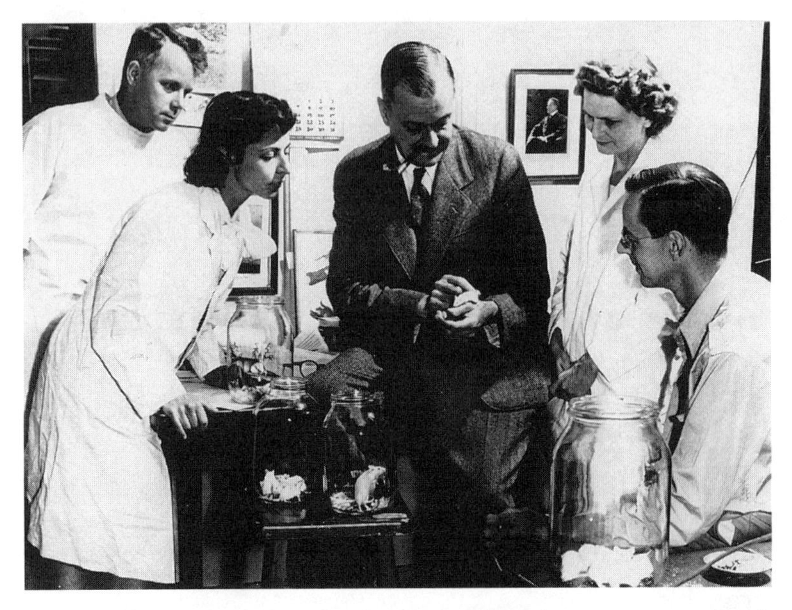

6.1. JAX publicity photo (note portrait of Jackson hanging in background), featuring , left to right, George Woolley, Liane Brauch [later, Russell], C. C. Little, unknown, and W. L. Russell, c. mid-1940s [Source Credit: Jackson Laboratory Archives].

Russell's departure from JAX marked the beginning of new institutional and experimental era for mammalian genetics research, as well as for the development and circulation of inbred mice. While entertaining some academic job offers, Russell got a letter from biologist Alexander Hollaender, asking him to come to Oak Ridge and set up a new Division of Biology in what was the old X-10 plutonium production building. In the immediate postwar biomedical research policy climate, public and scientific apprehension about the health risks posed by atomic radiation swamped other health concerns. Accordingly, the federal government began to make unprecedented investments in radiation biology research, especially in the question of the genetic effects of mutagenesis.[2] As a

Russell are available at from the Oak Ridge National Laboratory Library. Cf. W. L. Russell, "An Analysis of the Action of Color Genes in the Guinea Pig by Means of the Dopa Reaction," Ph.D. Thesis, University Of Chicago, 1937; W. L. Russell and J. G. Hurst, "Pure Strain Mice Born to Hybrid Mothers Following Ovarian Transplantation," *Proceedings of the National Academy of Sciences* 31 (1945): 267–73; Russell, "Inbred and Hybrid Animals."

[2] John Beatty and Angela Creager are both currently working on histories of the federal government's involvement in radiation biology research. Beatty's work focuses on genetics, and Creager's on the radioisotope program. The best published accounts to date are John Beatty, "Genetics in the Atomic Age: The Atomic Bomb Casualty Commission, 1947–1956," in *The Expansion of American Biology*, ed. K. Benson, J. Maienschein, and

result, even while inbred mouse materials continued to be produced at JAX for the U.S. government's chemotherapy screening program, radiation genetics displaced cancer as the public health problem for which both the discipline of genetics and the research materials of inbred mice were thought most suitable.

Russell married Liane Brauch, and in 1948 the couple moved to Oak Ridge, Tennessee, where scientific work started out small but grew even more quickly beyond the bounds of their work at JAX than either one of them had imagined. Liane finished her thesis on X-ray-induced developmental abnormalities in the mouse,[3] and Bill planned to re-create his lost ovarian transplantation studies in strain 129. Reconstituting this stock presented some challenges. Strain 129 was now a congenic strain (like the ones George Snell had developed) with two coat color markers, one allele of which they had to retrieve from a mouse in the possession of Bill's former graduate student worker at JAX. In the first few months Bill managed to do only a few experiments in the single room Hollaender gave him for mice. Even then, he remembered, "we were already overflowing that and having to put mice into our offices." But within five years the Russells had moved into wartime building Y-12 and developed one of the largest inbred mouse colonies in the country. It was not unusual for ten thousand first-generation animals to be analyzed in one of their experimental protocols (fig. 6.2). Within ten years, they were being visited by J. F. Loutit, head of the Medical Research Council's Radiobiology Division, and asked to send strains of mice to the MRC Harwell Unit, so that their experiments on inducing mutations with radiation in mice could be replicated by British scientists.[4]

The Russells' scientific work at Oak Ridge combined with the policymaking vision of Division of Biology head Alexander Hollaender to transform mammalian genetics from a discipline focused on routine theoretical analysis of rare mutants to a field whose results were scrutinized by state departments and foreign ministries in the United States and Europe. Such

R. Rainger (New Brunswick: Rutgers University Press, 1986), pp. 284–324; and M. Susan Lindee, *Suffering Made Real* (Chicago: University of Chicago Press, 1994). The Atomic Energy Commission offered to have the Jackson Lab reestablish itself at Brookhaven after the 1947 fire, but Little and the Board of Trustees preferred to remain in Maine. See Press Release no. 66, 29 October 1947, in *U.S. Atomic Energy Commission Releases, April 19, 1947–Jan. 1, 1948* (incomplete, held by Princeton University Firestone Library); many thanks to Angela Creager for this reference.

[3] Liane B. Russell, "X-Ray Induced Developmental Abnormalities in the Mouse and Their Use in the Analysis of Embryological Pattern," Ph.D. dissertation, University of Chicago, 1950.

[4] WLR-OH; on MRC labs obtaining strains from Russell's lab, see Earl Green and T.H. Roderick, "Radiation Genetics, in *The Biology of the Laboratory Mouse*, 2d ed., ed., Earl Green (New York: McGraw-Hill, 1966), pp. 165–85, quote on p. 165.

6.2. Liane and William Russell, in their mouse room at Oak Ridge, c. 1950 [Source Credit: Ed Westcott; Department of Energy, provided by Oak Ridge National Laboratory].

developments secured the inbred mouse's place as a genetic model organism in postwar biomedical research laboratories. Written off by some researchers as too chromosomally cumbersome to provide useful results (recall that mice have twenty-one chromosomes, compared with *Drosophila*'s four), inbred mice came to play a central role in the conception, construction, and execution of a federally sanctioned radiation risk policy in the years immediately following World War II. While contemporary biological scientists were working to evaluate radiation's genetic threat to humanity—as well as debating how best to define and investigate this threat—Hollaender was building what he envisioned as a new mode of biological Big Science: academic-quality research done at national laboratories. Hollaender charged Russell with the task of conceiving and carrying out a series of experiments that would accommodate competing perspectives on the genetic risk of radiation. Russell devised what came to be known as the "megamouse" study. By exposing hundreds of thousands of specially engineered inbred mice to radiation and measuring the induction of mutations in their offspring, the so-called Specific Locus Test (SLT) at the cornerstone of this work transformed large-scale mammalian genetics breeding, both in scale and in scope, and created a legacy of mutant materials that is still being used today.

The inbred mouse's early development as a radiation genetics test organism proceeded differently from the incorporation of this same animal in earlier American cancer research. By the mid-1930s, cancer research policy was administered according to government-sanctioned values of centralization and coordination in all experiments. Postwar radiation genetics, by contrast, proceeded according to mission-oriented imperatives, even while Hollaender simultaneously strove to establish Oak Ridge Laboratory as an institution where openness and a tolerance for theoretical exploration were also valued. In addition, for three decades of mouse cancer research (1933–55), JAX Lab produced and sold many inbred strains that were used as experimental tools or models in different ways by different researchers, and all these heterogeneous modes of practice coexisted more or less peacefully. But in mouse radiation work, one mouse model—the SLT—quickly emerged as the dominant experimental system for radiation testing in laboratory mammals. The SLT's monopoly over the field made consensus over its use necessary for settling scientific and policy debates over radiation risk, but this consensus was neither straightforward nor easily achieved.

In heated discussions with the genetic group of the National Academy of Science's Biological Effects of Atomic Radiation (BEAR) committee, Bill Russell strove to apply data from the SLT to real-world scenarios: namely, to address the problem of radiation and human genetic disease. A reduction in permissible human radiation dose initially resulted from this process, and two years later, Russell's observation that radiation dose rate (that is, the time over which radiation was administered) affected mutation rate transformed both policy recommendations and the field of theoretical genetics. Ultimately, as in cancer research, the SLT inbred mouse became an acceptable human stand-in for radiation studies. But in the charged arena of cold war politics, the combined efforts of scientists and policymakers to manage mouse models in their intended policy context raised just as many social and methodological issues as they solved.

GENETICS, RADIATION, AND RISK IN AMERICAN SCIENCE AND CULTURE, C. 1946

The biological and medical effects of radiation were subjects of American social concern and scientific investigation long before 1945. The pathological effects of radiation were first documented in 1908, and over the next twenty years, genetic investigators hailing from Paris to New York irradiated everything from fungi and plants to grasshoppers and even mice. All of this work suggested that radiation's heritable effects occurred across the biological world. Such effects included altered sex-ratio, major and minor phenotypic variations, reduced fertility, and reduced lifespan.[5] But it was a particular constellation of historical events that conspired to transform mouse radiation genetics from a small subdiscipline into a major research initiative, at the fore of the AEC's institutional agenda.

The discipline of radiation genetics is most often cited as beginning with H. J. Muller's 1927 paper, which reported that X-rays could induce sex-linked recessive lethal gene mutations in *Drosophila* germ cells.[6] Indeed, Muller's work, Muller himself, and the changing political climate of the interwar period all combined to provide impetus for more concerted scientific and policy actions regarding radiation's effects in the decades that followed this milestone. In 1928 the U.S. National Research Council appointed a Committee on the Effects of Radiation on Living Organisms (CERLO) to study these phenomena more thoroughly. From 1933 to 1935, working in Muller's laboratory, mouse geneticist George Snell produced some of the first suggestive findings in an experimental mammal. Snell reported that X-rays induced reciprocal chromosomal translocations in the spermatogonial (i.e., immature spermatozoa) cells of male mice and that litters sired by these spermatogonially affected males consequently expressed a genetically dominant lethal condition he called "semisterility"—the litters were half the size of the control litters.[7]

Subsequently, during World War II, Manhattan Project researchers monitored and studied the consequences of radiation for bomb-worker safety, while from 1943 to 1950, the U.S. government contracted with Donald Charles and his co-workers at the University of Rochester for a series of mouse genetic experiments. These experiments promised to shed light on the "types of genetic change which would be significant in human populations," specifically those exposed to a range of chronic low-level radiation doses (0–10 roentgens). Charles estimated a dominant mutation rate resulting from irradiation, based on nearly fourteen thousand F1 mouse offspring, of 3.85×10^{-5} roentgens per genetic unit, and he also presented data suggesting that his exposures led to reduced litter size,

[5] Lindee, *Suffering Made Real*, pp. 61–65; Nickolai Timofeef-Ressovsky, "The Experimental Production of Mutations," *Biological Reviews* 9 (1934): 411–57; for mouse work, see Bagg and Little, "Four Inheritable Morphological Variations in Mice."

[6] H. J. Muller, "Artificial Transmutation of the Gene," *Science* 66 (1927): 125–34.

[7] Oral History interview with George Snell by author, 28 June 1993, Bar Harbor, JLOH-KR. On Snell's training and career, see "George Snell," *Current Biography*. George Snell, "Genetic Changes in Mice Induced by X-rays," *American Naturalist* 67 (1933): 24; "X-ray Sterility in the Male House Mouse," *Journal of Experimental Zoology* 65 (1933): 421–41; "The Production of Translocation and Mutations in Mice by Means of X-rays," *American Naturalist* 20 (1935): 545–67. Paula Hertwig's lab in Germany was a center of mouse radiation genetics throughout the 1930s and into the 1940s: for a review, see W. L. Russell, "Genetic Effects of Radiation in Mammals," in *Radiation Biology*, vol. 1, ed. A. Hollaender (New York: McGraw Hill, 1954), pp. 826–47.

growth, weight, and postweaning mortality. But largely because of its secret status, and its failure to demonstrate clear or new trends, this work contributed little to contemporary public (or even semipublic) policy debates over human risk.[8] Still, by 1941, both Britain and the United States were beginning to enforce (however ineffectively, as Russell Olwell has shown)[9] their own crude existing limits on human exposure (0.1 roentgen/day) within government facilities.[10]

Almost immediately after the war ended, broader public concern about radiation began to grow, spurred on by media reports on the sufferings of Japanese bomb survivors. Through the world's fascination with these people, radiation began its journey—in J. Samuel Walker's words—"from the rarefied realms of scientific and medical discourse to the front page . . . in news reports, magazine stories, political campaigns, and congressional hearings."[11] As John Beatty has documented, the most frightening idea was that radiation's worst effects—damaging heritable mutations—might not be manifest until it was too late to prevent the suffering of the next generation.[12] Consequently, then, when AEC set up the Atomic Bomb Ca-

[8] Charles's first true publication of the results was in a medical, rather than a genetic, journal. See D. R. Charles, "Radiation-induced Mutations in Mammals," *Radiology*, 55 (1950): 579–81; and, more comprehensively, Donald Charles et al., "Genetic Effects of Chronic X-irradiation Exposure in Mice," U.S. AEC Commission Research and Development Report UR-565, RRT-UT. I have found no precise dollar figure for Charles's AEC contract at the University of Rochester, but Rochester was the second highest Manhattan Project Biological Research–funded institution—netting 1.7 million dollars in 1945–46 alone—and the Charles project represented a large part of that initiative: see "Proposed Medical Research Program, 1946–47," DOE #051094-A-157, DOE Archives (Germantown, MD, but also in part formerly housed at the Advisory Committee on Human Radiation, Washington, DC: now, "DOE-Hum" and "DOE-Ger"). Cf. Lenoir and Hays, "The Manhattan Project for Biomedicine."

[9] Russell Olwell, "Radiation Protection for Workers at Oak Ridge: Scientific Debate and Workplace Practice, 1942–1950," paper Presented at the History of Science Society Meeting, Atlanta, 1996.

[10] See "ORG:NAS:Com on the Effect of Radiation on Living Organisms: Beginning of Program," in the NAS-DC; Robley Evans (MIT) obituary in the *Boston Globe*, from 3-4 January 1995; Gilbert F. Whittemore, "The National Committee on Radiation Protection, 1928–60: From Professional Guidelines to Government Regulation," Ph.D. dissertation, Harvard University, 1986; Catherine Caufield, *Multiple Exposures: Chronicles of the Radiation Age* (Chicago: University of Chicago Press, 1989); Barton Hacker, *The Dragon's Tale: Radiation Safety on the Manhattan Project, 1942–1946* (Berkeley: University of California Press, 1987).

[11] J. Samuel Walker, "The Atomic Energy Commission and the Politics of Radiation Protection, 1967–1971," *Isis* 85 (1994): 57–78 (esp. p. 58), and "The Controversy over Radiation Safety: A Historical Overview," *Journal of the American Medical Association* 262 (1989): 664–68. For a more wide-ranging historical view, see Spencer Weart, *Nuclear Fear: A History of Images* (Cambridge: Harvard University Press, 1988).

[12] Beatty, "Genetics in the Atomic Age."

sualty Commission (ABCC) in 1945 to study the bomb survivors, this organization's genetics project attracted a great deal of public and scientific attention.

From its very first scientific advisory meeting in 1947, the ABCC genetics project was plagued with political and methodological controversies. A quick overview suggests that a spectrum of competing views co-existed regarding both how to define human radiation risk and how to investigate it. Susan Lindee has argued that the AEC began the ABCC human genetics study on Japanese survivors with the understanding that it had low potential for providing the kinds of conclusions about the fundamental nature of genetic risk in human material that the experimental geneticists had already generated in their work with lower organisms. To begin with, the "human experiment" could not, practically and ethically, be repeated, and researchers had no good dosimetry estimates for the amount of radiation received by Japanese survivors. Nevertheless, a key issue remained: how valuable would it be to investigate the basic phenomena in lower organisms and simply extrapolate to humans? ABCC genetics project director James Neel (a *Drosophila* geneticist turned human geneticist) had argued that any human data, even if only "detailed family studies" that were largely epidemiological, would be extremely valuable.[13]

Notably, though, nearly every other scientist at a 1947 genetics conference officially endorsed animal genetics as a much-needed experimental project to supplement the human one. Neurospora geneticist George Beadle suggested that animal studies were necessary from a "purely scientific viewpoint," since the human material was so uncontrolled that it could not produce reliable data on any front. The 1955 ad hoc ABCC evaluation committee echoed this criticism, noting that overall, the program was lacking in the "experimental viewpoint."[14] *Drosophila* geneticist Alfred Sturtevant took an even stronger position. He opened an influential 1954 *Science* article on the contemporary social implications of genetics with the statement: "Man is the most unsatisfactory of all organisms for genetic study," arguing that the only work that was needed could be done on flies and bacteria.[15]

[13] Lindee, *Suffering Made Real*, pp. 73–74. Ultimately, Lindee suggests that the ABCC genetics project survived for nearly ten years partly because of its unique research materials; the work also won the U.S. government political points for ongoing concern over the human consequences of nuclear war, regardless of whether it ever produced conclusive results.

[14] "Francis Report," pp. 21–23, 27 November 1955, NAS Central Files: "Medical Sciences: Committee on Atomic Casualties: Conference to Review Reports on Survey of ABCC, Ad Hoc," NAS-DC.

[15] Alfred Sturtevant, "The Social Implications of the Genetics of Man," *Science* 120 (1954): 60; see also coverage of Sturtevant's views in the *New York Times*, 11 November 1954.

Controversies nevertheless persisted over how to determine what events counted as demonstrating "real" genetic risk. In public and scientific forums, for example, Muller himself repeatedly pushed a strict genotypic basis for risk determination: he stressed the danger of overall increased mutation rate—no matter how small or functionally ineffectual each mutation (e.g., eye or hair color changes)—and he preached the necessity of carefully monitoring what he called the human's race's total "load of mutations."[16] While existing *Drosophila* data were suggestive, Muller argued, new mouse studies were needed to determine the frequency of recessive mutations—a category of genetic damage that even Neel reluctantly agreed would be important from a policy standpoint but virtually undetectable in the first generation of the human survivors' offspring.[17]

Alternatively, the ABCC survivor study interpreted that risk should be determined by a phenotypically visible distressful change within a specific human cultural context. Neel and his colleagues tracked parameters in the Japanese population that were intended to reflect a spectrum of mutational events, from dramatic phenotypic abnormalities (such as minor malformation and stillbirth) to "fitness indicators" (like birth weight and growth of children at nine-months); these occurrences were presumed to be the cumulative effect of minor deleterious, but not disastrous, genetic changes. But in the end, as Lindee has shown, these geneticists dismissed those signs of mutation that they believed would have little social cost in terms of true human suffering. These included reduced fertility (e.g., semisterility) and even abnormalities such as malformed ears and partial albinism.[18]

Pressure to reconcile these competing views of risk and its genetics study arose in both internally and in public relations circles. In 1947 the AEC's existing biological and medical research program underwent its first internal review, which sternly concluded that the AEC had funded too much basic research "directed to the furtherance of knowledge regarding cellular

[16] Cf. Diane Paul. "Our Load of Mutations Revisited," *Journal of the History of Biology* 20 (Fall 1987): 3–20. On Muller's mutation studies as politics by other means, see also Devora Kamrat-Lang, "Science as Political Activism: The Mutation Research of H.J. Muller, 1918–1927," paper presented at the Joint Atlantic Seminar in the History of Biology, Cold Spring Harbor Laboratory, Winter 1995.

[17] Conference report: "Concerning the Study of the Genetic Effects of the Atomic Bombs in Hiroshima and Nagasaki," 30 June 1947, Dr. 19—"Genetics—1947, #1"; minutes of the Committee on Atomic Casualties, 26 June 1947; both in NAS-DC. Ironically, mouse geneticist Don Charles was the only member of the conference who had more enthusiasm and methodological suggestions for the human study than he had to say about experimental mouse work. Muller and Neel were more interested in using his data to extrapolate to humans than he was. See Lindee, *Suffering Made Real*, p. 74.

[18] M. Susan Lindee, "What Is a Mutation? Identifying Heritable Change in the Offspring of Survivors at Hiroshima and Nagasaki," *Journal of the History of Biology* 25 (Summer 1992): 231–55; see also *Suffering Made Real*, chap. 9.

physiology as such" and in the process had shirked its "primary responsibility" as a promoter of research "of the programmatic, applied type"[19]—specifically, toward understanding the most-feared biological effects of atomic radiation for humanity. As the AEC put it in another policy document: "The need for answers to this question is quite urgent because the accepted dosage of 0.1 r per day for human exposure may not be as safe as we would expect from the usual statements." Now was the time, they wrote, for a "concerted attack" on this "major Public Health problem."[20] By the mid-1950s, in the wake of widespread media coverage of the "Bravo" test explosion in the Pacific, the public appears to have come to the same conclusion. As a contemporary *New Yorker* reporter concluded,[21] the world was now "fallout conscious," and internal memos reveal that the AEC was clearly aware of the problem. Americans were demanding answers that would reassure them not only about the safety of the Japanese "guinea pigs of atomic warfare" (as they were often called in press reports[22]), but also about the potential risk to their own descendants.

HOLLAENDER AND "BIG BIOLOGY" AT OAK RIDGE AND BEYOND

Because the Manhattan Project already incorporated a significant biomedical research program, it is not surprising that the AEC extended its commitment to biological work after the war. Sponsorship of genetics research alone, for example, increased nearly sevenfold during the period from 1945 to 1960. Nevertheless, after the war, important structural questions about AEC-sponsored biological research remained very much open: would the new biological work be secret and mission oriented (as was the earliest Manhattan Project biological research at the University of

[19] Robert Loeb to David Lillenthal, plus attached report by AEC Board of Review, 20 June 1947, MS 1067, Folder 2, RRC-UT; cf. also George Beadle to Hollaender, 15 December 1946: "In our meeting, I held out for as much basic work as possible . . . [but] I appreciate that some compromise between all basic work and all programmatic research must be made," from MS-652, Box 10, Folder 55, RRC-UT.

[20] "Why in the Hell Should the Hills of Tennessee Have a Good Biology Division? Conversations with Alexander Hollaender," compiled by Ida C. Miller, in MS 1709, Folder 2, RRC-UT; "Outline of Research of the Biology Division, A. Hollaender, Director," 1 December 1946, DOE-051094-A-456, DOE-Hum.

[21] Daniel Lang, "Fallout," originally published in the *New Yorker*, reprinted in *From Hiroshima To The Moon: Chronicles of Life in the Atomic Age* (New York: Simon and Schuster, 1959), p. 369.

[22] E.g., "A-Bomb Story Not Yet Told," *Science News Letter* 51 (1947): 219. Beatty makes this same point in "Genetics in the Atomic Age." Despite the prevalence of this expression, little has been written on guinea pigs as actual research organisms.

Rochester)?[23] Or open and dictated by the emerging interests of the scientists themselves? And should it address only applied or also basic research questions? Thus in the immediate postwar era of federal funding for biological research, the AEC faced critical decisions—similar to those faced by the private foundations (biology's key prewar sponsors) nearly sixty years earlier and the National Institutes of Health in 1937—about what features should constitute their new system of biological research.[24]

Manhattan Project Medical Director Stafford Warren held expansive ideas of social reach of postwar biomedicine, but for practical advice, the AEC called on NIH radiobiologist Alexander Hollaender. Hollaender's plan for biology and genetic research in the postwar world guided AEC initiatives at Oak Ridge, where he created and presided over the Division of Biology for nearly two decades (1947–66). Hollaender's vision was at once entrepreneurial and system-oriented, because it took as its starting point the need to achieve a genuine reconciliation between AEC priorities and the scientific inclinations of individual biological researchers. The biological research realized under his tenure made use of the unique resources provided by the AEC and by Oak Ridge's national laboratory setting, while at the same time created a model for postwar biomedical research practices more generally.[25]

By 1940 Hollaender had built a unique reputation among radiation biology researchers and policymakers, since he embodied an interdisciplinary perspective as well as a rigorous experimental approach. In 1934 the Rockefeller Foundation's Warren Weaver asked him to spend the next nine months making a comprehensive survey of all research on the biological effects of radiation. As Hollaender recalls: "I said 'Although nobody is interested in it, I want to work on it.' And they said 'You go ahead and do this for us, and then you tell us who we should support." After finishing the survey, Hollaender returned to Wisconsin as professor of biology and published a series of important papers in this field. The most famous of these was a debunking of mitogenetic rays, but his interest in the problem of mutation and the chemical components of biological systems also led him to explore the ability of certain bacterial cells to recover from, or

[23] This growth estimate is based on numbers of contracts, which grew from eight (primarily at Rochester, during and immediately after the war) to fifty-five in 1959. See U.S. Atomic Energy Commission, *Genetics Research Program of the Division of Biology and Medicine* (Oakridge, TN, January 1960) (published report #TID-4041).

[24] On foundation sponsorship of biological research, see Kohler, *Partners in Science*; Kay, *The Molecular Vision of Life*.

[25] For a more biographical treatment of some of the points raised in this section, see Karen A. Rader, "Alexander Hollaender's Postwar Vision for Biology: Oak Ridge and Beyond," paper presented at the "Master Builders" Workshop, Johns Hopkins University, April 1999.

to repair, radiation damage. To focus exclusively on this latter research, he accepted a U.S. Public Health Service appointment at the Washington Biophysical Institute, a short-lived interdisciplinary arm of the NIH. During this time, he also collaborated with several geneticists—Miloslav Demerec and members of the phage group on mutations and nucleic acids in *Drosophila*—and it was through this project that Vannevar Bush became aware of his work and granted him an appointment as an associate of the Carnegie Institution.[26]

In 1946 the strength of Hollaender's commitments to radiation biology and the stature of his existing independent research prompted the surgeon general to offer him the prospect of developing biology at postwar Oak Ridge. "I was asked," Hollaender recalls, "if I would like to go to Oak Ridge and I see what I could do with biology down there. If I developed a good group, I could bring it to Washington and set up an Institute of Radiation Biology for the NIH." Hollaender believed the key problem to attack first was the effect of radiation on genetic systems. For him, colleagues have suggested, genetics was "at the core, with the rest of the biological sciences radiating outward."[27]

From the AEC's perspective, the Biology Division's position at Oak Ridge reflected the growing ambivalence about the proper relationship between basic and applied research at the national labs. During the war, the lab's Health Division was charged with monitoring the conditions under which plant employees work. After the war, the government's contract with Monsanto Chemical Company was not renewed, and the University of Chicago, the proposed academic operator, failed to assemble a management team; this resulted in the selection of a new industrial contractor, Union Carbide. Of this decision, historians Richard Hewlett and Francis Duncan wrote, "Few at Oak Ridge, not even the indomitable Alvin M. Weinberg, had much faith in Carbide's ability to build a new . . . laboratory on the ruins of Clinton." Wartime Health Division Director John Wirth consequently left, to return to the National Cancer Institute, in September 1946. Oak Ridge Research Director Eugene Wigner

[26] "Why in the Hell"—section entitled "Highlights of Work in the 1930s" (document pages not numbered); Alexander Hollaender, "The Problem of Mitogenic Rays," in *Biological Effects of Radiation*, vol. 1, ed. B. M. Duggar (New York: McGraw Hill, 1936), pp. 919–60. For example, A. Hollaender and J. T. Curtis, "The Effects of Sublethal Doses of Monochromatic Ultraviolet Radiation on Bacteria in Liquid Suspensions," *Proc. Soc. Exp. Biol. Med.* 33 (1935): 61–62; F. S. Brackett to AH, 10 August 1937, MS 652, Box 10, Folder 81, Alexander Hollaender Papers, RRC-UT; V. Bush to AH, 20 January 1942, MS 1261, Box 4, Folder 5, RRC-UT.

[27] "Why in the Hell"; cf. AH to James H. Lum, 27 August 1946, MS 6521, Box 10, Folder 53, RRC-UT; R. C. von Borstel and Charles M. Steinberg, "Alexander Hollaender: Myth and Mensch," *Genetics* 143 (1995): 1054–56, quote on p. 1054.

then split the division into two sections: Biology and Health Physics. Health Physics quickly assumed a service function, monitoring the Oak Ridge facilities and conducting studies designed primarily to develop better methods of detecting radiation. But with internally focused research needs now tended to by Health Physics, the pressing question became how the Biology Division could—and should—do basic research in ways that would contribute to the AEC's mandate.[28]

When Hollaender began his tenure in 1947, the AEC's policy dictates had crystallized around the notion of "areas of availability." The AEC's director of research, James B. Fisk, openly declared his commitment to developing the national laboratories as the centers of research. But because of pending legislation on the National Science Foundation, Fisk felt hampered from acting programmatically to promote basic research. Where and how, he asked his advisors, would the AEC justifiably draw the line about which projects to fund in the absence of any existing rationalized criteria? Fisk's own short-term answer was to fund those areas that could make use of the unique materials, facilities, and information that were under government control and would occasionally be available in excess quantities "for fundamental research." But Fisk's cautious policy combined with Union Carbide's shoddy management to ensure that those remaining in biology at Oak Ridge (in the words of one historian) "would have little more to work with than the obsolete X-10 reactor . . . and the crumbling temporary buildings from the wartime project."[29]

Paradoxically, Hollaender maintained that these circumstances would be the ideal starting point for a new biology program and for developing the living research tools it would require. Where others saw programmatic compromise and material scarcity, he saw an easy opportunity to please the AEC while still countering its stultifying established wartime norms. Carbide manager Clark Center, with whom Hollaender developed a good professional rapport, explained (as Hollaender remembers): "he had a 'white elephant' sitting at Y-12 and asked if I would take the build-

[28] Richard Hewlett and Francis Duncan, *Atomic Shield: A History of the Atomic Energy Commission, 1947–52*, vol. 2 (Berkeley: University of California Press, 1990), p. 224. On the related controversy over the relationship between the military and civilian scientific communities in postwar defense research, see Daniel Kevles, "Scientists, the Military, and the Control of Postwar Defense Research: The Case of the Research Board for National Security, 1944–46," *Technology and Culture* 16 (1975): 20–47. Robert Seidel has argued that the AEC differentiated between "laboratories primarily devoted to programmatic research within limited basic research programs, and laboratories primarily devoted to basic research with limited applied research programs," but it is not clear to me whether biology programs at the national labs were approached this categorically. See Robert Seidel, "Accelerating Science: The Postwar Transformation of the Lawrence Radiation Laboratory," *Historical Studies in the Physical Sciences* 13 (1983): 375–400.

[29] On areas of availability, see Hewlett and Duncan, *Atomic Shield*, pp. 111, 223.

ings off his hands. They were scared of a congressional investigation. He would do anything to help me get these buildings fixed up for biological laboratories." In Hollaender's view, this translated into the possibility of a truly independent and exploratory program of basic biological research, even if it meant losing some existing personnel: "At first, the health physicists . . . wanted to come with me, but they were all scared to move away from the established laboratory. I said, 'I will love it!' because I could independently develop something that was different from the main laboratory. At that time, they did war-related work."[30] With regard to radiation studies, Hollaender understood that, as he wrote to physicist Edward Teller years later, "the answers to many of the questions . . . could be obtained, but they cannot be gotten overnight since the long term implications, even in mice, take a minimum of three years to be observed."[31]

Few members of the larger biological research community shared Hollaender's initial enthusiasm about the fate of the program, and nearly all worried about the peripheral nature of Oak Ridge's geographic location. "I tried to get my friends to come with me," he remembered, "but not a single one had enough confidence to come to this 'God-forsaken' place. They didn't even have sidewalks at the time, but they had started to pave the streets." But Hollaender's own optimism drew strength from his ability to connect successfully his long-term scientific vision with both biologists' and AEC policymakers' goals—in short, to combine mission-oriented and discipline-oriented biological research.

In turn, Hollaender recognized basic mammalian genetics as a program that would simultaneously mediate the political, material, and scientific concerns governing debates over biological research in postwar America. Shortly after he came to Oak Ridge, while the ABCC studies on Japanese survivors were still ongoing, he decided that "in the long run, it was absolutely essential that we prove whatever we found on mammals which are close in comparison to man." Such a project would be extremely resource intensive in terms of the space required to breed animals and the time required to get meaningful results, but both of these things fell within the AEC's more narrow "areas of availability" provision. Hollaender supported the inbred mouse genetics research that the Russells developed over the next ten years in part because it reflected his theoretical interests, but also because it was a good fit with both existing policy dictates and his own more expansive view of how the postwar life sciences should be developed.

[30] Hollaender, "Why in the Hell."

[31] Hollaender to Edward Teller, 3 July 1957, MS 1261, Box 4, Folder 22, RRC-UT.

THE RUSSELLS AND THE MEGA-MOUSE PROJECT: MODEL DESIGN, SETUP, AND INITIAL RESULTS

From the start, the "mega-mouse project" was embedded with public policy implications. As Liane Russell would later reflect: "the basic amount of money that was available came from the worry of the public and their offspring, and I think the government was determined to get more data on that." The seven-recessive SLT mouse, the model system at its center, was thus designed to be the "right tool" for all the jobs it needed to do. The SLT mouse promised to produce experimental data that could be used to make an information-based determination of risk assessment, regardless of how geneticists eventually decided to define that risk.

As a former JAX scientist, Bill Russell was understandably skeptical of benefits of industrial-style research, but Hollaender's persistence and assurances won him over. Both Walter Heston at JAX and *Journal of Heredity* editor Robert Cook forwarded Russell's name to Hollaender, who asked them for likely candidates for the Oak Ridge job, and Heston vouched for Russell's superior-quality research. Russell later remembered that he did not reply to Hollaender's initial letter because it had "Monsanto Chemical Company," on the letterhead: "I just didn't even bother to answer because I didn't want to be involved in that. It wasn't demanding. . . . And so I just didn't answer. And then I got another phone call from him saying; 'why the hell haven't you answered my letter?' He was very furious about it and that was my first introduction to Hollaender." Neither Bill nor Liane had previously worked in industry, or done radiation genetics research. But while Liane wrote up her dissertation work, Bill developed the SLT experimental protocol and pitched it to Hollaender at their initial meeting.[32]

Thus far, Russell argued, mammalian radiation genetics had missed its calling: mouse workers had focused on the induction of dominant mutations or chromosomal aberrations, but these were poor measures of human genetic risk. Instead, Russell—like Muller—believed that recessive gene mutation rates would provide more fundamental information, precisely because they could be used as a kind of midpoint in extrapolation from rigorous *Drosophila* data to the human case. Unlike Muller, but like many other geneticists, Russell preferred to remain cautious. In his initial project proposal, Russell reminded the AEC that, the ABCC

[32] WLR-OH; see also R. Cook to A. Hollaender, 24 July 1947; A. Hollaender to R. Cook, 13 August 1947; W. E. Heston to AH, 6 August 1947; AH to W. E. Heston, 13 August 1947; all Alexander Hollaender Papers, MS 652, Box 10, Folder 55 (July–December 1947), RRC-UT.

investigation aside, a strict scientific evaluation of the genetics hazards of radiation to man "is not possible at the present time." Nevertheless, in a final section of the proposal entitled "Importance to Human Genetics," he presented both the practical and programmatic rationales for developing mouse models:

> Even if we knew accurately how the rate of mutation in mammals compared with that of *Drosophila*, we could not, without the type of experiments outlined here, estimate the effect of a given induced mutation rate on the morphological and physiological health of mammalian populations in successive generations. The one big disadvantage of working with mammals is the expense. [But] it is solely the expense that has prevented the accumulation of data on mammals. The information is now of vital importance.[33]

To obtain this information, Russell suggested a method modeled after Muller's work on *Drosophila*.[34] First, he would expose a wild-type male mouse to radiation, and then he would mate that mouse with a female animal from an inbred stock engineered to be homozygous for seven recessive genes (or, as Russell put it, "as many as can be practicably managed"). All of the loci specified easily identifiable phenotypic characteristics, such as coat color or ear shape, and were designated with simple alphabetical notation: agouti (or a, the recessive nonagouti allele), brown (b), albino (c), dilute (d), short ear (se), pink-eyed dilution (p), and piebald (s). Thus in the absence of a mutation, the offspring would not express any of the seven recessive phenotypes visible in the mother. But any offspring derived from a treated germ cell that had mutated would resemble the characteristic of a particular pair of recessive trait or traits. As Hollaender recognized, such an undertaking would mean first spending a lot of time, money, and space breeding the seven-recessive strain of mice, without any guarantees that the method would actually produce mutants. But the payoff if it worked would be great, for it would represent the industrialization of mammalian genetics research: a quantitative test for mutations that (as Russell put it) "can be carried out so easily that one person, in one hour, without any instrumentation or knowledge of mouse physiol-

[33] Hollaender to J. B. Fisk, 21 October 1947, MS 652, Box 10, Folder 20, RRC-UT; Russell, "Reminiscences of a Mouse Specific-Locus Test Addict," *Environmental and Molecular Mutagenesis*, Supplement 14, 16 (1989): 17; Hollaender, "Why in the Hell"; "Mammalian Genetics Program, Revised Nov. 27, 1947 and Drawn Up after a Consultation with Wright, Muller, and Hollaender," DOE-051094-A-555, DOE-Hum.

[34] J. I. Valencia and H. J. Muller, "The Mutational Potentialities of Some Individual Loci in *Drosophila*," *Proceedings of the Eighth International Congress of Genetics*, 1949, pp. 681–83.

ogy, can score 2,000 loci" by simple visual inspection.[35] In October 1947 Hollaender presented Russell's "Specific Locus Test" plan to the AEC's director of research as a positive and necessary step in the government's research plan. It was, he wrote, an "effort to obtain information on the possible genetical [*sic*] implication of bomb explosions, and . . . we believe this information can be obtained."

Before committing AEC resources to the SLT mouse production, however, Hollaender brought Russell back for a consultation with Muller and population geneticist Sewall Wright. This meeting was explicitly intended to assimilate the new SLT procedure with the received understanding of genetic risk. As a fellow mammalian geneticist, Wright had no hesitation that Russell could get the SLT method to work. Instead, he objected that if the results were to be used for evaluating human genetic hazards, then measuring a wide range of phenotypic effects over many generations would be a better approach because this would "capture" unknown but potentially more consequential recessive mutations—"great enough to be important to man." To address these concerns, Russell agreed to extend the SLT experiments over at least five years, making extensive diagnostic and autopsy examinations of all offspring, and he added a clause in his AEC proposal: "*only as a group*, will [my projects] make possible a reasonably complete evaluation of the genetic hazards to man."

Muller, on the other hand, did not contest overall mutation rate as a useful measure for risk assessment—indeed, that was his chosen definition—but he did question whether Russell could get significant results with seven nonrandomly chosen loci. He suggested that Russell follow the drosophilists' methods even more closely by scoring mutations over a larger part of the genome. At first, Russell responded with frustration. He argued that what was doable in *Drosophila* in one generation of breeding would take three in mice. Furthermore, since there was no accurate estimate then available for the number of genes in mice, having information on more genes was not going to be useful without the specific loci measurements. Eventually Muller conceded—a move Russell later remembered as motivated by the Nobelist's "powerful social conscience"—but he also told Hollaender to try and get an animal facility three times the size of Russell's initial request, presumably to better ensure that the specific-locus experiments would be done on a scale large enough to produce the desired mutations.[36]

[35] For descriptions of the SLT method, see W.L. Russell, "X-Ray-Induced Mutations in Mice," *Cold Spring Harbor Symposium of Quantitative Biology* 16 (1951): 327–36, quote on 327; Russell, "Reminiscences"; Hollaender, "Why in the Hell."

[36] "Mammalian Genetics Program, Revised Nov. 27, 1947 and drawn up after a consultation with Wright, Muller, and Hollaender," DOE-051094-A-555, DOE-Hum.

Ultimately, then, the SLT model was configured as a laboratory technology that would mediate existing controversies at several levels. It was an acceptable middle ground, for both scientists and policymakers, between experimental studies of flies and bacteria and the study of Japanese survivors. Among radiation geneticists, its design accounted for both short-term, mutation-based definitions of risk and long-term, phenotype-based "social distress" definitions of risk. Perhaps not surprisingly, Shields Warren, director of the Division of Biology and Medicine's Advisory Committee, immediately approved the project for a minimum of $300,000 over three years.[37]

In terms of producing results, the mega-mouse project got off to a shaky start. The first mice were irradiated at Oak Ridge in March 1949, and a month later, the first litter from the SLT pilot experiment was born—without any mutants. In fact, it was not until the eighth litter that the first mutant appeared (at the *d*, dilute locus), and a second mutant (at *c*, albino) did not show up until another (by their own description) "nail-biting" seven months. Apparently, though, the scientists decided the problem was not one of model design but (as Muller had predicted) one of scale; sometime in early 1950—when the Y-12 animal quarters were ready—they began the main experiment, using far larger numbers of mice.[38]

By the summer of 1951, Bill Russell remembered, "things had gone well for us"—well enough to draw some preliminary conclusions about both the scientific and programmatic work of the project. He was invited to give a paper to his genetics colleagues at the prestigious 1951 Cold Spring Harbor Symposium on "Genes and Mutations," and there he first reported the mutation rate results based on nearly 100,000 mouse offspring. Russell was upfront about the applied aspects of his research. "The immediate practical purpose of the work reported here," he said, "was obtaining data that would be useful in the estimation of the genetic hazards of ionizing radiation in man." Yet he was also careful to distinguish the research question at hand ("What is the increase in mutation per dose?") from the "human values" policy-oriented question, "how great an increase can be tolerated?" This rhetorical distinction aside, he reported the mutation rate data—25×10^{-5} mutations per locus, per roentgen in mice, or roughly one order of magnitude (or fifteen times) higher than in comparable *Drosophila* studies, and he tentatively concluded that

[37] "Minutes of the Division of Biology and Medicine, 14 February 1948," p. 17, MS 1067, Folder 2, RRC-UT.

[38] Hollaender seems to have been a bit nervous about Russell's slow start; in August 1949, Muller was once more called down to Oak Ridge "for consultation and advice in connection with our Mouse Genetics Experiment." Hollaender to C. N. Rucker, 2 August 1949, in MS 652, Box 9, Folder 3, RRC-UT.

Muller and the drosophilists had essentially been right in their concerns about policy. "Estimates of human hazards based on *Drosophila* mutation rates may be too low," Russell wrote."[39] The SLT work met with great enthusiasm at the Cold Spring Harbor meeting, especially from Muller himself.[40]

Russell's early results also generated an apparent shift of attitude toward the overall importance of inbred mouse work in general AEC policy circles, as well as some competition from his former employer, JAX Lab. In 1949 the AEC convened a committee of biologists and policymakers to determine whether it should expand the Russells' operation at Oak Ridge. As Russell remembered, first there were practical concerns about the mouse colonies, as well as structural concern about the Oak Ridge group having a monopoly on mammalian radiation genetics: "There was a big discussion going on that this was such an important project that we shouldn't risk having a diseases break out here . . . and that it might be better, as it expanded , to set another group up somewhere else. And I think there was some competition from the Jackson Lab to want to do that." After months of deliberation, geneticist Bentley Glass boldly suggested that the SLT was already obsolete. Russell recalled him arguing, "Well, we're getting all these results with tissue culture—we probably won't need the specific locus test anymore, so we probably don't need to expand at all." Renowned *Neurospora* biologist George Beadle nixed the idea that tissue culture was an appropriate substitute for whole organisms right away and ultimately rejected the notion that SLT mouse production should be done at the institution that originated large-scale mouse inbreeding. The committee decided to give the Russells an additional floor of the Y-12 building for their animals.[41]

Bill Russell greatly appreciated this support, even though it bound him to additional experiments that he did not find very important from the perspective of genetics. In the summer of 1951, experiments with mice were incorporated into the medical and biological research done at AEC bomb testing sites in the Pacific (specifically, Eniwetok). Bill Russell was

[39] Russell's first SLT publication/presentation was at a 1950 NRC Symposium on Radiobiology ("Mammalian Radiation Genetics," *Symposium on Radiobiology*, Oberlin, 1950, [New York: Wiley and Sons, 1950, pp. 427–40]), but by then he had not obtained results from the main experiment; Russell, "X-Ray Induced Mutations in Mice," pp. 327–28, 334–35.

[40] Muller lauded the Russell's work in his Kimber Award acceptance speech as "a significant attack on the problem of how many mutations are produced by a given test. . .utilizing reasonably uniform biological material in precisely controlled crosses." Muller, "The Genetic Damage Produced by Radiation," 25 April 1955, p. 3 of typescript, Folder "Background Data: H.J. Muller," NAS-DC.

[41] Interview with Liane and Bill Russell, December 1995, JLOH-KR.

asked to design an elaborate experiment for a neutron bomb test on U.S. soil: Operation Upshot-Knothole, conducted in the spring of 1953 at the Nevada Proving Grounds. Scientifically, Russell remembered, this experiment was "most uninteresting" because its main objective was to measure the effects of time: to determine if the mutational effects of a short neutron blast were the same as or different from the mutational effects of the inbred mouse's typical minutes-long exposure in the cyclotron at Oak Ridge. Liane Russell also wanted to examine how pregnant mice responded to the blast.

Technically, however, Bill Russell remembered this mission-oriented project proudly as "our most exotic experiment!" and his experiences conducting it once again suggest the importance of inbred mouse work relative to the AEC's goals for its biology program. To hold the mice (strains *S* and *Ba* [later, BALB/c])[42] and keep them alive in the desert before the bomb went off, the AEC worked with the Army Corp of Engineers to construct special air-conditioned lead hemispheres (run by backup batteries, in the event that the bomb itself destroyed the power supply). Loading these containers with mice could be accomplished only "by lying on your belly and working through a tiny hole." After the test, which Russell witnessed, George Beadle himself came to the site—"because we needed an extra truck driver"—and all the experimental animals were quickly put in special racks in the back seat and driven away to the airport. Two days later, after Russell and his mice survived a harrowing storm in a DC-3 over Texas, Russell returned to his lab only to find it labeled "Radiation Hazard." When he asked why, he was told that "it turned out . . . the bomb we had been at had a big fallout over Oak Ridge and they were afraid that people would pick it up, carry it in [to the lab] and irradiate the control mice."[43]

Following Hollaender's dictate to embrace an ethos of openness about their work, both Russells spoke about the results of these experiments at

[42] *S* mice appear to have come from medical researcher R. G. Schott, who was using them to work on the history of disease resistance (R. G. Schott, "The Inheritance of Resistance to Salmonella in Various Strains of Mice," *Genetics* 17 (1932): 203–29; *Ba* mice were some substrain of Halsey Bagg's mammary cancer-prone albinos and probably came from JAX-based strains (Heston, "Genetics of Mammary Tumors in Mice").

[43] For Eniwetok, see materials in "DOE ARCHIVES" folder. Based on the number of deaths one week after the experiment, strain *S* mice were described (in language that resonated with earlier inbred mouse cancer studies) as "more resistant" to radiation than strain *Ba*. For a description of the 1953 bomb test and a full presentation of Russell's data, see "Operation Upshot-Knothole, Division of Biology and Medicine Report WT-820, DOE Archive; cf. W. L. Russell, Liane B. Russell, and A. W. Kimball, "The Relative Effectiveness of Neutrons from a Nuclear Detonation and from a Cyclotron in Inducing Dominant Lethals in the Mouse," *American Naturalist* 88 (1954): 269–86; quotes from William Russell, Oral History interview with Bill and Liane Russell, JLOH-KR.

the 1955 Geneva Conference on the Peaceful Uses of Atomic Energy. Bill argued that "recent advances in radiation genetics of mice make possible a reestimation of the genetic hazards of ionizing radiation in man." The 1953 test showed that, contrary to what had been observed in lower organisms, the rate of mutation observed in the offspring of these irradiated mice was not dependent on the length of the interval between irradiation and mating. He was clear about the implications for humans: "In the exceptional cases, where the gonads of a man are exposed to a considerable dose of radiation within a short time, it is recommended that he abstain from procreation for the few weeks required for the elimination of the germ cells that were irradiated in the postspermatagonial stages. . . . Unfortunately, even this risk is far from negligible . . . [since] in man, offspring conceived long after irradiation are just as likely to inherit induced mutations as those conceived just a few weeks after irradiation."

Liane Russell's results were equally discouraging from a policy standpoint: "doses high enough to produce developmental abnormalities do not necessarily cause abortion or prenatal death, as has been claimed by some medical authorities." The one piece of good news seemed to be that most mouse mutations were only deleterious (that is, caused death) in the homozygous condition—as opposed to *Drosophila* experiments by Curt Stern and others, which had shown that even heterozygous mutations could be lethal.[44] AEC leaders recommended that the results be sent on to the National Academy of Sciences, to determine if a formal policy recommendation should be made.

Significantly, not all mouse radiation work reported at this time received this stamp of approval. In August 1951, for example, the AEC responded harshly to the predictions made by mouse geneticist Arnold Grobman in his popular book *Our Atomic Heritage*. Grobman, an associate of Donald Charles at Rochester, claimed that extrapolation from wartime data obtained in the Rochester mouse genetic experiments indicated that exposures to long-term, low-level radiation would produce generations of deformed and diseased human children.[45] Although both Russell and Grobman were supported by government funds, following the accepted norms of science appears to have made a big difference to the AEC.

[44] Text of Russell's speech ("Radiation in Mice: The Genetic Effects and Their Implications for Man") appears in *Bulletin of the Atomic Scientists* 12, 1 (January 1956): 19–20. Cf. quotes from abstracts of the "O+M Peacetime Uses of Atomic Energy 1955," RG 326 (AEC), DBM Collection, Box 3367, DOE-MD.

[45] At Rochester, Grobman worked with Curt Stern, whose low-level radiation work with *Drosophila* did not distinguish between spontaneous and induced mutations and therefore came to the same conclusion: "There is no threshold below which radiation fails to induce mutations." See D. E. Uphoff and Curt Stern, "Genetic Effects of Low Intensity Irradiation," *Science* 109 (1949):609–10.

Russell published his findings in a peer-reviewed journal and did not go directly to the public, and his work—even while potentially damaging in its implications for the AEC—was taken seriously. Grobman's book generated little response among policymakers, with the exception of a dismissive press statement.[46]

PUBLIC RECOMMENDATIONS, LINGERING QUESTIONS: ANIMAL MODELS IN THE BEAR COMMITTEE DEBATES

The widely acknowledged scientific success of the Russells' work after 1951, however, did not necessarily mean that the policy uses of this work would be unproblematically achieved on behalf of any of the interested parties: Muller; other radiation geneticists, the AEC policy makers, or the American public. In fact, the results generated from Russell's mouse experiments proved to be an unexpectedly blunt instrument for settling the very scientific and policy debates that were literally their raison d'être. Nowhere is this point better demonstrated than in the machinations of the genetics subcommittee of the NAS's Committee on the Biological Effects of Atomic Radiation (the BEAR committee).

The BEAR committee was officially called into existence in April 1955, as a cooperative effort between the AEC and the National Academy. Fully funded by the Rockefeller Foundation, it was to be made up of several expert scientific panels, each charged with "undertak[ing] a broad appraisal of present knowledge about the effects of atomic radiation on living organisms [and] . . . identify[ing] questions upon which further intensive research is urgently needed." Essentially, though, AEC chairman Lewis Strauss's hand had been forced.[47] In the wake of the Bravo and 1953 bomb tests, the BEAR genetic subcommittee was to be a science policy proving ground. The committee's official mission was to develop radiation protection standards that took into account all the new genetic studies.[48]

[46] Arnold B. Grobman, *Our Atomic Heritage* (Gainsville: University of Florida Press, 1951); "Statement on the Genetic Effects of Radiation on Human Beings," 28 August 1951, DOE Archives Collection 326: U.S. Atomic Energy Commission, Division of Biology and Medicine, Box 3354, Folder 26, DOE-Hum.

[47] Cf. reference to Lewis Strauss's 1954 press release re: fallout, in W. Weaver to George Beadle, 24 January 1957 (ORG:NAS:Coms on BEAR:Genetic:General); BEAR Committee NAS Press Release, 8 April 1955 and excerpt (pp. 2–3) from NAS Council Meeting minutes, 18 June 1955 (both in ORG:NAS:Coms on BEAR:Beginning of Program); all NAS; "Editorial: Genetics in Geneva," *Bulletin of the Atomic Scientists* 11 (December 1955): 314–16, 343.

[48] C. I. Campbell to S. D. Cornell re: John Bugher's chosen topics for BEAR, 19 October 1955 (ORG:NAS: Coms on BEAR, Beginning of Program); Agenda for the Conference on Genetic Effects of Radiation, 20–22 November 1955, Princeton, NJ, and "Remarks by Pres-

Transcripts of the committee's initial 1956 meeting reveal that whenever Russell or his mouse data were called upon to scientifically reconcile competing views of genetic risk, they proved practically ineffective for compelling group members to adopt a common epistemological rationale for that recommendation. First, especially given the individual members' diverse research backgrounds, the issue of extrapolation was still not easily resolved just by pointing to Russell's mouse model as closer to the human case. Muller argued, for example, that Russell's increased mutation rate estimates in mice made determining permissible dose for humans merely a mathematical problem of scale translation—from *Drosophila* to mice to humans. Both Neel and Wright objected: "When does an estimate get so large," Neel challenged, "that it loses its usefulness and defeats its own purpose?"[49] Indeed, Neel's inquiries were complicated even further by Russell's own admission that the seven loci on which the specific locus test focused were "highly biased"—that is, likely to give artificially high rates of mutation because they were all loci at which spontaneous mutations had already occurred many times in the past.[50] "It is possible," he told the group, "that the spectrum of relative damage may be greater or less," and this experimental uncertainty, he argued, must be seriously weighed against the certain detrimental effects that a low permissible dose would have on national defense or atomic energy development.[51]

Beyond the issue of extrapolation, however, Russell's model also proved troublesome for those wanting some objective indication of how serious a problem the increased mutational effects of radiation might be for mammals. Repeatedly, for example, group members turned to Russell for advice on what kinds of mutations counted the most in terms of long-term risk incurred by a mouse population. At points, Russell seemed not to want to draw any conclusions at all, such as in an early exchange with James Crow, who believed Russell's results were too preliminary for extrapolation:

> *Dr. Crow*: Bill, wouldn't it be fair to say that the logical consequence of your statements is that the Oak Ridge experiment is of no consequence for this problem?

ident Bronk, Study Group on Genetics, 20 November 1955," typescript, p. 6, which refers to genetic effects as of "prime importance" (both ORG: NAS: COMs on BEAR: Genetic Meeting); all NAS-DC.

[49] Transcribed Proceedings, Conference on Genetics, 5–6 February 1956, Chicago (hereafter, referred to as BEAR Genetics Proceedings), p. 280, in "ORG:NAS: Comm. on BEAR:Genetics Meeting:Transcripts, Feb. 1956," NAS, NAS-DC.

[50] Ibid., p. 134.

[51] Ibid., pp. 139 and 61, respectively. Cf. the following exchange between Russell and human geneticist Jim Crow (p. 137).

Dr. Russell: That is right. I have never claimed that it was and I don't now.
Dr Crow: Well, I disagree.

But when pushed by others present, Russell argued that, because the observable mutations in any SLT experimental test would (by design) be those that were genetically recessive, no laboratory work could illuminate the dominant mutation effect by which "the higher proportion of the risk will be borne by the [first] offspring of the exposed individuals."[52] Later, when asked by Wright, Chairman Warren Weaver, and bacterial geneticist Tracy Sonneborn to give "measures . . . of definiteness" indicating what such long-term dominant effects might be, Russell pointed to his observed data along these lines: a reduction in gene loci in three-week-old mice, and a 5 percent reduction in lifespan.[53] A spirited debate then ensued about whether the group's risk calculations should be based on total numbers of increased mutations or only on those mutations that eventually express themselves as deleterious in the first generation. Was a reduction of genetic loci truly a long-term problem? And what about those mutations that express themselves prenatally—for example, recessive lethal conditions that killed an embryo before it even had a chance to be born. Should they be considered deleterious if, as Wright put it, "nobody will ever know anything about them"? For his part, Weaver implied that human beings might want any unnecessary death counted in a risk calculation—a view supported not only by other BEAR Committee members but also in one widely read popular treatment of mouse radiation experiments. Charlotte Auerbach, a former student of Muller's now working at the MRC Radiobiology Unit in Harwell, England, chose to illustrate the phenomenon of hidden but lethal recessive mutations in her popular genetics textbook with a picture of mouse "parents" crying over their radiation-induced "empty nest" (fig. 6.3).[54]

Ultimately, the BEAR Committee members did all agree to a permissible lifetime dose recommendation of no more than ten cumulative roentgens, but this consensus was not reached by appeal to the Russells' mouse data. Indeed, Russell himself appears to have sanctioned the figure for impressionistic, as well as political, reasons. As he summed up his position for the group: "If you really want me to say something, I am ready to say something but it won't mean anything. . . . My feeling that this is reason-

[52] BEAR Genetics Committee Transcript, p. 75, NAS-DC.

[53] Ibid., pp. 263–65.

[54] Ibid., pp. 264 and 260–70 inclusive; cf. James F. Crow, "How Well Can We Assess Genetic Risk? Not Very," lecture no. 5 in the Lauriston Taylor Lecture Series in Radiation Protection and Measurements (Washington, DC: National Council on Radiation Protection Measurements, 1981); Charlotte Auerbach, *Genetics in the Atomic Age*, with illustrations by I. G. Auerbach (New York: Oxford University Press, 1965), p. 81.

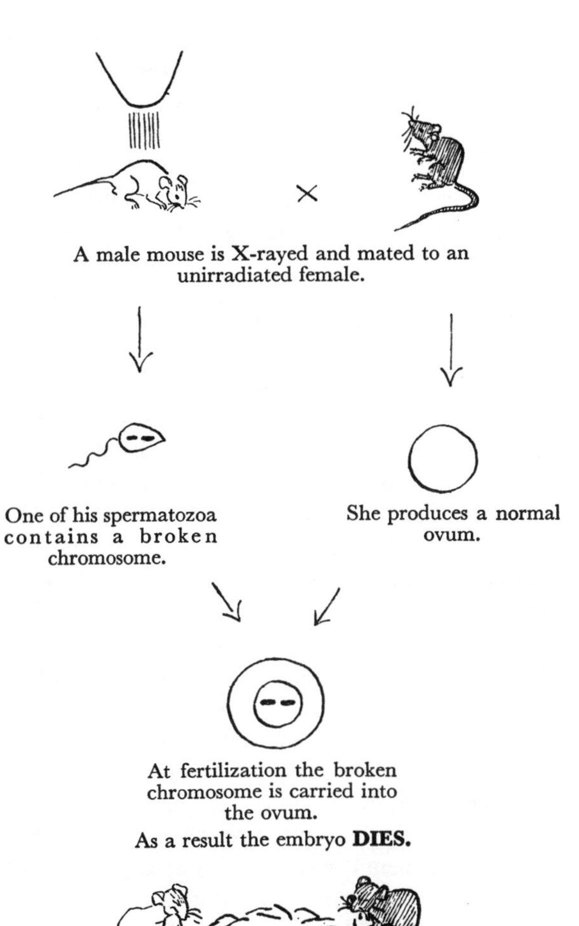

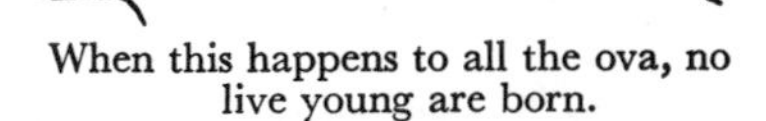

6.3. Crying mice, from Charlotte Auerbach, *Genetics in the Atomic Age*.

able figure . . . is also based partly on my laboratory experience that all-spring mice that receive much higher doses get along pretty well, and this is a considerably lower dose. . . . Also, I think that we can arrive at a guess that seems reasonable genetically and if we can have it within a limit that does not upset present operations, I think it is desirable."[55] Even in the

[55] BEAR Genetics Proceedings, pp. 142–43, NAS-DC.

published report of the committee, Weaver noted that, "although [there were] no disagreements as to fundamental conclusions," the group's decision was made on an assumption that made many members uncomfortable: "risk on neither side"—i.e., scientific/theoretical or political/social—is completely visible "at the present time.[56] SLT mice, it seemed, had fulfilled their mission, but even then their applications were not as transparent as their well-known pedigrees would lead biologists to believe.

MANAGING MICE, MANAGING POLICY

In 1954, when English scientists arrived at Oak Ridge and took back strains of the SLT mice to their radiobiology laboratories at Harwell Laboratories, two developments were clear to all involved: radiation genetics had become a problem of political interest to the entire Western world, and Russell's mouse model had become the standard animal in its study. John Loutit—the Medical Research Council's equivalent of Hollaender—"had early realized the importance of genetics effects in radiation damage" and sought to make his MRC radiobiology unit "particularly strong in this area." Thus he pushed for independent confirmation of Russell's results. Bill Russell remembered that "there was a lot of cooperation, especially at the start," between the groups at Oak Ridge and the new MRC's Harwell unit, whose members now included Tobe Carter (a former poultry geneticist) and cytogeneticist Mary Lyon.[57]

Two years later, however, the Russells published additional results from the 1953 test, and new scientific controversies arose. Contrary to the received wisdom of experimental work on lower organisms, they showed that the mutation rate in mouse spermatagonia exhibited a dose-rate effect. It was lower when the radiation was protracted (as in the cyclotron) than when delivered quickly (as in a bomb). Bill Russell suggested that this observed effect might be due to intracellular repair of mutational or premuta-

[56] "Report of the Genetic Effects of Atomic Radiation," in Summary Reports: The Biological Effects of Atomic Radiation, National Academy of Sciences—National Research Council, June 1956. See also the "Report to the Public," a version of the group's conclusion that has been remarkably purged of conflict; both versions were available to any U.S. citizen, group, or institution that requested them.

[57] Bill Russell quote from Oral History interview with Bill and Liane Russell, JLOH-KR. On Loutit's goals, see Mary Lyon and P. L. Mollison, "John Freeman Loutit," *Biographical Memoirs of the Fellows of the Royal Society* 40 (1994): 239–52. On the prehistory of the MRC radiobiology group, see Donald Falconer, "Quantitative Genetics in Edinburgh: 1947–1980," *Genetics* 133 (February 1992): 137–42, esp. 138–41. Cf. Toby Carter, "Radiation Induced Gene Mutation in Adult Female and Foetal Male Mice," British *Journal of Radiology* 31 (1958): 407–11; Toby Carter, Mary Lyon, and R.J.S. Phillips, "Genetic Hazards of Ionizing Radiation," *Nature* 182 (1958): 409.

tional damage. Gene repair is a widely accepted idea today, but at the time it was a heresy. Nevertheless, this finding sparked both additional research and further policy debate: if mutations can be reversed, how? And was there some public health recommendation that can be made to encourage gene repair in cases where radiation exposure has already occurred?[58]

Geneticists have argued that the biggest scientific legacy of the mega-mouse project was its mutant material, since its large colonies and spinoff projects (such as Bill Russell's subsequent work in the 1970s with ENU (ethylnitrosourea) as a chemical mutagen) greatly increased the probability that spontaneous variations would occur.[59] Indeed, many of the earliest mouse genes mapped arose in Oak Ridge colonies or in colonies descended from them (especially in the 129 strain).[60] But beyond the genetics laboratory, what does the inbred mouse's development for radiobiological studies teach us about the relations among cultural understandings of risk, the experiments that scientists conduct to model that risk, and policy formulations that eventually get made?

First, the story reveals that inbred mice as research tools—even when they are standardized and explicitly policy-informed in their creation—are a lot easier to manage in the laboratory than in a discussion of public policy. As Brian Wynne has argued, quantifiable pictures of risk, reduced (or at least reducible) to a single dimension (e.g., gene mutation rate in the mouse), promise an essentially naturalistic basis for policy choices and, by definition, seek to separate risk from its social context.[61] But sci-

[58] W. L. Russell, L. B. Russell, and E. M. Kelly, "Radiation Dose Rate and Mutation Frequency," *Science* 128 (1958): 1546–50; L. B. Russell et. al. "Influence of Dose Rate on the Radiation Effect on Fertility of Female Mice," *Proceedings of the Society of Experimental Biology and Medicine* 102 (1959): 471–79. On the policy effects, see "W. L. Russell—Research Summary," typescript from Oak Ridge Library. The Russells' contribution to the field of DNA repair has been largely ignored, for reasons suggested by biochemist Bernard Strauss to researcher Errol Friedberg: "The radiation biologists did killing curves and tried to understand what was going on. I believe that until there was some biochemistry attached to those curves none of the molecular biologists engaged in working out the details of the genetic code . . . paid the slightest attention;" see Errol Friedberg, *Correcting the Blueprint of Life: An Historical Account of the Discovery of DNA Repair Mechanisms* (Plainview, NY: Cold Spring Harbor Press, 1997), pp. 66–67.

[59] Allan Peter Davis and Monica J. Justice, "An Oak Ridge Legacy: The Specific Locus test and Its Role in Mouse Mutagenesis," in *Perspectives on Genetics*, ed. James F. Crow and William F. Dove (Madison: University of Wisconsin Press, 2000), pp. 638–43. Originally appeared in *Genetics*.

[60] Margaret Green, "Mutant Genes and Linkages," in *The Biology of the Labortory Mouse*, ed. Earl Green (New York: McGraw-Hill, 1966); Murial Davisson and Tom Roderick, "Linkage Map," in *Genetic Variants and Strains of the Laboratory Mouse*, ed. Mary Lyon and A. G. Searle (Oxford UP, 1989): 416–427.

[61] Brian Wynne, *Rationality and Ritual: The Windscale Inquiry and Nuclear Decisions in Britain* (St. Giles Chalfont: History of Science Society, 1982), p. 144.

ence-based policymaking—the preferred strategy of the United States since World War II—is a more complicated project. It is, as Sheila Jasanoff puts it, a process of "co-production" of scientific knowledge and social order.[62] Thus at the BEAR Committee meetings, Russell's mouse experiments by themselves were literally unusable for determining acceptable risk. Social order had to be added in order for the data to translate into effective policy. Consensus over the extrapolation of work with experimental mammals to the human case emerged relatively quickly within the scientific community and was not challenged by either policymakers or the lay public. But before SLT mice could be viewed as acceptable stand-ins for humans exposed to radiation, a political consensus needed to be reached about what kinds of biological events were detrimental enough to constitute "risk," as well as what level of risk was acceptable.

For most American citizens, the Russells and their SLT inbred mice are mere footnotes to a decade of "nuclear fear," but in the history of experimental biology and medicine, the mega-mouse project represents a more significant transition. Anthropologist Mary Douglas has suggested that any analytically workable model of risk perception must consider the relationship between power and knowledge by "tak[ing] account of culturally distinct attitudes towards authority."[63] Accordingly, the development and use of inbred mouse models for understanding radiation risk reveals a great deal about changing contours of power and knowledge in postwar America. By the mid-1950s, the success of the mega-mouse project suggests that experimental biologists more generally—and geneticists in particular—had emerged as key scientific and social authorities of biomedicine. In retrospect, then, the large-scale nature of the Russells' mouse breeding operation could be seen as practically necessary on more than just the biological level. Unlike JAX's production of inbred mice for cancer research, which eventually came to tolerate and even encourage heterogeneous experimental uses for its animals, the Russells' breeding enterprise was singularly geared toward posing the problem of radiation risk as a problem of Mendelian genetics. Also, their SLT animal models circulated more rhetorically than literally. These mice were used as levers for shifting political understandings of responsibility or blame—away from deadly government-sponsored radiation sources, like bombs—toward life-affirming "physics for life" research done in the name of public health. Like JAX mice, however, inbred SLT mice were also resonant boundary objects: they embodied Hollaender's expansive vision for the

[62] Sheila Jasanoff, "Beyond Epistemology: Relativism and Engagement in the Politics of Science," *Social Studies of Science* 26, 2 (May 1996): 393–418, quote on 397.

[63] Weart, *Nuclear Fear*; Mary Douglas, *Risk and Blame: Essays in Cultural Theory* (London: Routledge, 1992), p. 44.

Oak Ridge Biology Division as a premiere center for postwar research, as well as the AEC's belief in the value of mission-oriented consensus for shaping a rational public policy. Those geneticists like the Russells who eagerly sought their share of expanding federal funding for work in radiation biology held the power to mediate the knowledge produced by their experimental models for policymakers and the lay public. But what these scientists could neither predict nor contain was the expansion of genetic modeling of health and disease that would result from their experimental material's successful negotiation of these social worlds. In the decades that followed, inbred mice became the leading technological edge in a platform of genetic research that would eventually change the topography of biomedical practice, both inside and outside the laboratory.[64]

[64] Cf. Peter Keating, "Biomedical Platforms," *Configurations* 8 (2000): 337–87.

EPILOGUE

ANIMALS AND THE NEW BIOLOGY

Oncomouse™ and Beyond

In 1960, on the occasion of his fiftieth Harvard reunion, C. C. Little composed a touching ode to the inbred mouse for the Class Notes. Little summarized the creature's scientific achievements and then gave credit to the animal's "work": "Sometimes I have difficulty in believing what has actually happened over the years. The mouse has indeed labored and brought forth a mountain."[1] The history of science is full of legends about significant coincidences in the deaths of celebrated scientists. For example, Newton was born in the same year that Galileo died, and Copernicus saw the printed edition of his *De Revolutionibus* only on his deathbed. Soon after Little wrote these words, the parallel histories of inbred mouse use and medical research would also claim an entry in this pantheon. C. C. Little, the man who labored his whole life for both these causes, died of a heart attack on 22 December 1971—only one day before U.S. President Richard Nixon officially signed the 1971 National Cancer Act into law.[2]

The "ecology of knowledge"[3] for the inbred laboratory mouse began as medical and remains primarily so. To build the Jackson Lab, C. C.

[1] Entry on "Clarence Cook Little," written by Little in the Harvard College Class of 1910 Fiftieth Anniversary Class Notes, pp. 283–87, esp. p. 287, Harvard University Archives, Pusey Library, Harvard University, Cambridge, MA.

[2] See Strickland, *Politics, Science, and Dread Disease*, p. 289, n. 50.

[3] On the historiographic use of this concept, see Susan Leigh Star, *Ecologies of Knowledge: Work and Politics in Science and Technology* (Albany: SUNY Press, 1995), pp. 1–38.

Little first promoted this standardized mammal a tool for understanding the genetics of cancer. More than sixty-five years after its founding, the institution is still going strong, and mouse genetics has undergone a dazzling renaissance, culminating in the complete sequencing of the mouse genome in December 2002. The story I have told suggests that all of this would have been quite unexpected in the 1950s. But technical and scientific developments involving mouse breeding and genetic analysis over the last fifty years have been staggering, in both quality and quantity. In conclusion, then, I want to reflect on how JAX's subsequent history and that of its primary product together highlight the ways in which both animal material standardization and animal research in biomedicine remain contingent and deeply historically situated achievements, even in this contemporary time of triumph.

The legacy of the inbred mouse as a "model organism" today draws momentum from the choices made by Little and other early twentieth-century geneticists. In turn, these choices both shaped and were shaped by broader trends (made retrospectively even more salient) in the structure of biological research—specifically, shifts in sources of patronage and the commercialization of scientific infrastructure, including the mass production and specialized production of lab animals. Current academic work in "animal studies" provides a useful framework for thinking about how cultural assumptions about mice inform the development and use of engineered animals in research. Seen in this light, then, the laboratory mouse's historical story calls for a re-engagement with questions of animal and human integrity raised by contemporary mouse use.

JAX AFTER LITTLE

Little began thinking about a professional life beyond JAX in the early 1950s. When Little applied for a new cancer research grant from the Rockefeller Foundation in 1951, Warren Weaver told his old friend that he was worried about the implications of a shift in leadership at his institution, defined as much by its director as by its research materials: "Dr. Little was born in 1888. That means he will be 63 this fall. Everyone realizes to what a remarkable and magnificent extent the success of the Jackson Memorial Laboratory has depended upon Dr. Little's energy, imagination, scientific competence, and enthusiasm. . . . I am sure that you are most sensitively aware of this problem." Yet Little sent no reassuring response, and JAX did not get the grant. L. C. Dunn later told Little: "I don't disagree with Weaver's basic position. . . . It cannot come as a surprise that a foundation seeking to invest its funds . . . should seek for organizations equipped to carry forward the foundation's purposes." In turn, the lab's board officially named a "successor committee" in early 1954, consisting of inside scientists (including Little), trustees, and members of the scientific Board of Directors.[4]

In 1956 Little retired from JAX Lab, but he did not leave genetic work behind. The year of his retirement, Little published his final book on dog coat color genetics, an interest he had nurtured since his days at Harvard and redeveloped through J. P. Scott's project on behavioral genetics at JAX (1945–56). Ironically, Castle, who had initially discouraged him from dog work, sent Little a supportive note: "Hurrah and hearty congratulations!"[5] That same year, Little also took a controversial position as the scientific advisor to the Tobacco Industry Research Council, a cigarette company lobbying group. Because of JAX's interest in cancer research, the Board of Trustees was openly uncomfortable with this new alliance. Nevertheless, Little was allowed to stay and given lab space for his new research. Before his retirement, Little had told Warren Weaver that current institutional circumstances made him pessimistic about the future of mouse genetics. JAX could not get research grants in this field, and Little believed it was "still an open question whether the continuity of mammalian genetics can be and will be maintained at any university."[6]

In retrospect, Little's pessimism was unwarranted. During the late 1960s and early 1970s, new chromosome identification techniques were perfected that enabled researchers to distinguish the finer points of mouse chromosome structure with a degree of precision previously unattainable. These developments proved to be a boon to mouse genetic mapping, and no institution was better equipped to deal with this scientific windfall than the Jackson Lab. Existing mutants, often identified by caretakers with a good eye for variance in a strain, now became a new resource for those seeking genetic knowledge of this mammalian species. In 1972 *Science* magazine called the Jackson Laboratory a "Mecca to anyone tuned into mammalian genetics."[7]

In 1955, however, no precise mammalian gene manipulation techniques existed, and in the immediate wake of Watson and Crick's announcement of the structure of DNA, the future of biological research looked like it was going to be in the molecular genetics of smaller organisms. Thus

[4] CCL to Weaver, 7 June 1951, and Weaver to CCL, 15 June 1951; LCD to CCL, 12 July 1951; all in RF, RG 1.1, Series 200D, Box 144, Folder 1780.

[5] Castle to CCL, 30 September 1956, Box 740, CCL-UMO. Cf. Paul, "The Rockefeller Foundation."

[6] See the file on "TIRC" at JLA-BH; see also "Tobacco Men's Cancer Detective," *Milwaukee Journal*, 4 November 1954, in JLA-BH (attached to letter from Harry Sonneborn, 9 November 1954, Box 12/Little correspondence).

[7] Barbara J. Culliton, "The Jackson Laboratory: 'Mice Are Our Most Important Product,' " *Science* 177 (1 September 1972): 871–74. I am indebted to Patrick Catt for this reference.

Little's first choice for a new JAX director was not a "mouse man," but someone who worked on bread mold, a biochemically simpler and more pliant organism. However, George Beadle of Caltech, a *Neurospora* researcher to whom the Board of Directors offered the job in July 1955, turned it down.[8] Ultimately, the job went to a man who, like Little, had made his career in administration as well as mouse research: Earl Green of Ohio State University.

Contemporary observers agreed that Little and his successor could not have been more opposite in terms of personality and temperament. Little had a dramatic, visionary style. In the words of Elizabeth Russell, "he thought of the ideas as more important than the details"—while Green was "very, very systematic" and "very serious-minded in an idealistic type of way." In short, Little operated "very much ad hoc" as an administrator, whereas Green possessed an almost technocratic style.[9] These different approaches became evident in the day-to-day operation of JAX. For example, the leadership transition affected mouse production: in 1957, 470,000 animals were shipped to other laboratories, but the JAX system was unable to fill all extant orders. Green moved quickly to improve the situation by rationalizing and consolidating mouse production and sales.[10]

Many older JAX staffers insist that Green's keen business sense kept the laboratory afloat during the 1960s. In August 1957 he had Little's initial "JAX Mouse" trademark extended to include any reference whatsoever to "JAX." Simultaneously, he instituted new, stringent mouse husbandry rules to prevent the spread of disease in the JAX colonies.[11] Two years later, Green sought another Rockefeller grant—this one from the Rockefeller Family—to further the development of the mouse production buildings known as Morell Park. Green emphasized the sound business practices of JAX, rather than the link between mouse production and research science that, by the judgment of the time, Little had failed to achieve. Green noted that JAX was the largest single employer on Mount Desert Island and "a major force in the island's total economy."[12]

Green's style was thereafter thoroughly imprinted on the JAX institution, as well as on several of its subsequent scientific contributions, but

[8] Harold Bozell to CCL, 5 November 1954, JLA-BH; Frank Adair (president of the JAX Board of Trustees) to Beadle, 29 July 1955; Beadle to CCL, 15 October 1958; both Box 740, CCL-UMO.

[9] Interview with Elizabeth Russell, June 1993, and interview with Tom Roderick, June 1993; both JLOH-KR.

[10] *RBJ Laboratory Alumni Bulletin* (an internal, seemingly sporadic publication) 7 (December 1957), JLA-BH.

[11] Trade-Mark 650,295, registered 20 August 1957 and first used 8 October 1956.

[12] "Physical Plant" and "Economic Impact on the Economy of Mount Desert Island," HOMES/Rockefeller Family Collection, RG III/2I, Box 133, Folder "Roscoe B. Jackson Lab, May 1959," RAC-NY.

JAX's institutional role in mouse research did not drastically change. During the 1960s JAX further consolidated the literary technologies surrounding laboratory mouse use. In 1962 JAX supplemented the already existing *Biology of the Laboratory Mouse* with the first *Handbook on Genetically-Standardized JAX Mice.* Reprinted in 1964, with a second edition in 1968, this new booklet was a thorough and even more formalized catalogue of JAX mice and their biological specifications than Little's earlier mouse sales listings. In a compact, user-friendly reference format, it presented everything a researcher would want to know about inbred mice, as well as information on how to order conveniently these organisms from JAX. Mouse features such as "breeding performance" were further transformed into quantifiable, objectively observable items such as "age of dam at first litter" and "percent of fertile matings." Green still did not list JAX mouse prices—these were enclosed on a separate sheet.[13]

The Biology of the Laboratory Mouse went through a second edition in 1966, and its programmatic focus on genetics remained, even while editorial changes suggest that JAX's new scientific leader understood that the market for mice had expanded beyond one particular set of biomedical problems. Earl Green's preface noted that the new book reflected three developments: "the increased uses of inbred strains, an increase in the number and use of stocks carrying named mutant genes, and an increase in awareness on the part of most biologists of the importance of hereditary factors in determining the characteristics of organisms." Of the thirty-three chapters, six cover aspects of standard animal husbandry and nomenclature (as compared to one in the first edition), six cover physiology and biochemistry (as compared to one in the first edition), and the experimental embryology of the mouse was omitted entirely.[14] Also, sometime in the 1950s, JAX librarian Joan Staats began to compile a comprehensive bibliography of mouse work. Staats constructed a punch card system for keeping track of the fast-growing body of mouse literature, and she made her project known as the "Classified Mouse Bibliography" in brief articles for science journals. JAX then charged researchers who

[13] Earl Green, ed., *Handbook of Genetically Standardized JAX Mice* (Bar Harbor: Bar Harbor Times Publishing Corporation, 1962). Reprinted in 1964; second edition in 1968, JLA-BH.

[14] Earl Green, ed. *The Biology of The Laboratory Mouse*, 2d edition (New York: McGraw-Hill, 1966), quote from Green's preface, p. vii. On the historical split between embroyology and genetics, see Scott Gilbert, "Cellular Politics: Just, Goldschmidt, and the Attempts to Reconcile Embryology and Genetics," in *The American Development of Biology*, ed. R. Rainger, K. Benson, and J. Maienschein (Philadelphia: University of Pennsylvania Press, 1988), pp. 311–46; Gilbert, "The Synthesis of Embryology and Human Genetics: Paradigms Regained," *American Journal Human Genetics* 51 (1992): 211–15.

used the service a fee, which depended on the specificity or comprehensiveness of the search.[15]

The mercantile aspects of many of these developments remain striking, although such elements of JAX's institutional identity did not originate under the Green directorship. Alongside the mouse sales venture that Little successfully fostered, there were also incidents when the line between mouse science and mouse business did not seem as clear to the outside observer as it had to JAX's founder. Little, for example, was confronted in 1952 by a Maine state tax assessor about the propriety of selling mice under JAX's not-for-profit incorporation agreement.[16] Ten years later, the carefully balanced tension between the scientific goals and commercial activities of the Jackson Lab was dramatically highlighted in what has since become known as "The JAX Tax Case." In March 1962, Kendall Young, a private resident of Maine's Salisbury Cove, decided he had heard one too many lighthearted tales about the millions of dollars that the "nonprofit" Jackson Laboratory was bringing into Bar Harbor from inbred mouse sales, while he and other neighbors bore all the tax burden for the town's expensive oceanfront property. Young brought suit against the State of Maine's Tax Assessor's Office, attempting to force it to make JAX admit that its scientist-employees were more involved in the cutthroat business of selling mice than in the noble vocation of research. Production and sales of JAX inbred mice were then in excess of one million animals per year and brought in an annual revenue of nearly $850,000. Clearly, Young argued, such "sale of mice on the open market" was a highly commercialized, industrial venture, and the lab's property should be taxed accordingly.[17]

JAX scientists balked at this suggestion, and the first among them to speak out in court was the man who had first decided to sell JAX mice nearly thirty years earlier. Now seventy-four years old, Little confronted the central issue of the trial by admitting that JAX Lab could have easily allowed more mercenary values to rule the exchange of its living commodities. Industrial production of inbred mice, however, was not the main purpose of the institution. As Little told the court: "Had we tried at the

[15] Interview with Joan Staats, June 1993, JLOH-KR; cf. Staats, "The Classified Bibliography of Inbred Strains of Mice," *Science* 119 (26 February 1954): 295–96. Information on later developments in Staats's project can be found in Box 1–8, JLA-BH.

[16] Little got Maine's assistant attorney general to write a "sympathetic" letter to the state tax assessor on behalf of the laboratory: see CCL to Ernest Johnson (Maine tax assessor) and CCL to Boyd Bailey (Maine assistant attorney general), 13 June 1952, Box 740, CCL-UMO.

[17] Earl L. Green, "TJL: The Tax Case," typescript dated 15 April 1993, furnished by the author, gives the background on the lawsuit. The quoted phrase from Young is from the court transcript of the case: see Earl L. Green to JAX Lab Trustees and Staff, "Extracts from Transcript of Hearings on the Tax Case," 31 January 1964, JLA-BH.

start of the Jackson Laboratory to make money, we could have monopolized these mice. We could have sent out only mice of one sex of a strain, for experiments, and of another sex from another strain, but . . . we said, "We shall share these mice with everybody that we can.". . .This has been a very great service. We have not gotten rich on it. . . . There has been no business. There has been no industry."[18]

JAX Lab's former director thus painted a picture of a scientific activity defined by the quintessential value of objectivity and governed by classic Mertonian norms.[19] The goal of his institution was to contribute to the communal production of knowledge, a pursuit that by definition spurned personal gain as a motive. Inbred mice were biomedical research tools, whose production, availability, and sales were determined solely by researchers' needs. Little and his Jackson Lab were mere facilitators of the knowledge-making process.

The judge who presided over the 1962 JAX Tax Case ruled squarely according to Little's formulation of the dispute. The fact that the Jackson Laboratory produced and sold mice, he argued in his decision, cast no aspersions on its status as a charitably minded research institution. In less politically charged contexts, Little himself acknowledged that JAX mice had come to hold a more complicated place in the social and intellectual ecology of animal mouse research (see fig. I.5). But the local judge's opinion reflects the more widespread public belief that JAX was to be lauded for the service it provided other scientific workers.[20] In the ensuing decades—especially after George Snell's 1980 Nobel Prize for mouse genetic histocompatibility research—the lab's own research in cancer and genetics underwent a revival, and the "service" argument became even easier for JAX to sustain in the burnished glow of its own acknowledged scientific successes.

NEW BIOLOGICAL LANDSCAPES: DISCIPLINARY, INSTITUTIONAL, CULTURAL

Changes at the Jackson Lab, however, represent a small cross-section of broader shifts in American biomedical research since the 1950s. During this time, mouse breeders and biomedical scientists working with inbred

[18] Green, "TJL: The Tax Case," excerpted from sections 281–83.

[19] Robert Merton, "The Normative Structure of Science," in *Sociology of Science: Theoretical and Empirical Investigations* (Chicago: University of Chicago Press, 1973), pp. 267–80.

[20] Much to the chagrin of Young, the presiding judge considered inbred mouse sales to be so irrelevant to JAX Lab's tax status that he refused to address that line of questioning in his ruling. Young appealed the decision, to no avail. See Green, "TJL: The Tax Case."

mice have been forced to confront the ways in which new modes of production permeate their research practices. As I have shown, the utility of inbred mice as reliable mammalian stand-ins for humans was more or less established in various biomedical arenas by the decade immediately following World War II. As genetic explanations for cancer were on the wane in biomedical circles, the development of a mouse model for radiation damage gave mice a new relevance to both genetics and issues of public policy. But the extent to which these arenas themselves would grow and prosper has been shaped by larger disciplinary and structural trends in biological research during the last five decades.

One important trend amplified is the commercialization of the infrastructure of biological research. Sales of JAX mice to outside researchers now exceed two million organisms annually (fig. E.1),[21] but this figure no longer guarantees JAX the largest share of the scientific mouse market. Mouse breeding became a private money-making venture in the 1960s when several large, multinational "mouse manufacturers"—such as Charles River Laboratories and Harlan Sprague Dawley—were founded in response to the need for mice for the National Cancer Institute's massive chemotherapy drug screening program (CCNSC). Proprietary information on exact numbers of animals these companies once produced for the CCNSC is difficult to come by, but by best estimates, they currently provide somewhere between one-third and one-half of the mice used by all U.S. laboratories. Their market edge, summarized Ralph Dell, director of the Institute for Lab Animal Research at the National Academy of Sciences, is that they offer "quicker turnaround for limited numbers of popular mutant material."[22] As with other instruments of biological research, inbred mouse production has been largely "deskilled" and removed from the preserve of "doing science."[23] JAX as an institution now attempts to capitalize on the fact that it does both, although the physical arrangement of the place reinforces the divide: since the mid-1960s, these two domains have been housed in different buildings at the lab.

Another transition, begun in the 1940s and continuing to shape the more contemporary uses of inbred mice, is the shift in patronage away

[21] Personal communication, JAX Public Information Office, June 1992. See also Jackson Laboratory Annual Report, 1991.

[22] Because of increasing animal rights activism in the United States, specific, up-to-date information about the number of mice produced and sold is difficult to obtain. The figures I cite are from 1980 stock market analyses and Institute of Laboratory Animal Resources (ILAR) reports: see Rowan, *Of Mice, Models and Men* (Albany: SUNY Press, 1984), pp. 70–71; cf. Henry L. Foster, "The History of Commercial Production of Laboratory Rodents," *Laboratory Animal Science* (30) (1980): 793–98. Dell quoted in Ellen Barry, "Mouse Shortage Hinders ALS Research," *Boston Globe*, 26 January 2000, p. A1.

[23] Deborah Fitzgerald, "Farmers Deskilled: Hybrid Corn and Farmer's Work," *Technology and Culture* 34 (1993): 324–43.

E.1. JAX mice readied for shipping, c. 1980 [Source Credit: Jackson Laboratory Archives].

from private foundation grants for individual researchers and toward federal government grants for targeted biomedical research areas. As of 2000 the budget of the NIH still made up less than 1 percent of the total federal budget, but this money funded a significantly greater pool of biological researchers relative to the total pool.[24] Susan Wright has argued that this shift brought with it increasing demands for accountability, both to one's peers in science (who evaluate grant applications and make funding decisions based on internal criteria of research productivity) and to the public (who have increasingly challenged the effects and direction of the science and technology they pay for with their tax dollars).[25] Both recombinant DNA technology and the Human Genome Project emerged out of this milieu in the 1970s and 1980s, and collectively they offered molecular biologists a way to serve two masters: science and its new public.

Basic knowledge that had accrued about inbred laboratory mice and the accumulation of diverse mutant strains positioned the mouse to become the leading vertebrate for mammalian genetics. The mouse was the first mammal chosen as a "model organism" for the human genome, and its use has been described as a "Rosetta stone" for the genome's biomedical interpretation. The field of mouse genetics by some accounts has become crowded, with public and private mouse genome projects competing for resources. Molecular biologists began embracing such organismal genetics with a vengeance, in part because advances in gene manipulation techniques made possible molecular mapping and sequencing in mammals on par with *Drosophila* and *E.coli*. But their scientific conversion was also rooted in a belief that C. C. Little first articulated fifty years earlier: these creatures represented the key to understanding more complex problems of heredity applicable to humans. "A dog may be man's best friend," wrote a *Nature* reporter upon publication of the mouse genome in 2002, "but the humble mouse, *mus musculus*, is certainly our greatest ally."[26]

New genetic manipulations of mice gave rise to new issues of scientific and animal integrity. In 1985 Harvard University scientists developed a

[24] Cf. "2003 Federal Budget: Historical Tables," w3.access.gpo.gov/usbudget/fy2003/maindown.html (accessed December 2002).

[25] Wright, *Molecular Politics*.

[26] On mouse as "Rosetta stone," see Shirley Tilghman's comments in "The Mouse Sequence: A Rosetta Stone" and also the Oak Ridge Museum, as discussed in Rader, "The Mouse People." Alison Abbot, "Drive for More Genomes Threatens Mouse Sequence," *Nature* 405 (8 June 2000): 602; Nicolas Wade, "Mouse Genome is New Battleground for Project Rivals," *New York Times*, 7 May 2002, p. A1. On mouse mapping techniques, see L. C. Dunn "Oral History"; cf. Richard Saltus, "Scientists Finish DNA Map Project," *Boston Globe*, 14 March 1996, p. A7. Tom Clarke, "Mice Make Medical History," *Nature: Science Update*, 5 December 2002, p. 1.

completely new mouse tool: a "transgenic mouse," engineered to contain a piece of human DNA.[27] Since December 1988, when Harvard was awarded a U.S. patent for Oncomouse (so named because the gene inserted causes breast cancer), scientists and policymakers alike have debated how to insure "fair" access to research materials without neglecting or, to the other extreme, overcrediting the efforts of the creator scientist. Often this filled the coffers of biotechnology companies at the expense of slowed research. Given the verdict of the 1962 JAX trial, it is in many ways not surprising that both researchers and science administrators have looked to JAX for guidance in settling such matters.[28] But the Canadian Supreme Court's 2002 ruling against Dupont's original patent suggests that the internal needs and wants of the biomedical community regarding experimental materials will not always trump the ethical sensibilities of the broader public.[29]

Likewise, boundary work[30] has been taken up by scientists and the American popular press alike around what might be called the "designer" mutant mice. In media representations such creatures sometimes appear as a generic aggregate—"genes-for-X mice," where one can substitute for X medical conditions as diverse as Alzheimer's disease, male patterned baldness, obesity, and aggression (to name only a few).[31] But occasionally such mice emerge as individual *faits accomplis*, with unique names and commercial applications attached to their identities. One of the most widely reported of the latter was called "Doogie, the Smart Mouse." In mid-September 1999, Princeton University neurobiologist Joe Tsien created this animal and named him after the television child prodigy "Doogie Houser, MD." Doogie is a so-called smart mouse: his memory has been enhanced through genetic manipulation of NR2B, a gene whose product pairs with another gene's product, NR1, to open what biologists believe to be the physical mechanism of memory in the brain. Doogie's forebrain

[27] Cf. David L. Wheeler, "Biologists Discuss Ways to 'Share' Genetically Engineered Mice," *Chronicle of Higher Education*, 7 April 1993, p. A14.

[28] Ibid., cf. John R. Wilke, "Furry Freaks: Mutant Mice Maker Is Serving Science," *Wall Street Journal*, 4 February 1994, p. 1 (I am indebted to Judy Johns Schloegel for this reference).

[29] Daniel J. Kevles, "Of Mice and Money: The Story of the World's First Animal Patent,' *Daedalus* (Spring 2002): 78–88. For a comparative international perspective regarding the patenting of the Harvard mouse, see Soren Germer, "German Experiments in the Science and Politics of Evolution," Ph.D. dissertation, University of California, Berkeley, 1997.

[30] Thomas F. Gieryn, "The Boundaries of Science," in *Handbook of Science and Technology Studies*, ed. S. Jasanoff et al. (Thousand Oaks, CA: Sage, 1995).

[31] Nicholas Wade, "Of Mice, Men, and a Gene That Jump-Starts Follicles." *New York Times*, 5 October 1999, p. F1; Sharon Begley, "The Mice That Roared," *Newsweek*, 4 December 1995, p. 76; "New Breed of Mouse Aids Alzheimer's Work," *New York Times*, 4 October 1996, p. A22.

now produces some extra NR2B product, so his memory mechanism stays open an extra 150 thousandths of a second. This is enough time, Tsien and his colleagues say, for the creature to outperform other normal mice its age on "standard" tests of rodent intelligence. *Time* magazine proclaimed that Tsien's work "sheds lights on how memory works and raises questions about whether we should use genetics to make people brainier."[32] Meanwhile, Princeton has filed for a use patent on the NR2B gene, which would give the institution the right to develop drugs to enhance NR2B production in humans. As a material incarnation of the practical and ideological values of genetic-based experimental biomedicine, then, designer mutant mice like Doogie are true cyborgs. Doogie is part nonhuman animal, insofar as scientists have altered his genes and experimented on his body in ways ethically unthinkable for members of our own species, but bearing a human name and performing functions significant only within our culture.[33]

Critics of genetic engineering seek to mobilize these contradictions to challenge to current biomedical research values and assumptions. The environmental advocacy organization Turning Point Project, for example, ran an advertisement in the front-page section of the *New York Times* in October 1999 featuring a photo of "the mouse with a human ear on its back." This creature was produced in 1995 in the lab of Charles Vacanti. Vacanti collaborated with a chemical engineer to create an earlike scaffolding made of biodegradable polyester fabric, to distribute human cartilage cells onto the form, and to implant it onto the back of a hairless (or "nude") mouse that lacks the immune system to reject human tissue.[34] The resulting ear-shaped human cartilage was then removed, and the mouse lived. At the time of Vacanti's initial publication, *Associated Press* reported: "Someday, ears and noses for cosmetic surgery will be grown in a test tube using the patient's own cells and custom-designed scaffolding." But the Turning Point Project asked readers to reject this formulation: "Whether you give credit to God or to Nature . . . there is a *boundary between life forms* that gives each its integrity and identity. . . . Biotech companies are blithely removing components of human beings (and other creatures) and treating us all like auto parts at a swap meet. . . . Some of these experiments may save lives, but so far, there are few successes and an equal or greater chance for terribly negative consequences" (emphasis

[32] Wade, "Of Smart Mice"; Michael D. Lemonick, "Smart Genes?" *Time* 154, 11 (September 13, 1999): 54–59. On possible drug applications, see also Catherine Arnst, "Building a Smarter Mouse," *Business Week* 3646 (September 13, 1999): 103.

[33] Cf. Donna Haraway, *Modest Witness @ Second_Millenium.FemaleMan(c)Meets Oncomouse™: Feminism and Technoscience* (New York: Routledge, 1997).

[34] K. T. Paige et al., "Injectable Cartilage," *Plastic & Reconstructive Surgery* 96 (1995): 1390–1400.

in original). In this formulation, then, cultural dissonance springs from biomedicine's misguided intrusion on what might be called "natural mousehood." No knowledge obtained from these mice is natural, and the creatures themselves are artifacts of a dangerous arrangement between science and society.

Likewise, persistent biotechnology critic Jeremy Rifkin has collaborated with developmental biologist Stuart Newman to create a series of laboratory protocols for making a human-mouse chimera they call Humouse. One method, for example, merges an eight-cell human blastula (the structure resulting from a fertilized egg) with a mouse embryo at the same stage of development. Another introduces human embryonic stem cells into early stage mouse development, then implants the resulting embryo into the uterus of a mouse to gestate. Newman argues that such procedures would "take only a day of laboratory work" to complete successfully, but neither method has been tried. Instead, Rifkin and Newman submitted the protocols to the U.S. Patent Office in 1997, in an attempt to force the legislative hand of the federal government on the question of patenting of human life. In April 1998 the patent commissioner broke a long-term policy of not commenting on pending application to tell reporters: "There will be no patents on monsters, at least not while I'm commissioner."[35]

The lay public has for the most part tolerated the tensions of scientific mouse and scientific human co-existence, and even at times embraced them with a noticeable sense of humor. A 1948 *Boston Herald* cartoon imagined a "strain of neurotic mice" developed by "a Bar Harbor, ME scientist"—these animals were "so afraid of women that they will jump on a chair" to avoid them (fig. E.2). By contrast, in 1953 the staff of the Vincent Memorial Hospital donated an advertisement for their institution to the local Boston theater's *Playbill* (fig. E.3). The ad features a Jackson Lab mouse standing on its hind quarters and holding a syringe (filled presumably with cancerous tumor cells or cancer-inducing chemicals). Embodying the wish and the contradiction of nonhuman animal agency in science, the creature appears to be dancing its way through a laboratory protocol that will surely result in the loss of its own life.[36]

More recently, the 1999 NSF Science Indicators Survey indicated that public approval of the use of mice in research was approximately 66 percent—significantly higher than the use of animal model generally. At the

[35] Dashka Slater, "Humouse™," *Legal Affairs* 1, 4 (November/December 2002): 1+; Aaron Zitner, "Patently Provoking a Debate," *Los Angeles Times*, 12 May 2002, p. A1+.

[36] Both photos appear in the last "Clipping File" from the C. C. Little Papers (CCL-UMO), which suggests that Little carefully followed the media images of JAX and of laboratory mice.

E.2. "Smelling Salts" cartoon from the *Boston Herald*, February 13, 1948 [Source Credit: C. C. Little Papers, University of Maine].

same time that the Minnesota-based Alternatives Research and Development Foundation took on the biomedical research industry by mounting a challenge to the Animal Welfare Act, forcing the USDA (its administering institution) to include previously excluded mice, rats, and birds under its more stringent laboratory care provisions, the Web-based magazine *Fast Company* conducted a survey indicating that its readers rated the development of the laboratory mouse as "the best of the best innovations" in technology during the twentieth century.[37]

Incongruous images and beliefs about the standardized lab mouse, then, remain powerful as much for their ability to bring us together under an umbrella of common humanity served by these animals, as to tear us apart in periodic waves of social and ethical conflict over the meaning of their creation and use. Bryan Crockett's large and powerful sculptures of

[37] On 1999 NSF Science and Engineering Indicators Report, see *Biomedical Communication* (San Diego: Academic Press, 2001); Meredith Wadman, "U.S. Lab Animals May Win in Lawsuit," *Nature* 407 (2000): 549. Cf. "Mighty Mice," at www.fastcompany.com/online/47/agendaitems.html (accessed August 2002).

I come from

THE ROSCOE B. JACKSON
MEMORIAL LABORATORY
BAR HARBOR, MAINE

The oldest and largest research institution in the world devoted principally to the study of heredity of cancer and allied diseases.

I serve at the

VINCENT MEMORIAL LABORATORY

Contributed by the Staff

E.3. Boston theater *Playbill* advertisement from "Pass the Salt, Mr. Jones," March 1953 [Source Credit: C. C. Little Papers, University of Maine].

laboratory mice—especially *Ecce Homo* or "Monument to Man" (2000)—engage these incongruities. On the one hand, they are grotesque and absurd caricatures, but on the other (as Ronald Bailey noted), they "wryly honor humanity's growing biological ingenuity and would be the perfect complement to the lobby of some brash, self-confident biotech laboratory that wants a work of art to symbolize its creative activities."[38] Nowhere has this tension been better expressed than in the climactic scene of Zadie Smith's novel *White Teeth*, which features the public launch of celebrated (fictional) scientist Marcus Chalfen's FutureMouse(c). FutureMouse(c) is a transgenic mouse in which fatal genes have been engineered to be "turned on" and expressed along a predictable timetable. This animal, Chalfen claims, is the "site for an experiment" into the aging of cells, the ultimate embodiment of human agency in the process of scientific development: "FutureMouse(c), he tells the crowd, holds out the tantalizing promise of a new phase in human history where we are not the victims of the random but instead directors and arbitrators of our own fate." For Chalfen's middle- and working-class teenage children and their friends, however, FutureMouse(c) is a metaphor for the degree to which their father's personal affections for them (and, by extension, his concern for the problems affecting contemporary Western, especially British, society) have been distracted by science and other bourgeois values. Still, when asked by a group of animal activists to liberate the animal, Chalfen's oldest son Josh (with whom Marcus has an especially tense relationship) surprises himself by resisting his friends and expressing what he holds to be a certain fatalism about this mouse's development: "This isn't like the other animals you bust out. It won't make any difference. The damage is done."[39]

To the extent that history can inform the present, my story of the development of the standardized laboratory mouse demonstrates the importance of contingency in science in several ways. The first is practical, for both scientists and science studies scholars. The current utility of inbred mice was determined through a multivalent consensus among benchtop scientists, policymakers, and various public constituencies over the values that should govern American biomedical research—namely, standardization, coordination, efficiency, and, later, minimal extrapolation from animal data to the most important human disease threats. However, these values were not predetermined but contextually defined through the cir-

[38] Ronald Bailey, "Pink Mice and Petri Dishes: Artists Contemplate Biotechnology," in *Reason Online* (December 2000), reason.com/0012/cr.rb.pink.shtm (accessed 11 June 2003).

[39] Zadie Smith, *White Teeth: A Novel* (New York: Random House, 2000), pp. 401, 357. I am grateful to Julie Abraham for pointing me toward this book.

cumstances in which inbred mice were actually used. In other words, use determines utility, and research materials are more often the consequence of, rather than the reason for, laboratory developments—so the stories we tell ourselves (and others) about science should reflect this insight.

The second is cultural, for anyone who seeks to understand how the production of natural and technological knowledge structures nearly every facet of our contemporary Western world—commerce, industry, healing, literacy, and entertainment, to name just a few. In one gesture—his letter to Walt Disney—C. C. Little demonstrated that he grasped an important tenet from the anthropology of science: science can shape culture, just as culture shapes science.[40] What he perhaps failed to realize is something that seems obvious in hindsight: all animals, both human and nonhuman, are contested actors in this process. Remarking on the genetic similarities between mice and humans, yeast geneticist Ira Herkowitz has said: "I don't consider the mouse a model organism. The mouse is just a cuter version of a human, a pocket-sized human." But if we are to take the study of science and society seriously, then we must acknowledge that this biomedical translation goes both ways. If (as Jean Baudrillard infamously declared) "Disneyland exists to conceal the fact that America is Disneyland," then the laboratory mouse exists to conceal the fact that man is also a mouse. In the introduction to *Representing Animals*, Nigel Rothfels summarized this central notion from recent scholarship in "animal studies," which sheds analytic light on this insight: "The way animals are understood is bound in time and place, and . . . the careful scrutiny of that understanding reveals not only important limits to our knowledge of animals but important limits to our knowledge of ourselves."[41]

Today's inbred laboratory mice are in many ways different animals from the mice of Little's time, and their use in research has permanently transformed the mouse's cultural legacy. Whether we assent or not, the laboratory mouse's very existence engages us in a complicated process of technical, cultural, and political formation, but at the same time, its historical situatedness provides a tool kit for intervening in this process. This acknowledgment is not an open invitation to practice counterfactual history: envisioning the "what ifs" of biomedicine's past is a serious project that can be taken on only by those who appreciate the degree to which

[40] Pickering, *Science as Practice and Culture*.

[41] Hershkowitz quoted in "The Mouse Sequence: A Rosetta Stone," *The Genes We Share*, Howard Hughes Medical Institute, accessed at www.hhmi.org/genesweshare, February 2003. On Baudrillard and mice, see Marc Holthof, "To the Realm of Fables: The Animal Fables from Mesopotamia to Disneyland," in *Zoology on (Post) Modern Animals*, ed. Bart Verschaffel and Marc Vermick, trans. Milt Papatheophanes (Dublin: Lilliput Press, 1993): pp. 37–55, quote on p. 55. Nigel Rothfels, "Introduction," in *Representing Animals*, ed. Rothfels (Bloomington: Indiana University Press, 2002) xii.

E.4. Drawing of a mouse in the announcement of the JAX Workshop in Mouse Neurogenetics, September 1999 [CREDIT: Jackson Laboratory Archives].

they have a stake in its future. Rather, I am suggesting that this history invites us to view ourselves like the laboratory mouse pictured in a 1999 Jackson Lab summer program announcement (fig. E.4): humbled and perhaps even a little confused by the knowledge that confronts us and what we should do with it, but nevertheless poised to evaluate, measure, cut, paste—in short, to engage with the questions of animal and human integrity that now define biomedicine and, in the process, define us.

BIBLIOGRAPHY

Abbot, Alison. "Drive for More Genomes Threatens Mouse Sequence," *Nature*, 405 (8 June 2000): p. 602.

Abir-Am, Pnina G. "Converging Failures: Science Police, Historiography and Social Theory of Early Molecular Biology." In *Scientific Failure*. Edited by Tamara Horowitz and Allen Janis. Lanham, MD: Rowmand & Littlefield, 1994.

———. "The Discourse of Physical Power and Biological Knowledge in the 1930s: A Reappraisal of the Rockefeller Foundation's 'Policy' in Molecular Biology." *Social Studies of Science* 12 (1982): 341–82.

"A-Bomb Story Not Yet Told." *Science News Letter* 51 (1947): 219.

Ahrweiler, Jacques, ed. *Écrits sur L'hérédité*. Paris: Editions Seghers, 1964.

Alder, Kenneth. *Engineering the Revolution*. Princeton: Princeton University Press, 1997.

———. "A Revolution to Measure: The Political Economy of the Metric System in France." *The Values of Precision*. Edited by N. Wise, pp. 39–71. Princeton: Princeton University Press, 1995.

Allen, Garland. "Aaron Franklin Shull." *In Dictionary of Scientific Biography*. Vol. 12. Edited by C. Gillispie, pp. 416–18. New York: Scribners, 1970.

———. "The Eugenics Record Office at Cold Spring Harbor: An Essay in Institutional History." *Osiris* 2 (1986): 225–64.

———. "William Ernest Castle." In *Dictionary of Scientific Biography*. Vol. 3. Edited by C. Gillispie, pp. 120–24. New York: Scribner's, 1970.

Andervont, H. B. "The Influence of Foster Nursing on Incidence of Spontaneous Mammary Tumor Cancer in Resistant and Susceptible Mice." *Journal of the National Cancer Institute* 1 (1949): 147–53.

———. "J. W. Schereschewsky: An Appreciation." *Journal of the National Cancer Institute* 19.2 (August 1957): 331–33.

———. "Spontaneous Tumors in a Subline of Strain C3H Mice." *Journal of the National Cancer Institute* 1 (June 1941): 737–43.

———. "Studies on Immunity Induced by Mouse Sarcoma 180." *Public Health Reports* 47, 37 (9 September 1932): 1859–77.

Ankeny, Rachel Allyson. "The Conqueror Worm: An Historical and Philosophical Examination of the Use of the Nematode *Caenorhabditis Elegans* as a Model Organism." Ph.D. diss., University of Pittsburgh, 1997.

Arluke, Arnold, and Clifford Sander. *Regarding Animals*. Philadelphia: Temple University Press, 1996.

Arnst, Catherine. "Building a Smarter Mouse." *Business Week* 3646 (September 13, 1999): 103.

Ascheim, Selmar, and Bernhard Zondek. "Die Schwangerschaftsdiagnose aus dem Harn durch Nachweis des Hypophysenvorderlappenhormons." *Klinische Wochenschrift* (Berlin) 7 (1928): 8–9, 1404–11, 1453–57.

Auerbach, Charlotte. *Genetics in the Atomic Age*. With illustrations by I. G. Auerbach. New York: Oxford University Press, 1965.

Bagg, Halsey. "Further Studies on the Relation of Functional Activity to Mammary Carconoma in Mice." *American Journal of Cancer* (1936): 542+.

Bagg, Halsey., and J. Jacksen. "The Value of a 'Functional Test in Selecting Material for Genetic Study of Mammary Tumors in Mice and Rats.' " *American Journal of Cancer* 30 (1937): 539–48.

———. "Obituary." *New York Times*, 15 April 1947.

——— and C.C. Little, "The Occurrence of Four Inheritable Morphological Variations in Mice and Their Possible Relation to Treatment with X-rays," *Journal Of Experimental Zoology*, 1924, 41: 45–91.

Bagg, Halsey, and C. C. Little. "Hereditary and Structural Defects in the Descendants of Mice Treated with Roentgen-ray Irradiation." *American Journal of Anatomy* 33 (1924): 119–45.

Barry, Ellen. "Mouse Shortage Hinders ALS Research." *Boston Globe*, 26 January 2000, p. A1.

Bartlett, Arthur. "The Big Mouse Man of Cancer Research." *Coronet* 26 (August 1949): 161–62.

Beatty, John. "Genetics in the Atomic Age: The Atomic Bomb Casualty Commission, 1947–1956." In *The Expansion of American Biology*. Edited by K. Benson, J. Maienschein, and R. Rainger, pp. 284–324. New Brunswick: Rutgers University Press, 1986.

Begley, Sharon. "The Mice That Roared." *Newsweek*, 4 December 1995, p. 76.

"Birth Control Urged for Improving the Race." *New York Times*, 21 November 1929, p. 56.

Bittner, John J. "Breast Cancer in Mice." *American Journal of Cancer* 36 (1939): 44.

———. "Changes in the Incidence of Mammary Carcinoma in Mice of the A Stock." *Cancer Research* 1 (1941): 113–14.

———. "A Color Mosaic in the Mouse." *Journal of Heredity* 23 (1932): 421–22.

———. "Foster Nursing and Genetic Susceptibility for Tumors of the Breast in Mice." *Cancer Research* 1 (1941): 793–94.

———. "A Genetic Study of the Transplantation of Tumors Arising in Hybrid Mice." *American Journal of Cancer* 15 (1931): 2202–47.

———. "The Influence of Transplanted Normal Tissue on Breast Cancer Ratios in Mice." *Public Health Reports* 54 (1939): 1827–31.

———. "Linkage in Transplantable Tumors." *Journal of Genetics* 29 (1934): 17–27.

———. "Mammary Cancer in Mice." In *Lectures on Genetics, Cancer, Growth, and Social Behavior*, pp. 51–57. Bar Harbor: Bar Harbor Times, 1949.

———. "Mammary Tumors in Mice in Relation to Nursing." *American Journal of Cancer* 30 (1937): 530–38.

———. "Possible Types of Mammary Gland Tumors in Mice." *Cancer Research* (November 1942): 755–58.

———. "Relation of Nursing to Extrachromosomal Theory of Breast Cancer in Mice." *American Journal of Cancer* 35 (1939): 90.

———. "A Review of Genetic Studies in the Transplantation of Tumors." *Journal of Genetics* 31 (1935): 471–87.

———. "Some Enigmas Associated with the Genesis of Mammary Cancer in Mice." *Cancer Research* 8 (December 1948): 625–39.

———. "Transplantability of Mammary Cancer in Mice Associated with the Source of the Mammary Tumor Milk Agent." *Cancer Research* 7 (1947): 741–45.

———. "Tumor Incidence in Reciprocal F1 Hybrid Mice—A X D High Tumor Stocks." *Proceedings of the Society for Experimental Biology and Medicine* 34 (1936): 42–48.

Boguski, Mark S. "The Mouse That Roared." *Nature* 420 (5 December 2002): 515–16.

Borrel, A. "Parasitisme et tumeurs." *Annales de l'Institute Pasteur* 24 (1910): 778.

Bowker, Geoffrey, and Susan Leigh Star. *Sorting Things Out: Classification and Its Consequences*. Cambridge: MIT Press, 1999.

Brand, Steward. *How Buildings Learn: What Happens to Them After They're Built*. New York: Penguin, 1995.

Brent, Leslie. *History of Transplantation Immunology*. San Diego: Academic Press, 1997.

Brinkley, Alan. *Liberalism and Its Discontents*. Cambridge: Harvard University Press, 1998.

Bud, Robert. "Strategies in American Cancer Research after World War Two: A Case Study." *Social Studies of Science* 8, 4 (November 1978): 425–59.

Bud, Robert, and Deborah Jean Warner. *Instruments of Science: An Historical Encyclopedia*. New York: Science Museum, London, and National Museum of American History, Smithsonian Institution, in association with Garland Publishing, 1998. "Mouse" entry on pp. 403–4.

Bugos, Glenn E., and Daniel J. Kevles. "Plants as Intellectual Property: American Practice, Law and Policy in World Context," *Osiris* 7 (1992): 75–104.

Callon, Michel. "Some Elements of a Sociology of Translation: Domestication of the Scallops and the Fisherman of St. Brieuc Bay." In *Power, Action, Belief: A New Sociology of Knowledge?* Edited by John Law, pp. 196–229. (New York: Routledge and Kegan Paul, 1986).

Cambrosio, Albert, and Peter Keating. "The New Genetics and Cancer: Contributions of Clinical Medicine in an Era of Biomedicine." *Journal of the History of Medicine and Allied Sciences* 56, 4 (October 2001): 321–52.

"Cancer Army." *Time* 29 (1937): 40–41.

"Cancer Files, Mice Lost at Institute." *New York Times*, 25 October 1947.

"Cancer Heredity Laid to Single Gene." *New York Times*, 15 July 1933.

"Cancer: National Research Center Explores Its Nature and Cause." *Life* 8 (17 June 1940): 35–38.

"Cancer Prevented in Some Animals." *New York Times*, 28 June 1930, p. 18.

"Cancer Research Laboratory Is to Be Established This Summer." *Bar Harbor Times*, 8 May 1929.

"Cancer: The Great Darkness." *Fortune* 15 (March 1937): 112, 114, 162.

Capecchi, Mario. "Altering The Genome by Homologous Recombination." *Science* 244 (1988):1288–92.

———. "The New Mouse Genetics: Altering the Genome by Gene Targeting." *Trends in Genetics* 5 (1989):70–76.

Capshew, James, and Karen Rader. "Big Science: Price to the Present." *Osiris* 7 (1992): 3–25.

Carter, Toby. "Radiation Induced Gene Mutation in Adult Female and Foetal Male Mice." British *Journal of Radiology* 31 (1958): 407–11.

Carter, Toby, Mary Lyon, and R.J.S. Phillips. "Genetic Hazards of Ionizing Radiation." *Nature* 182 (1958): 409.

Castle, W. E. *Mammalian Genetics*. Cambridge: Harvard University Press, 1940.

———. "The Part of Mammalian Genetics in Founding the Jackson Memorial Laboratory." *Journal of the National Cancer Institute* 15 (1954): 593–605.

———. "Piebald Rats and Selection: A Correction." *American Naturalist* 53 (1919): 370–75.

———. "Pure Lines and Selection." *Journal of Heredity* 5 (1914): 93–97.

Castle, W. E., et al. "The Effects of Inbreeding, Cross-Breeding, and Selection upon the Fertility and Variability of Drosophila." *Proceedings of the American Academy of Arts and Sciences* 41 (1906): 729–86.

Castle, W. E., and C. C. Little. "The Peculiar Inheritance of Pink-eyes among Colored Mice." *Science* 30 (1909): 313–15.

Caufield, Catherine. *Multiple Exposures: Chronicles of the Radiation Age*. Chicago: University of Chicago Press, 1989.

Charles, Donald R. "Radiation-induced Mutations in Mammals." *Radiology* 55 (1950): 579–81.

Charles, Donald, et al. "Genetic Effects of Chronic X-irradiation Exposure in Mice." U.S. AEC Commission Research and Development Report UR-565, RRT-UT.

"Clarence Cook Little." *Who Was Who in America*. Vol. 5 (1969–73), pp. 435–36.

Clark, Charles E. *Maine: A Bicentennial History*. New York: W. W. Norton, 1977.

Clark, Roberta. "The Social Uses of Scientific Knowledge: Eugenics in the Career of Clarence Cook Little, 1919–1954." M.A. thesis, University of Maine, Orono, 1986.

Clarke, Adele E. "Research Materials and Reproductive Science in the United States, 1910–1940." In *Physiology in the American Context, 1850–1940*. Edited by Gerald L. Geison. Bethesda: American Physiological Society, 1987.

Clarke, Adele E., and Joan H. Fujimura. *The Right Tools for the Job: Materials, Instruments, Techniques and Work Organization in Twentieth Century Life Sciences*. Princeton: Princeton University Press, 1992.

Clarke, Tom. "Mice Make Medical History." *Nature: Science Update*, 5 December 2002, p.1.

Clause, Bonnie Tocher. "The Wistar Rat as a Right Choice: Establishing Mammalian Standards and the Ideal of a Standardized Mammal." *Journal of the History of Biology* 26 (Summer 1993): 329–49.

Cloudman, Arthur. "A Comparative Study of Transplantability of Mammary Tumor Arising in Inbred Mice," *American Journal of Cancer* 16 (1932): 568–630.

———. "Gross and Microscopic Diagnoses in Mouse Tumors at the Site of the Mammary Glands. *AJC* 27 (1936): 510–12.

———. "Organophilic Tendencies of Two Transplantable Tumors of the Mouse." *Cancer Research* (September 1947): 585+.

Coburn, Charles A. "Heredity of Wildness and Savageness in Mice." *Behavioral Monographs* 4 (1921–22): 1–32.

Cochrane, Rexmond. *The National Academy of Sciences: The First Hundred Years, 1863–1963*. Washington, DC: The Academy, 1978.

Cook, Robert. "Front Line against Cancer." *Collier's Weekly* 107 (8 February 1941): 21+.

———. "Heredity and Cancer." *Journal of Heredity* 23 (1932): 160. (A journalist's summary of H. Gideon Wells, "The Nature and Etiology of Cancer," *American Journal of Cancer* 15 [July 1931]: 1919.)

Corner, George W. *A History of the Rockefeller Institute, 1901–1953: Origins and Growth*. New York: Rockefeller Institute Press, 1964.

Cowan, Ruth Schwarz. "The Consumption Junction: A Proposal for Research Strategies in the Sociology of Technology." In *The Social Construction of Technological Systems: New Directions in the Sociology and History of Technology*. Edited by Wiebe E. Bijker, Thomas P. Hughes, and Trevor J. Pinch, pp. 261–80. Cambridge: MIT Press, 1987.

Creager, Angela N. M. *The Life of a Virus: TMV as an Experimental Model, 1930–1965*. Chicago: University of Chicago Press, 2001.

Crist, Eileen. *Images of Animals: Anthropomorphism and the Animal Mind*. Philadelphia: Temple University Press, 1999.

Cross, F. C. "What Animals Know about Medicine." *Popular Science* 127 (September 1935): 24+.

Crow, James. "A Century of Mammalian Genetics and Cancer: Where Are We at Mid-Passage?" In *Mammalian Genetics and Cancer: The Jackson Laboratory Fiftieth Anniversary Symposium*. Edited by Elizabeth Russell, pp. 309–24. New York: A. R. Liss, 1981.

Crow, James. "How Well Can We Assess Genetic Risk? Not Very." Lecture no. 5 in the Lauriston Taylor Lecture Series in Radiation Protection and Measurements. Washington, DC: National Council on Radiation Protection Measurements, 1981.

Culliton, Barbara J. "The Jackson Laboratory: 'Mice Are Our Most Important Product.' " *Science* 177 (September 1972): 871–74.

Davies, C. J. *Fancy Mice*. 5th edition. London: L. Upcott Gill, 1912.

Davis, Allan Peter, and Monica J. Justice. "An Oak Ridge Legacy: The Specific Locus Test and Its Role in Mouse Mutagenesis." In *Perspectives on Genetics*. Originally appeared in *Genetics* (January 1998). Edited by James F. Crow and William F. Dove, pp. 638–43. Madison: University of Wisconsin Press, 2000).

Davisson, Muriel, and Tom Roderick. "Linkage Map." In *Genetic Variants and Strains of the Laboratory Mouse*. Edited by Mary Lyon and A. G. Searle. New York: Oxford University Press, 1989), pp. 416–27.

De Chadarevian, Soraya. "Of Worms and Programmes: 'Caenorhabditis elegans' and the Study of Development." *Studies in the History and Philosophy of the Biological and Biomedical Science* 29 (1998): 81–105.

Dewey, John. "The Ethics of Animal Experimentation." *Atlantic Monthly* 138 (September 1926): 343–46.

Dickie, Margaret. "The Expanding Knowledge of the Genome of the Mouse." *Journal of the National Cancer Institute* 15 (1954): 679+.

Donnelly, Walter A., et al., eds. *The University of Michigan: An Encyclopedic Survey* Vol. 4, pp. 1538–39. Ann Arbor: University of Michigan Press, 1958.

Douglas, Mary. *Risk and Blame: Essays in Cultural Theory*. London: Routledge, 1992.

"Dr. Flexner Denies Cruel Vivisection." *New York Times*, 17 January 1910, p. A1.

"Dr. Little Explains Resignation." *New York Times*, 16 February 1929, p. 36.

"Dr. Little Welcomed Here." *New York Times*, 11 October 1929, p. 30.

Dunn, L. C. "Analysis of a Case of Mosaicism in the House Mouse." *Journal of Genetics* 29 (1934): 317–26.

———. "A New Series of Allelomorphs in Mice." *Nature* 129 (1932): 130.

Dunn, L. C. Oral History. Interview by Saul Benison. New York: Columbia University Oral History Project. 1960.

———. *A Short History of Genetics*. New York: McGraw-Hill, 1965.

———. "Testing Mendel's Rules." In *A Short History of Genetics*. New York: McGraw-Hill, 1965.

———. "William Ernest Castle." *Biographical Memoirs, National Academy of Sciences* 38 (1962): 31–80.

Dupree, A. Hunter. "The Great Instauration of 1940: The Organization of Scientific Research for the War." In *Twentieth Century Sciences: Studies in the Biography of Ideas*. Edited by Gerald Holton. New York: Norton, 1977.

Durham, Florence M. "Further Experiments on the Inheritance of Coat Color in Mice." *Journal of Genetics* 1 (1911): 159–68.

"Editorial: Genetics in Geneva." *Bulletin of the Atomic Scientists* 11 (December 1955): 314–16, 343.

Endicott, Kenneth. "The Chemotherapy Program." *Journal of the National Cancer Institute* 19 (1957): 275–93.

"The English Craze for Mice." *Reader's Digest* 30 (March 1937): 19.

Ewing, James. "The Public and the Cancer Problem." *Science* 87 (6 May 1938): 399–407.

Falconer, Donald. "Quantitative Genetics in Edinburgh: 1947–1980." *Genetics* 133 (February 1992): 137–2.

Fekete, Elizabeth, and Charles Green. "The Influence of Complete Blockage of the Nipple on the Incidence and Location of Spontaneous Mammary Tumors in Mice." *AJC* 27 (1936): 513–14.

Fekete, Elizabeth, and C. C. Little. "Observations on the Mammary Tumor Incidence of Mice Born from Transferred Ova." *Cancer Research* 2 (August 1942): 525–26.

Festing, Michael, and Elizabeth Fisher. "Mighty Mice." *Nature* 404, 6780 (20 April 2000): 815.

Fissell, Mary. "Imagining Vermin in Early Modern England." *History Workshop Journal* 47 (1999): 1–29.

Fitzgerald, Deborah. "Farmers Deskilled: Hybrid Corn and Farmer's Work." *Technology and Culture* 34 (1993): 324–43.

Foster, Henry L. "The History of Commercial Production of Laboratory Rodents." *Laboratory Animal Science* 30 (1980): 793–98.

Fox, Daniel. "Abraham Flexner's Unpublished Report: Foundations and Medical Education, 1909–1928." *Bulletin of the History of Medicine* 54 (Winter 1980): 475–96.

———. "The Politics of the NIH Extramural Program, 1937–1950." *Journal of the History of Medicine and Allied Sciences* 42 (1987): 447–66.

Fox, Richard, and T. J. Jackson Lears. *The Culture of Consumption*. New York: Pantheon, 1983.

Fox Keller, Evelyn. "Critical Silences in Scientific Discourses: Problems of Form and Re-Form." In *Secrets of Life, Secrets of Death*, pp. 73–92. New York: Routledge, 1992.

Fraser, Steven, and Gary Gerstle, eds. *Rise and Fall of the New Deal Order, 1930–1980*. Princeton: Princeton University Press, 1989.

Friedberg, Errol. *Correcting the Blueprint of Life: An Historical Account of the Discovery of DNA Repair Mechanisms*. Plainview, NY: Cold Spring Harbor Press, 1997.

Fujimura, Joan. *Crafting Science: A Sociohistory of the Quest for the Genetics of Cancer*. Cambridge: Harvard University Press, 1996.

"Fundamental Cancer Research: Report of a Committee Appointed by the Surgeon General." *Public Health Reports* 53 (1938): 2121–30.

Gates, William, and Elizabeth Lord. "Shaker: A New Mutation in the House Mouse." *American Naturalist* 63 (1929): 435.

Gaudillière, Jean Paul. "Circulating Mice and Viruses: The Jackson Memorial Laboratory, the National Cancer Institute, and the Genetics of Breast Cancer, 1930–1965." In *Practices of Human Genetics*. Edited by Michael Fortun and Everett Mendelsohn, pp. 89–124. Dordrecht: Kluwer, 1999.

Gaudillière, Jean Paul. "Rockefeller Strategies for Scientific Medicine: Molecular Machines, Viruses, and Vaccines." *Studies in the History and Philosophy of Biology and the Biomedical Sciences* 31, 3 (2000): 491–509.

———. "Taking Mice for Men: The Production and Uses of Animal Models in Postwar Biomedical Research." Paper presented at the Davis Center Workshop of Mice and Men: Animals and Medical Models, 14 March 1997.

Geison, Gerald, ed. *Physiology in the American Context, 1850–1940*. Baltimore: American Physiological Society, distributed by Williams & Wilkins, 1987.

Genes, Mice, and Men: A Quarter-Century of Progress at the Roscoe B. Jackson Memorial Laboratory. Bar Harbor: The Jackson Laboratory, 1954.

"George Snell." *Current Biography.* (New York: H. W. Wilson, May 1986), pp. 40–43.

Germer, Soren. "German Experiments in the Science and Politics of Evolution." Ph. D. dissertation, University of California, Berkeley, 1997.

Gest, Howard. "Arabidopsis to Zebrafish: A Commentary on the 'Rosetta Stone' Model Systems in the Biological Sciences." *Perspectives in Biology and Medicine* (Fall 1995) 39: 77–85.

Gieryn, Thomas F., "The Boundaries of Science." In *Handbook of Science and Technology Studies*. Edited by S. Jasanoff et al. Thousand Oaks, CA: Sage, 1995.

Gilbert, Scott, "Cellular Politics: Just, Goldschmidt, and the Attempts to Reconcile Embryology and Genetics." In *The American Development of Biology.* Edited by R. Rainger, K. Benson, and J. Maienschein, pp. 311–46. Philadelphia: University of Pennsylvania Press, 1988.

———. "The Synthesis of Embryology and Human Genetics: Paradigms Regained." *American Journal of Human Genetics 51* (1992): 211–15.

Green, C. V. "On the Nature of Size Factors in Mice." *American Naturalist* 65 (1931): 407–16.

Green, Earl, ed. *The Biology of The Laboratory Mouse*. 2d edition. New York: McGraw-Hill, 1966.

———. *Handbook of Genetically Standardized JAX Mice*. Bar Harbor: Bar Harbor Times Publishing Corporation, 1962. Reprinted 1968. JLA-BH.

Green, Earl, and Thomas H. Roderick. "Radiation Genetics." In *The Biology of the Laboratory Mouse*. 2d edition. Edited by Earl Green, pp. 165–85. New York: McGraw-Hill, 1966.

Green, Elizabeth Ufford. "On the Occurrence of Crystalline Material in the Lungs of Normal and Cancerous Swiss Mice." *Cancer Research* 2 (March 1942): 210+.

Green, Margaret. "Mutant Genes and Linkages." In *Biology of the Labortory Mouse*. Edited by Earl Green, pp. 87–150. New York: McGraw-Hill, 1966.

Griffith, Frederick. "The Influence of Immune Serum on the Biological Properties of Pneumococci," *Reports on Public Health and Medical Subjects*, no. 18. London: His Majesty's Stationery Office, 1928.

Grobman, Arnold B. *Our Atomic Heritage*. Gainsville: University of Florida Press, 1951.

Gross, Ludwig. "Is Cancer a Communicable Disease?" *Cancer Research* 4 (May 1944): 293–303.

Gruneberg, Hans. *The Genetics of the Mouse.* Cambridge: Cambridge University Press, 1943.

Guerrini, Anita. *Animal and Human Experimentation: An Introductory History.* Baltimore: Johns Hopkins University Press, 2002.

———. "Animal Tragedies: The Moral Theater of Anatomy in Early Modern Europe." Paper presented at the HSS Annual Meeting. Santa Fe, NM, 1993.

Hacker, Barton. *The Dragon's Tale: Radiation Safety on the Manhattan Project, 1942–1946.* Berkeley: University of California Press, 1987.

Hacking, Ian. "Weapons Research and the Form of Scientific Knowledge," *Canadian Journal of Philosophy,* supp. vol. 12 (1987): 237–60.

Haldane, J.B.S. "The Genetics of Cancer." *Nature* 132, 19 August 1933: 265–67.

Haldane, J. B. S., A. D. Sprunt, and N. M. Haldane. "Reduplication in Mice." *Journal of Genetics* 5 (1915): 133–35.

Hanau, A. "Erfolgreiche experimentelle Ubertragung von Carcinom." *Forschung die Medizinisch* 7 (1889): 321.

Haraway, Donna. *Modest Witness@Second_Millenium.FemaleMan(c)Meets Oncomouse™: Feminism and Technoscience.* New York: Routledge, 1997.

"Harvard College v. Canada (Commissioner of Patents)." 2002 Supreme Court of Canada 76, File No. 28155; 2002: May 21; 2002: December 5.

Hays, Samuel P., in collaboration with Barbara D. Hays. *Beauty, Health, and Permanence: Environmental Politics in the U.S.* New York: Cambridge University Press, 1987.

———. *Conservation and the Gospel of Efficiency: The Progressive Conservation Movement, 1890–1920* (New York, Atheneum, 1959).

"H. B. Andervont," *Annual Obituary.* New York: St. Martin's Press, 1981, pp. 171–72.

Heller, J. R. "The National Cancer Institute: A Twenty-Year Retrospective." *Journal of the National Cancer Institute* 19, 2 (August 1949): 147–90.

Heston, W. E., M. K. Deringer, and H. B. Andervont. "Gene-Milk Agent Relationship in Mammary Tumor Development." *Journal of the National Cancer Institute* 5 (1945): 289–307.

Heston, Walter. "Genetics of Mammary Tumors in Mice." In *A Symposium on the Mammary Tumors of Mice.* Edited by F. R. Moulton, pp. 55–84. Washington, DC: AAAS, 1945.

Hewlett, Richard, and Francis Duncan. *Atomic Shield: A History of the Atomic Energy Commission, 1947–52.* Volume 2. Berkeley: University of California Press, 1990.

Hogle, Linda. "Standardization Across Non-Standard Domains: The Case of Organ Procurement." *Science, Technology, and Human Values* 20, 4 (1995): 482–501.

Hollaender, Alexander. "The Problem of Mitogenic Rays." In *Biological Effects of Radiation.* Vol. 1. Edited by B.M. Duggar, pp. 919–60. New York: McGraw Hill, 1936.

Hollaender, Alexander, and J. T. Curtis, "The Effects of Sublethal Doses of Monochromatic Ultraviolet Radiation on Bacteria in Liquid Suspensions." *Proceedings of the Society of Experimental Biology and Medicine* 33 (1935): 61–62.

Holstein, Jean. *The First Fifty Years at the Jackson Laboratory, 1929–1979*. Bar Harbor: The Jackson Laboratory, 1979.

Holthof, Marc. "To the Realm of Fables: The Animal Fables from Mesopotamia to Disneyland. In *Zoology on (Post) Modern Animals*. Edited by Bart Verschaffel and Marc Vermick. Translated by Milt Papatheophanes. Dublin, 1993.

Hoyt, Janet. *Wings of Wax*. New York: J. H. Sears, 1929.

Hurst, Marsha, and Jane Nusbaum. "Advocating for Women's Health: Breast Cancer and Models of Advocacy in the Mid-Twentieth Century." Paper presented at the Berkshire Conference on the History of Women, Chapel Hill, NC, June 1996.

Ibsen, Herman, and Emil Steigleder. "Evidence for the Death in Utero of the Homozygous Yellow Mouse." *American Naturalist* 53 (1919): 185–87.

Jackson, Gardner. "Michigan Takes Live Wire from Maine in Clarence Cook Little." *Detroit News*, 3 July 1925.

James, Henry. Letter to W. E. Castle. 8 November 1917. Rockefeller University Archives, Record Group 210.2, Box 1, Folder "Abderhalden—Councilman," RAC-NY.

Jasanoff, Sheila. "Beyond Epistemology: Relativism and Engagement in the Politics of Science." *Social Studies of Science* 26, 2: (May 1996): 393–418.

Jennings, Ann. "The Social Construction of Measurement: Three Vignettes from Recent Events and Labor Economics." *Journal of Economic Issues* 35, 2 (2001): 365–71.

Jennings, H. S. "Formulae for the Results of Inbreeding." *American Naturalist* 48 (1914): 693–96.

———. "The Production of Pure Homozygotic Organisms from Heterozygotes by Self-Fertilization." *American Naturalist* 46 (1912): 487–91.

Johannsen, Wilhelm. *Elements de Exakten Erblichkeitslehre*. Jena: G. Fischer, 1909.

Kamrat-Lang, Devora. "Healing Society: Medical Language in American Eugenics." *Science in Context* 8 (1995):175–96.

———. "Science as Political Activism: The Mutation Research of H. J. Muller, 1918–1927." Paper presented at the Joint Atlantic Seminar in the History of Biology, Cold Spring Harbor Laboratory, 1995.

Kanigel, Robert. *The One Best Way: Frederick Winslow Taylor and the Enigma of Efficiency*. New York: Viking, 1997.

Kay, Lily. "Laboratory Technology and Biological Knowledge: The Tiselius Electrophoresis Apparatus, 1930–1945." *History and Philosophy of the Life* Sciences 10, 1 (1988): 51–72.

———. *The Molecular Vision of Life: Caltech, the Rockefeller Foundation and the Rise of the New Biology*. New York: Oxford University Press, 1993.

Keating, Paul, and Alberto Cambrosio. "The New Genetics and Cancer: The Contributions of Clinical Medicine in the Era of Biomedicine." *Journal of the History of Medicine* 56 (October 2001): 321–52.

Keating, Peter. "Biomedical Platforms." *Configurations* 8 (2000): 337–87.

Keeler, Clyde. "How It Began." In *The Origins of Inbred Mice*. Edited by Herbert S. Morse III, pp. 179–93. New York: Academic Press, 1978.

———. *The Laboratory Mouse: Its Origin, Heredity and Culture*. Cambridge: Harvard University Press, 1931.

———. "In Quest of Apollo's Sacred White Mice." *Scientific Monthly* 34 (January 1933): 48–53.

Keen, W. W. "Anti-vivisectionists' Methods." *Hygeia* 5 (1927): 36–38.

Kennedy, David M. *Birth Control in America: The Career of Margaret Sanger*. New Haven: Yale University Press, 1970.

Kevles, Daniel J. "Foundations, Universities, and Trends in Support for the Physical and Biological Sciences, 1900–1992." *Daedalus* 121, 4 (1992): 195–235.

———. "Of Mice and Money: The Story of the World's First Animal Patent." *Daedalus* (Spring 2002): 78–88.

———. *In the Name of Eugenics*. Berkeley: University of California Press, 1985.

———. "The National Science Foundation and the Debate over Postwar Research Policy, 1942–45." *Isis* 68 (1977): 5–26.

———. "Pursuing the Unpopular: A History of Courage, Viruses, and Cancer." In *Hidden Histories of Science*. Edited by Robert Silvers, pp. 69–114, New York: New York Review, 1995.

———. "Scientists, the Military, and the Control of Postwar Defense Research: The Case of the Research Board for National Security, 1944–46." *Technology and Culture* 16 (1975): 20–47.

Kevles, Daniel J., and Gerald L. Geison. "The Experimental Life Sciences in the Twentieth Century." *Osiris* 10 (1995): 97–121.

Kimmelman, Barbara. "Organisms and Interests: R. A. Emerson's Claims for the Unique Contributions of Agricultural Genetics." In *The Right Tools for the Job: At Work in 20th-Century Life Sciences*. Edited by Clarke and Fujimura, pp. 198–232. Princeton: Princeton University Press, 1992.

———. "A Progressive Era Discipline: Genetics at American Agricultural College and Experiment Stations, 1900–1920." Ph.D. dissertation, University of Pennsylvania, 1987.

Kirschbaum, A., and L. C. Strong. "Transplantation of Leukemia Arising in Hybrid Mice." *Cancer Research* 1 (1941): 785–86.

Klein, George. *The Atheist and The Holy City*. Translated by Theodore and Ingrid Friedman. Cambridge: MIT Press, 1990.

Kluckhohn, Frank L. "18 Dead, Damage $25,000,000, as Forest Fires Sweep on in Wide New England Area." *New York Times*, 25 October 1947, A1+.

———. "Much of Bar Harbor Razed as 4,300 Flee Forest Fire; Whole Maine Towns Gone." *New York Times*, 24 October 1947, A1+.

Kohler, Robert. *From Medical Chemistry to Biochemistry: The Making of a Biomedical Discipline*. New York: Cambridge University Press, 1982.

———. *Lords of the Fly: Drosophila and the Experimental Life*. Chicago: University of Chicago Press, 1994.

———. *Partners in Science: Foundations and Natural Scientists, 1900–1945*. Chicago: University of Chicago Press, 1991.

Kohler, Robert E. "Warren Weaver and the Rockefeller Foundation's Program in Molecular Biology: A Case Study in the Management of Science." In *Managing Medical Research in Europe: The Role of the Rockefeller Foundation (1920s–*

1950s). Edited by Giuliana Gemelli, Jean-Francois Picard, and William H. Schneider, pp. 51–90. Bologna: CLUEB, 1999.

Kolata, Gina. "A Star Is Born: Even a Lab Mouse Needs an Agent." *New York Times*, 26 January 1997, p. E5.

Kolmer, John A. "What Science Owes to Animals." *Hygeia* 6 (1928): 618–22.

Konkel, D. A., S. M. Tilghman, and P. Leder. "The Sequence of the Chromosomal Mouse Beta Globin Major Gene: Homologies in Capping, Splicing and PolyA Sites." *Cell* 15 (1978): 1125–32.

Korteweg, R. "On the Manner in Which the Disposition to Carcinoma of the Mammary Gland is Inherited in Mice." *Genetics* 18 (1936): 350.

Kuznick, Peter J. *Beyond the Laboratory: Scientists as Political Activists in 1930s America*. Chicago: University of Chicago Press, 1987.

Lacassagne, A. "Relationship of Hormones and Mammary Adenocarcinoma in the Mouse." *American Journal of Cancer* 37 (1939): 414–24.

Lang, Daniel. "Fallout." Originally published in the *New Yorker*. Reprinted in *From Hiroshima to the Moon: Chronicles of Life in the Atomic Age*. New York: Simon and Schuster, 1959.

Lauber, Patricia. *Of Man and Mouse: How House Mice Became Laboratory Mice*. New York: Viking Press, 1971.

Law, Lloyd. "Mouse Genetics News, Number 2." *Journal of Heredity* 39 (1948): 300–307.

Leach, William. *Land of Desire: Merchants, Power, and the Rise of a New American Culture*. New York: Vintage, 1993.

Lederer, Susan E. "The Controversy over Animal Experimentation in America, 1880–1914." In *Vivisection in Historical Perspective*. Edited by N. Rupke, pp. 236–55. New York: Routledge, 1987.

———. "Human Experimentation and Anti-Vivisection in the Turn-of-the-Century America." Ph.D. diss., University of Wisconsin-Madison, 1987.

———. "Moral Sensibility and Medical Science: Gender, Animal Experimentation, and the Doctor-Patient Relationship." In *The Empathic Practitioner: Empathy, Gender, and Medicine*. Edited by Ellen Singer More and Maureen A. Milligan, pp. 59–73. New Brunswick: Rutgers University Press, 1994.

———. "Political Animals: The Shaping of Biomedical Research Literature in Twentieth Century America." *Isis* 83 (1992): 61–79.

———. *Subjected to Science: Human Experimentation in American Before the Second World War*. Baltimore: Johns Hopkins University Press, 1995.

Lemonick, Michael D. "Smart Genes?" *Time* 154, 11 (September 13, 1999): 54–59.

Lenoir, Timothy, and Marguerite Hays. "The Manhattan Project for Biomedicine." In *Controlling Our Destinies*. Edited by Phillip R. Sloan, pp. 29–62. Notre Dame: University of Notre Dame Press, 1995.

Lerman, Nina, Arwen Palmer Mohun, and Ruth Oldenziel. "Versatile Tools: Gender Analysis and the History of Technology" and "The Shoulders We Stand on and the View from Here: Historiography and Directions for Research." *Technology and Culture* 38 (1997): 1–32.

Lewis, George. "The Maine That Never Was: The Construction of Popular Myth in Regional Culture." *Journal of American Culture* 16, 2 (Summer 1993): 91–99.

Lewis, Ricki. "A Stem Cell Legacy: Leroy Stevens." *The Scientist* 14 (5–6 March 2000): 19.

Lewis, Sinclair. *Arrowsmith*. New York: Harcourt Brace, 1925.

Lindee, M. Susan. *Suffering Made Real: American Science and the Survivors at Hiroshima*. Chicago: University of Chicago Press, 1994.

———. "What Is a Mutation? Identifying Heritable Change in the Offspring of Survivors at Hiroshima and Nagasaki." *Journal of the History of Biology* 25 (Summer 1992): 231–55.

Little, C. C. "Agents Modifying the Germ Plasm." *Surgery, Gynecology and Obstetrics* 46 (1928): 155–58.

———. "Alternative Explanations for Exceptional Color Classes in Doves and Canaries." *American Naturalist* 53 (1919): 186–87.

———. *The Awakening College*. New York: W.W. Norton, 1930.

———. "Cancer and Heredity." *Science* 42 (1915): 1076–77.

———. *Civilization Against Cancer.* New York: Farrar and Rinehart, 1939.

———. "The Conquest of Cancer." *Good Housekeeping* (June 1936): 78–79, 108, 110, 112.

———. "The 'Dilute' Forms of Yellow Mice." *Science* 33 (1911): 896–97.

———. "A Discussion of Certain Phases of Sterility." *Annals of Clinical Medicine* 5 (1926): 1–4.

———. "Education in Cancer." *American Journal of Cancer* 15 (1931): 280–83.

———. "Evidence That Cancer Is Not a Simple Mendelian Recessive." *Journal of Cancer Research* 12 (1928): 30–46.

———. *Experimental Studies of the Inheritance of Coat Color in Mice*. Carnegie Institute of Washington Publication no. 179. Washington, DC: Gibson Brothers Press, 1913.

———. "Extrachromosomal Influence in Relation to the Incidence of Mammary and Non-Mammary Tumors in Mice." *American Journal of Cancer* 27 (1936): 516–18.

———. "Further Studies on the Genetics of Abnormalities Appearing in Descendants of X-rayed Mice." *Genetics* 17 (1932): 674–88.

———. "Genetic Investigations and the Cancer Problem." *Commonwealth Review* 8 (1926): 130–36.

———. "Genetics in Relation to Carcinoma." *Proceedings of the Staff Meeting of the Mayo Clinic* 11 (1936): 782–83.

———. "The Genetics of Tissue Transplantation in Mammals." *Journal of Cancer Research* 8 (1924): 75–95.

———. "Halsey Joseph Bagg." *Anatomical Record* 100 (1948): 397.

Little, C. C., with J. M. Murray and W. T. Bovie. "Influence of Ultra-violet Light on Nutrition in Poultry." *Maine Agricultural Experiment Station Bulletin* 320 (1924): 141–64.

———. "The Inheritance of Cancer." *Science* 42 (1915): 494–95.

———. "Inheritance of a Predisposition to Cancer in Man." *Eugenics, Genetics and the Family* 1 (1923): 186–90.

Little, C. C., with J. M. Murray and W. T. Bovie. "A New Deal for Mice: Why Mice Are Used in Research on Human Diseases." *Scientific American* 152 (1935): 16–18.

———. "Not Dead but Sleeping." *Journal of Heredity* 24 (1933): 149.

———. "A Note on the Fate of Individuals Homozygous for Certain Color Factors in Mice." *American Naturalist* 53 (1919): 185–87.

———. "A Note on the Origin of Piebald Spotting in Dogs." *Journal of Heredity* 11 (1920): 12–15.

———. "The Occurrence of Three Recognized Color Mutations in Mice." *American Naturalist* 50 (1916): 335–49.

———. "Opportunities for Research in Mammalian Genetics." *Scientific Monthly* 26 (1928): 521–34.

———."A Possible Mendelian Explanation for a Type of Inheritance Apparently Non-Mendelian in Nature." *Science* 40 (1914): 904–906.

———. "A Preliminary Note on the Occurrence of a Sex-Limited Character in Cats." *Science* 35 (1912): 784–85.

———. "The Present Status of Our Knowledge of Heredity and Cancer." *Journal of the American Medical Association* 106 (27 June 1936): 2234–35.

———. "The Relation of Heredity to Cancer in Man." *Scientific Monthly* 3 (1916): 196–202.

———. "Relations Between Research in Human Heredity and Experimental Genetics. *Scientific Monthly* 14 (1922): 401–14.

———. "The Role of Heredity in Determining the Incidence and Growth of Cancer." *American Journal of Cancer* 15 (1931): 2780–89.

———. "Shall We Live Longer and Should We?" *Proceedings of the Third Race Betterment Conference*, 2–6 January 1928. Battle Creek: Race Betterment Foundation, 1928.

———. "Some Contributions of the Laboratory Rodent to Our Understanding of Human Biology." *American Naturalist* 73 (1939): 127–38.

———. "Symposium on Cancer: The Biology of Cancer." *Proceedings of the Annual Congress on Medical Education and Licensure*. Chicago: The Association, 1941.

———. "A University President [on] Sterilization: A Symposium." *Birth Control Review* 12, 3 (March 1928): 89.

———. "Unnatural Selection and Its Resulting Obligations." *Birth Control Review* 10 (August 1926): 243–44.

———. "Will Birth Control Promote Race Improvement?" *Birth Control Review* 13, 12 (December 1928): 343–44.

———. "Yellow and Agouti Factors in Mice." *Science* 38 (1913): 205.

"Little, Clarence Cook." In *Who Was in America*. Vol. 5 (1969–73). 1974, pp. 435–36.

Little, C. C., and R. A. Hicks. "The Blood Relationship of Four Strains of Mice." *Genetics* 16 (1931): 397–421.

Little, C. C. and Beatrice Johnson. "Experimental Breeding of Young Mammals." *Birth Control Review* 10 (September 1926): 267–68.

———. "The Inheritance of Susceptibility to Implants of Splenic Tissue in Mice. I. Japanese Waltzing Mice, Albinos and Their F1 Generation Hybrids." *Proceedings of the Society of Experimental Biology and Medicine* 19 (1922): 163–67.

Little, C. C., and B. W. McPheters. "The Incidence of Mammary Cancer in a Cross between Two Strains of Mice." *American Naturalist* 66 (1932): 568–71.

Little, C. C. and W. M. Murray. "Further Data on the Existence of Extra-Chromosomal Influence in the Incidence of Mammary Tumors in Mice." *Science* 82 (1935): 228–30.

———. "The Genetics of Mammary Tumor Incidence in Mice." *Genetics* (1935): 466–96.

Little, C. C., and John Phillips. "A Cross Involving Four Pairs of Mendelian Characters in Mice." *American Naturalist* 47 (1913): 760–62.

Little, C. C., and L. F. Strong. "Genetic Studies on the Transplantation of Two Adenocarcinomata." *Journal of Experimental Zoology* 41 (1924): 93–114.

Little, C. C., and E. E. Tyzzer. "Further Experimental Studies on the Inheritance of Susceptibility to a Transplantable Carcinoma (J.W.A) of the Japanese Waltzing Mouse." *Journal of Medical Research* 33 (1916): 393–427.

———. "Studies on the Inheritance of Susceptibility to a Transplantable Sarcoma (J.W.B.) of the Japanese Waltzing Mouse." *Journal of Cancer Research* 1 (1916): 387–89.

Little, C. C., and Emelia Vicari. " 'Lipid-Steroid' Fractions of Mouse Adrenal Lipids." *Proceedings of the Society of Experimental Biology and Medicine* 58 (1945): 59–60.

"Little Forced Out of Michigan University." *New York Telegram*, 11 February 1929.

Loeb, Leo. "Development of a Sarcoma and Carcinoma after the Inoculation of a Carcinomatous Tumor of the Submaxillary Gland of Japanese Mouse." *University of Pennsylvania Medical School Bulletin* 19 (July 1906): 113–16.

———. "Uber Enstanhug sines Sarcoms nachs Transplantation eines Adenocarcinoms einer japanischen Maus." *Zeitschrift auf Krebforschung* 7 (1908): 80.

Logan, Cheryl. "The Altered Rationale for the Choice of a Standard Animal in Experimental Psychology: Henry H. Donaldson, Adolf Meyer and 'the' Albino Rat." *History of Psychology* 2 (1999): 3–34.

———. "Before There Were Standards: The Role of Test Animals in the Production of Empirical Generality in Physiology." *Journal of the History of Biology* 35 (2002): 329–63.

Löwy, Ilana. *Between Bench and Bedside: Science, Healing, and Interleukin-2 in a Cancer Ward*. Cambridge: Harvard University Press, 1996.

Löwy, Ilana, and Jean-Paul Gaudillière, "Disciplining Cancer: Mice and the Practice of Genetic Purity." In *The Invisible Industrialist*, pp. 209–49. New York: Macmillan, 1998.

Ludmerer, Kenneth M. *Genetics and American Society: A Historical Appraisal*. Baltimore: Johns Hopkins University Press, 1972.

Lynch, Clara. "Strain Differences in Susceptibility to Tar-Induced Skin Tumors in Mice." *Proceedings of the Society of Experimental Biology* 31 (1933–34): 215.

Lyon, Mary F. "L. C. Dunn and Mouse Genetic Mapping." *Genetics* 125 (June 1990): 231–36.

Lyon, Mary F., and P. L. Mollison. "John Freeman Loutit." *Biographical Memoirs of the Fellows of the Royal Society* 40 (1994): 239–52.

McCoy, J. J. *The Cancer Lady.* Nashville: Thomas Nelson Publishers, 1977.

Macklin, Madge. "An Analysis of Tumors in Monozygous and Dizygous Twins." *Journal of Heredity* 31 (1940): 277–90.

"Mammary Cancer in Mice," in *Lectures on Genetics, Cancer, Growth, and Social Behavior* (Bar Harbor: Bar Harbor Times, 1949).

Marks, Harry. "Leviathan and the Clinic." Paper presented at the History of Science Society Annual Meeting, 27–30 December 2002.

———. *The Progress of Experiment: Science and Therapeutic Reform in the United States, 1900–1990.* New York: Cambridge University Press, 1997.

Marsh, M.C. "Evidence of Heredity among Mammary Tumor Mice." *Journal of Cancer Research* 8 (1924): 518.

———. "Simple Experimental Cancer Research." *American Journal of Cancer* 26 (1936): 181.

Marshino, Ora. "Administration of the National Cancer Institute Act, August 1937 to June 1943." *Journal of the National Cancer Institute* 4 (April 1944): 429–43.

Marx, Karl. *Outlines of the Critique of Political Economy.* Translated by Martin Liclaus. New York: Penguin Books, 1973.

"Mass Suicide Fate Conceived for Man: A Report on the AAAS Meeting." *New York Times*, 30 December 1936, p. 10L.

Mayr, Ernst. *The Growth of Biological Thought.* Cambridge: Belknap Press of Harvard University, 1982.

Megill, Allan, ed. *Rethinking Objectivity.* Durham: Duke University Press, 1994.

Merton, Robert. "The Normative Structure of Science." *Sociology of Science: Theoretical and Empirical Investigations*, pp. 267–80. Chicago: University of Chicago Press, 1973.

"Mice Beautiful." *Time* 30 (19 July 1937): 50.

"Mice Exposing Man's Ills." *Science Newsletter* 56 (3 September 1949): 146.

Midgley, Mary. *Animals and Why They Matter.* Athens: University of Georgia Press, 1983.

Miller, Howard. *Dollars for Research.* Seattle: University of Washington Press, 1970.

Monod, Jacques, and François Jacob. "General Conclusions: Teleonomic Mechanisms in Cellular Metabolism and Growth." *CSH Symposium on Quantitative Biology* 26 (1961): 389–401.

Moreau, Henri. "Recerches experimentales sur la transmissibilite de certains neoplasm." *Archives de Médecine: Experiementale et d'Anatomie Pathologique* 6 (1894): 677.

Morse, Herbert S. "Introduction." In *The Origins of Inbred Mice.* New York: Academic Press, 1980.

Morton, John J., et al. "The Effect of Visible Light on the Development of Tumors Induced by Benzopyrene in the Skin of Mice." *American Journal of Roentgenology* 43 (1940): 896–98.

"The Mouse Sequence: A Rosetta Stone." In *The Genes We Share*, Howard Hughes Medical Institute, at www.hhmi.org/genesweshare. Accessed February 2003.

"Mouse Show." *Newsweek* 9 (23 January 1937): 40.

Muller, H. J. "Artificial Transmutation of the Gene." *Science* 66 (1927): 125–34.

Murphy, James B. "An Analysis of Trends in Cancer Research." *Journal of the American Medical Association* 120 (1942): 107–11.

———. "Certain Etiological Factors in the Causation and Transmission of Malignant Tumors." *The American Naturalist* 60 (1926): 227–33.

National Science Foundation, "Science and Engineering Indicators Report." *Biomedical Communications*. Academic Press, 2001.

Neushul, Peter. "Science, Government, and the Mass Production of Penicillin." *Journal of the History of Medical Alllied Science* 48 (1993): 371–95.

Oberling, Charles. *The Riddle of Cancer.* New Haven: Yale University Press, 1944.

"Of Fat Mice and Men." *Newsweek* 40 (4 August 1952): 52–54.

"Of Mice and Men." *Newsweek* 68 (12 September 1966): 93.

Olwell, Russell. "Radiation Protection for Workers at Oak Ridge: Scientific Debate and Workplace Practice, 1942–1950." Paper presented at the History of Science Society Meeting, Atlanta, 1996.

Paige K. T., et al. "Injectable Cartilage." *Plastic & Reconstructive Surgery* 96 (1995):1390–1400.

Paigen, Kenneth. "Director's Message." JAX Annual Report 2001, pp. 5–6.

———. "Seventy-five Years of Mouse Genetics: Some Perspectives and Lessons," m.s., 1983.

Parascondola, John. "Maud Slye." In *Notable American Women: The Modern Period*. Edited by B. Sicherman and C. H. Green, pp. 651–52. Cambridge: Belknap Press of Harvard University, 1980.

Patterson, James T. *The Dread Disease: Cancer and Modern American Culture.* Cambridge: Harvard University Press, 1987.

Paul, Diane. *Controlling Human Heredity: 1865 to the Present*. New York: Prometheus Books, 1995.

———. "The Eugenic Origins of Medical Genetics." In *Politics of Heredity*, pp. 133–56. Albany: SUNY Press, 1998.

———. "Our Load of Mutations Revisited," *Journal of the History of Biology* 20 (Fall 1987): 3–20.

———. "The Rockefeller Foundation and the Origins of Behavioral Genetics." In *The Expansion of American Biology*, pp. 262–83. New Brunswick: Rutgers University Press, 1991.

Pauly, Phillip. "The Appearance of American Biology in Late Nineteenth-Century America." *Journal of the History of Biology* 17, 3 (Fall 1984): 369–97.

———. *Biologists and the Promise of American Life*. Princeton: Princeton University Press, 2000.

———. "Summer Resort and Scientific Discipline: Woods Hole and the Structure of American Biology, 1882–1925." In *The American Development of Biology.* Edited by R. Rainger et al., pp. 121–50. New Brunswick: Rutgers University Press, 1991.

Pearl, Raymond. "A Contribution Towards an Analysis of the Problems of Inbreeding." *American Naturalist* 47 (1913): 577–614.

Peckham, Howard. "President Little Embattled." In *The Making of the University of Michigan 1817–1967*, p. 157. Ann Arbor: University of Michigan Press.

"A Persistant Pioneer in Research," *Boston Globe*, 11 October 1980.

———. "On the Results of Inbreeding in a Mendelian Population." *American Naturalist* 48 (1914): 57–62.

Pickering, Andrew. *The Mangle of Practice: Time, Agency, and Science*. Chicago: University of Chicago Press, 1995.

———, ed. *Science as Practice and Culture*. Chicago: University of Chicago Press, 1992.

Podberscek, Anthony, Elizabeth Paul, and James Serpell, eds. *Companion Animals and Us: Exploring Relationships Between People and Pets*. New York: Cambridge University Press, 2000.

Podolsky, Scott, and Alfred Tauber. *The Generation of Diversity: Clonal Selection Theory and the Rise of Molecular Immunology*. Cambridge: Harvard University Press, 1977.

Porter, Theodore M. *Trust in Numbers: The Pursuit of Objectivity in Science and Public Life*. Princeton: Princeton University Press, 1995.

Proctor, Robert N. *Cancer Wars*. New York: Basic Books, 1995.

"The Production of Translocation and Mutations in Mice by Means of X-rays." *American Naturalist* 20 (1935): 545–67.

Provine, William B. *Sewall Wright and Evolutionary Biology*. Chicago: University of Chicago Press. 1986.

Rader, Karen A. "Alexander Hollaender's Postwar Vision for Biology: Oak Ridge and Beyond." Paper presented at the "Master Builders" Workshop, Johns Hopkins University, April 1999.

———. " 'The Mouse People': Murine Genetics Work at the Bussey Institution of Harvard, 1910–1936." *Journal of the History of Biology*, 31, 3 (Fall 1998): 327–54.

———. "The Origins of Mouse Genetics: Beyond the Bussey Institution, II. Defining the Problem of Mouse Supply." *Mammalian Genome* 12 (2002): 2–4.

Rae, John B. *American Automobile Manufacturers: The First Forty Years*. New York: Chilton Company, 1959.

Rainger, Ronald, Keith Benson, and Jane Maienschein. "Introduction." In *The American Development of Biology*, pp. 3–14. Philadephia: University of Pennsylvania Press, 1988.

Rasmussen, Nicolas. *Picture Control: The Electron Microscope and the Transformation of Biology in America, 1940–1960*. Stanford: Stanford University Press, 1997.

Ratcliff, J. D. "Of Important Mice and Men." *Colliers* 130 (8 November 1952): 13–15.

Reilly, Philip R. *The Surgical Solution*. Baltimore: Johns Hopkins University Press, 1991.

Reingold, Nathan. "Choosing the Future: The U.S. Research Community, 1944–46." *Historical Studies in the Physical and Biological Sciences* 25, 2 (1995): 301–28.

Reinhard, M. C., and C. G. Candee. "Influence of Sex and Heredity on the Development of Tar Tumors." *Journal of Cancer Research* (1930–31): 640–44.

Rettig, Richard A. *Cancer Crusade: The Story of the National Cancer Act of 1971*. Princeton: Princeton University Press, 1977.

"Report on Clinical Congress of American College of Surgeons." *New York Times*, October 1936, p. 36.

"Reports Dr. Little Will Quit Michigan." *New York Times*, 21 January 1929, p. 20.

"Resigns as President of Maine University: Dr. Clarence C. Little Plans to Accept Offer to Head University of Michigan." *New York Times*, 3 July 1925, p. 21.

Rheinberger, Hans-Jorg. *Towards a History of Epistemic Things: Synthesizing Proteins in a Test Tube*. Stanford: Stanford University Press, 1997.

Richmond, Marsha L. "Women in the Early History of Genetics: William Bateson and the Newnham College Mendelians, 1900–1910." *Isis* 92 (2001): 55–90.

Riddle, Oscar. "Any Hereditary Character and the Kinds of Things We Need to Know About It." *American Naturalist* 58 (1924): 41–425.

Ritvo, Harriet. *The Animal Estate: The English and Other Creatures in the Victorian Age*. Cambridge: Harvard University Press, 1987.

Robbins, Frank. "The Administration of Clarence Cook Little" and "The Administration of Marion LeRoy Burton." In *The University of Michigan: An Encyclopedic Survey*. vol. 1. Edited by Wilfred B. Shaw, pp. 81–87 and 88–98. Ann Arbor: University of Michigan Press, 1942.

Robinson, Judith. *Noble Conspirator: Florence S. Mahoney and the Rise of the National Institutes of Health*. Washington, DC: Francis Press, 2001.

Rodgers, Daniel. *Atlantic Crossings: Social Politics in a Progressive Age*. Cambridge: Belknap Press of Harvard University, 1998.

Rothfels, Nigel. "Introduction." In *Representing Animals*. Edited by Nigel Rothfels, p. xii. Bloomington: Indiana University Press, 2002.

Rowan, Andrew. *Of Mice, Models, and Men*. Albany: SUNY Press, 1984.

Russell, Elizabeth. "Inside the Inbred Mouse." *Bulletin for Medical Research* 9 (January–February 1955): 2–6.

———. "Mouse Phoenix Rose from the Ashes." In *Perspectives on Genetics Anecdotal, Historical, and Critical Commentaries, 1987–1998*. Edited by James F. Crow and William F. Dove, pp. 29–30. Madison: University of Wisconsin Press. Originally published in *Genetics* (October 1987).

———. Oral History Interview. Jackson Laboratory Oral History Collection, American Philosophical Society, p. 13.

Russell, J. C. and D. C. Secord. "Holy Dogs and the Laboratory." *Perspectives in Biology and Medicine* 28 (1985): 374–81.

Russell, Liane B. "X-Ray Induced Developmental Abnormalities in the Mouse and Their Use in the Analysis of Embryological Pattern," Ph.D. dissertation, University of Chicago, 1950.

Russell, Liane B., et al. "Influence of Dose Rate on the Radiation Effect on Fertility of Female Mice." *Proceedings of the Society of Experimental Biology and Medicine* 102 (1959): 471–79.

Russell, William L. "An Analysis of the Action of Color Genes in the Guinea Pig by Means of the DOPA Reaction," Ph.D. dissertation, University of Chicago, 1937.

———. "Inbred and Hybrid Animals and their Value in Research." In *Biology of the Laboratory Mouse*. Edited by George Snell, pp. 325–45. Philadelphia: Blakiston Company, 1941.

———. "Genetic Effects of Radiation in Mammals." In *Radiation Biology. vol. 1*. Edited by A. Hollaender, pp. 826–47. New York: McGraw Hill, 1954.

———. "Mammalian Radiation Genetics." *Symposium on Radiobiology* (Oberlin, 1950). New York: Wiley and Sons, 1950, pp. 427–440.

———. "Radiation in Mice: the Genetic Effects and Their Implications for Man." *Bulletin of the Atomic Scientists* 12, 1 (January 1956): 19–20.

———. "Reminiscences of a Mouse Specific-Locus Test Addict." *Environmental and Molecular Mutagenesis*, supplement 14, 16 (1989): 16–22.

———. "X-Ray-Induced Mutations in Mice." *Cold Spring Harbor Symposium of Quantitative Biology* 16 (1951): 327–36.

Russell, William L., and J. G. Hurst. "Pure Strain Mice Born to Hybrid Mothers Following Ovarian Transplantation." *Proceedings of the National Academy of Sciences* 31 (1945): 267–73.

Russell, W. L., L. B. Russell, and E. M. Kelly. "Radiation Dose Rate and Mutation Frequency." *Science* 128 (1958): 1546–50.

Russell, W. L., Liane B. Russell, and A. W. Kimball. "The Relative Effectiveness of Neutrons from a Nuclear Detonation and from a Cyclotron in Inducing Dominant Lethals in the Mouse." *American Naturalist* 88 (1954): 269–86.

Rx Mouse. Bar Harbor, ME: Jackson Lab, c. 1952.

Saltus, Richard. "Scientists Finish DNA Map Project." *Boston Globe*, 14 March 1996, p. A7.

Sapp, Jan. *Beyond the Gene: Cytoplasmic Inheritance and the Struggle for Authority in Genetics*. New York: Oxford University Press, 1987.

Sax, Karl. "The Bussey Institution: Harvard's Graduate School of Applied Biology, 1908–1936." *Journal of Heredity* 57 (1966): 175–78.

Schloegel, Judith Johns. "Intimate Biology: Herbert Spencer Jennings, Tracy Sonneborn, and the Career of American Protozoan Genetics." Ph.D. diss., Indiana University, in progress.

———. "Life Imitating Art, Art Imaging Life: Intimate Knowledge, Agency and the Organism as Aesthetic Object." *Videnskabsforskning* (Danish Newsletter for the Network of History and Philosophy of Science) 20 (1998): 2–18.

Schott, R. G. "The Inheritance of Resistance to Salmonella in Various Strains of Mice." *Genetics* 17 (1932): 203–29.

Scott, Pam, Eveleen Richards, and Brian Martin. "Captives of Controversy: The Myth of the Neutral Social Researcher in Contemporary Scientific Controversies."*Science, Technology, & Human Values* 15 (Fall 1990): 474–94.

Sealander, Judith. *Private Wealth and Public Life: Foundation Philanthropy and the Reshaping of American Social Policy from the Progressive Era to the New Deal*. Baltimore: Johns Hopkins University Press, 1997.

Secord, James A. "Darwin and the Breeders: A Social History." In *The Darwinian Heritage*. Edited by David Krohn, pp. 519–42. Princeton: Princeton University Press, 1986.

Seidel, Robert. "Accelerating Science: The Postwar Transformation of the Lawrence Radiation Laboratory." *Historical Studies in the Physical Sciences*. 13, 2 (1983): 375–400.

Serpell, James. *In the Company of Animals: A Study of Human-Animal Relationships*. New York: Basil Blackwell, 1986.

Sevari, Lucio, and Franz Helberg. "Obituary: John J. Bittner." *Nature* 197 (7 February 1963): 539–40.

Shapin, Steven. "Pump and Circumstance: Robert Boyle's Literary Technology." *Social Studies of Science* 14 (1984) 484–520.

Shaughnessy, Donald. "The Story of the American Cancer Society." Ph.D. diss., Columbia University, 1957.

Shear, Murray J. "Studies on the Chemical Treatment of Tumors." *Journal of Cancer Research* (1935): 66+.

Shimkin, Michael. "A.E.C. Lathrop: Mouse Woman of Granby (1868–1918)." *Cancer Research* 35 (June 1975).

———. *Contrary To Nature: Being an Illustrated Commentary on Some Persons and Events of Historical Importance in the Development of Knowledge Concerning Cancer.* Bethesda: U.S. Dept. of Health, Education, and Welfare, Public Health Service, National Institutes of Health; Washington, DC: U.S. Govt. Printing Office, 1977.

———. *As Memory Serves: Essays on a Personal Involvement with the National Cancer Institute*, 1938–1978. PHS/NIH: NIH Publication no. 83–2217, September 1983.

Shoenfield, Allen. "U' Pins Faith in Dr. Little," *The Detroit News*, 7 July 1925.

Shryock, Richard Harrison. *The Development of Modern Medicine*. Philadelphia: University of Pennsylvania Press, 1936.

Silver, Lee. "Mice as Experimental Organisms." *Encyclopedia of the Life Sciences*. Nature Publishing Group, 2001. Available at www.els.net.

———. *Mouse Genetics*. New York: Oxford University Press, 1995.

Silvers, W. K. *The Coat Colors of the Mouse*. New York: Springer-Verlag, 1979.

Slater, Dashka. "Humouse™." *Legal Affairs* 1, 4 (November/December 2002):, 1+.

Slaton, Amy E. *Reinforced Concrete and the Modernization of American Building, 1900–1930*. Baltimore: Johns Hopkins University Press, 2001.

Slye, Maud. "The Incidence and Inheritability of Spontaneous Cancer in Mice." *Journal of Medical Research* 32 (1915): 159–72.

———. "The Relation of Heredity to Cancer." *Journal of Cancer Research* 12 (1928): 83–133.

———. "A Reply to Dr. Little." *Science* 42 (1915): 246–48.

Smith, David C. *The First Century: A History of the University of Maine, 1865–1965*. Orono: University of Maine Press, 1979.

Smith, Zadie. *White Teeth: A Novel*. New York: Random House, 2000.

Snell, George D. "Clarence Cook Little." *Biographical Memoirs, National Academy of Sciences* 46 (1975): 240–63.

———. "Clarence Cook Little." *Dictionary of Scientific Bibliography.* Vol. 17. Edited by F. L. Holmes, pp. 563–64. New York: Scribner's 1990.

———. "Genetic Changes in Mice Induced by X-rays." *American Naturalist* 67 (1933): 24.

———. "Studies in Histocompatibility." *Nobel Lectures* 8 (December 1980): 645–60.

———. "The Production of Translocation and Mutations in Mice by Means of X-rays." *American Naturalist* 20 (1935): 545–67.

———. "X-Ray Sterility in the Male House Mouse." *Journal of Experimental Zoology* 65 (1933): 421–41.

Snell, George, J. Dausset, and S. Nathenson. *Histocompatibility.* New York: Academic Press, 1976.

Snell, George, and Sheldon Reed. "William Ernest Castle, Pioneer Mammalian Geneticist." *Genetics* 135 (April 1993): 751–53.

Southwick, Ron. "Senate Votes to Block Expansion of Lab-Animal Regulations." *Chronicle of Higher Education*, 1 March 2002, p. 25.

Spiegelman, Art. *Maus: A Survivor's Tale.* New York: Pantheon, 1986.

Staats, Joan. "The Classified Bibliography of Inbred Strains of Mice." *Science* 119 (26 February 1954): 295–96.

Staff of the Jackson Laboratory. "The Existence of Non-Chromosomal Influence in the Incidence of Mammary Tumors in Mice." *Science* 18 (November 1933): 465–66.

Star, Susan Leigh. *Ecologies of Knowledge: Work and Politics in Science and Technology* (Albany: SUNY Press, 1995).

Starr, Paul. *The Social Transformation of American Medicine.* New York: Basic Books, 1982.

Stevens, Leroy. "The Development of Transplantable Teratocarcinomas from the Intratesticular Grafts of Pre and Post-implantation Mouse Embryos." *Developmental Biology* 21, 3 (March 1970): 364–82.

Strickland, Stephen P. *Politics, Science, and Dread Disease: A Short History of United States Medical Research.* Cambridge: Harvard University Press, 1972.

Strong, L. C. "A Baconian in Cancer Research: Autobiographical Essay." *Cancer Research* 36 (October 1976): 3545–53.

———. "Inbred Mice in Science." In *Origins of Inbred Mice.* Edited by Herbert Morse, pp. 44–60. New York: Academic Press, 1978.

———. "The Origins of Some Inbred Mice." *Cancer Research* 3 (1942): 531–39.

———. "Transplantation Studies on Tumors Arising Spontaneously in Heterozygous Individuals." *Journal of Cancer Research* 13 (1929): 103–15.

Sturtevant, Alfred. "The Social Implications of the Genetics of Man." *Science* 120 (1954): 60.

Southwick, Ron. "Senate Votes to Block Expansion of Lab-Animal Regulations." *Chronicle of Higher Education*, 1 March 2002.

Swain, Donald C. "The Rise of a Research Empire: NIH, 1930 to 1950." *Science* 138, 3546 (14 December 1962): 1233–37.

Tetry, Andree. "Lucien Cuenot." In *Dictionary of Scientific Biography*, vol. 3. Edited by C. Gillispie, pp. 492–94. New York: Scribner's, 1970.

Thorkelson, H. J. "Interview: C. C. Little, 5 March 1926." General Education Board Collection, Record Group 2239, Box 641, Folder 6712, RAC-NY.

Thurman, Jr., Emmet. *Dental Pattern in the Mice of the Genus Peromyscus*. Ann Arbor: University of Michigan Press, 1957.

Tilghman, Shirley. "The Sins of the Fathers and Mothers: Genomic Imprinting in Mammalian Development." *Cell* 96 (1999): 185–93.

Timofeef-Ressovsky, Nickolai. "The Experimental Production of Mutations." *Biological Reviews* 9 (1934): 411–57.

"To Regulate Vivisection." *New York Times*, 3 February 1910, p. A1.

"To Finance Dr. Little in Studies of Cancer." *New York Times*, 8 May 1929, p. 2.

"To Begin Cancer Research." *New York Times*, 7 October 1929, p. 10.

Todes, Daniel. *Pavlov's Physiology Factory: Experiment, Interpretation, Laboratory Enterprise*. Baltimore: Johns Hopkins University Press, 2002.

Tokuda, Mitosi. "An Eighteenth Century Japanese Guide-Book on Mouse Breeding." *Journal of Heredity* 26 (1935): 481–84.

Triolo, Victor, and I. I. Riegel. "The American Association for Cancer Research, 1907–1940: Historical Reiew." *Cancer Research* 21 (1961): 137–67.

Tyzzer, E. E. "The Inoculable Tumors of Mice." *Journal of Medical Research* 18 (1907): 137–53.

———. "A Study of Heredity in Relation to Development of Tumors in Mice." *Journal of Medical Research* 17 (1907): 199–211.

———. "A Study of Inheritance in Mice with Reference to Their Susceptibility to Transplantable Tumors." *Journal of Medical Research* 21 (1909): 519–74.

"Unique Animals Aid Science." *Science Newsletter* 56 (2 July 1949): 10–11.

Uphoff, D. E., and Curt Stern. "Genetic Effects of Low Intensity Irradiation." *Science* 109 (1949): 609–10.

"Urges More Cancer Aid." *New York Times*, 8 December 1929, p. 24.

"U.S. Science Wars against Unknown Enemy: Cancer." *Life* 2 (March 1937): 11–17.

Valencia, J. I., and H. J. Muller. "The Mutational Potentialities of Some Individual Loci in *Drosophila*." *Proceedings of the Eighth International Congress of Genetics*. Lund University, 1949, pp. 681–83.

Von Borstel, R. C., and Charles M. Steinberg. "Alexander Hollaender: Myth and Mensch." *Genetics* 143 (1995): 1054–56.

Wade, Nicholas. "Of Mice, Men, and a Gene That Jump-Starts Follicles." *New York Times*, 5 October 1999, Section F, p. 1, col. 2.

———. "Mouse Genome Is New Battleground for Project Rivals." *New York Times*, 7 May 2002, p. A1.

———. "Of Smart Mice and an Even Smarter Man." *New York Times*, September 7, 1999, p. A1.

Wadman, Meredith. "U.S. Lab Animals May Win in Lawsuit." *Nature* 407 (2000): 549.

Walker, J. Samuel. "The Atomic Energy Commission and the Politics of Radiation Protection, 1967–1971." *Isis* 85 (1994): 57–78.

———. "The Controversy over Radiation Safety: A Historical Overview." *Journal of the American Medical Association* 262 (1989): 664–68.

Ward, Harold. "The Sciences and Society: Prejudices against Animal Vivisection," *The Living Age* 348 (August 1935): 549–52.

Warthin, A. S. "Heredity of Carcinoma in Man." *Annals of Internal Medicine*, 4 (January 1931).

Wassink, W. F. "Cancer er Hérédité." *Genetica*. 17 (1935): 103–44.

Weart, Spencer. *Nuclear Fear: A History of Images*. Cambridge: Harvard University Press, 1988.

"The Week in Science: Hereditary Cancer." *New York Times*, 12 July 1936.

Weindling, Paul, and Marcia Meldrum. "Johannes Andreas Grib Figinger." In *Nobel Laureates in Medicine or Physiology, Biographical Dictionary* pp. 177–81. (New York: Garland, 1990).

Weir, J. A. "Harvard, Agriculture and the Bussey Institution." *Genetics* 136 (April 1994): 1227–31.

Weller, Thomas H. "Ernest Edward Tyzzer." *Biographical Memoirs, National Academy of Sciences* 40 (1978): 353–73.

Wells, H. Gideon, "The Nature and Etiology of Cancer." *American Journal of Cancer* 15 (July 1931): 1919.

Westermann-Cicio, Mary L. "Of Mice and Medical Men: The Medical Profession's Response to the Vivisection Controversy in Turn of the Century America." Ph.D. diss., State University of New York, 2001.

Wheeler, David L. "Biologists Discuss Ways to 'Share' Genetically Engineered Mice." *Chronicle of Higher Education*, 7 April 1993, p. A14.

Whiting, P. W. "Propaganda Versus Basic Progress." *Birth Control Review* 14 (March 1930): 78–79.

Whittemore, Gilbert F. "The National Committee on Radiation Protection, 1928–60: From Professional Guidelines to Government Regulation." Ph.D. dissertation, Harvard University, 1986.

Wiebe, Robert H. *The Search for Order, 1877–1920*. New York: Hill and Wang, 1967.

Wilke, John R. "Furry Freaks: Mutant Mice Maker Is Serving Science." *Wall Street Journal*, 4 February 1994, p. 1.

"Will Be Memorial for R. B. Jackson." *Bar Harbor Times*, 8 May 1929.

Williams, Greer. *Virus Hunters*. New York: Knopf, 1961.

Wise, M. Norton. *The Values of Precision*. Princeton: Princeton University Press, 1995.

Woolley, George, Lloyd Law, and C. C. Little. "The Occurrence of Whole Blood of Material Influencing the Incidence of Mammary Carcinoma in Mice." *Cancer Research* (1941): 955–56.

Wright, Susan. *Molecular Politics: Developing American and British Regulatory Policy for Genetic Engineering, 1972–1982*. Chicago: University of Chicago Press, 1994.

Wynne, Brian. *Rationality and Ritual: The Windscale Inquiry and Nuclear Decisions in Britain*. St. Giles Chalfont: History of Science Society, 1982.

Yerkes, Robert. *The Dancing Mouse: A Study in Animal Behavior.* New York: Macmillan, 1907.

INDEX